Society of Economic Geologists Annual Conference (SEG 2023)

Resourcing the Green Transition

London, United Kingdom
26 - 29 August 2023

ISBN: 978-1-7138-7964-0

Printed from e-media with permission by:

Curran Associates, Inc.
57 Morehouse Lane
Red Hook, NY 12571

Some format issues inherent in the e-media version may also appear in this print version.

Society of Economic Geologists (SEG)
7811 Shaffer Parkway
Littleton, CO 80127-3732
USA

Phone: +1(720) 981-7882
Fax: +1(720) 981-7874

www.segweb.org

Additional copies of this publication are available from:

Curran Associates, Inc.
57 Morehouse Lane
Red Hook, NY 12571 USA
Phone: 845-758-0400
Fax: 845-758-2633
Email: curran@proceedings.com
Web: www.proceedings.com

Welcome to the SEG 2023 Conference!

Thank you for joining us as we explore this year's theme, *Resourcing the Green Transition*, which addresses the major challenges facing the energy transition and the demand for strategic metals. The conference theme also embraces the environmental, social, and governance (ESG) issues that are fundamental to responsible sourcing of these essential metals.

The technical program provides wide-ranging insights on the science and technology that are fundamental to meeting these challenges—reflecting the unique science-technology interface provided by the Society of Economic Geologists. We have a truly remarkable lineup of leading industry and academic speakers, supported by a wide range of contributions drawn from 45 countries—a global response to a global issue.

This in-person program of oral and poster presentations is accompanied by field trips to key resource areas and six workshops. A strong emphasis is placed on students and early career professionals. A new dimension is added with local financial and service organizations.

The venue is the Hilton London Metropole, located in London's West End—an ideal location from which to see the sights of this amazing city and then follow up with excursions elsewhere in the United Kingdom and throughout Europe.

Thank you for joining us as we network with old friends and forge new relationships—especially so this year as we welcome the participation of members of multiple affiliated major geoscientific societies around the world that share our vision!

Bob Foster
Chair, SEG 2023
Organizing Committee

Thank You, Exhibitors

Actlabs

GEO TEK

ARIANA RESOURCES plc

GOLDENEYE

IAGOD — International Association on the Genesis of Ore Deposits, 1964

B2GOLD

BARRICK

iCRAG
SFI RESEARCH CENTRE IN APPLIED GEOSCIENCES

BHP

Krux

BRUKER

mdru Mineral Deposit Research Unit
SCIENCE. EXPLORATION. DISCOVERY.

CODES

SciAps

corescan

Society of Economic Geologists — Founded 1920

EXPLORATION MAPPING

TERRACORE
GEOSPECTRAL IMAGING

EQUINOX GOLD

UNEXMIN
GEOROBOTICS

GEOSENSE
Natural Resources
Remote Sensing
Data Integration

VERACIO™

SEG 2023 Affiliates

Thank you to our SEG 2023 Affiliates. Affiliate organizations are professional societies and organizations with shared concerns and visions for the future of geology.

- Amira
- British Geological Survey (BGS)
- Critical Minerals Association
- European Federation of Geologists (EFG)
- The Geological Society of America
- Irish Association for Economic Geology (IAEG)
- Mineralogical Society of the UK and Ireland
- Society for Geology Applied to Mineral Deposits (SGA)
- Women in Mining UK (WIM UK)
- Young Mining Professionals (YMP)

Conference Information

Conference Organizing Committee

- **Conference Chair:** Bob Foster
- **Steering Group:** Richard Sillitoe, Karen Hanghøj, Richard Herrington, Laurence Robb
- **Technical Program:** Jamie Wilkinson, Dan Smith
- **Poster Program:** Katie McFall
- **Students/Early Career Professionals:** Alex Gorman, Huw Richards
- **Workshops:** Frances Wall, Lynnette Greyling
- **Field Trips:** Graham Brown, John Clifford
- **Sponsorships and Exhibitors:** Richard Herrington
- **Advisory Panel:** Edith Newton Wilson
- **UK Advisory Panel:** John Meyer
- **Europe Advisory Panel:** Nick Arndt, Maria Boni, Jens Gutzmer, Pasi Eilu, Lorena Ortega

SEG Contact Information

7811 Shaffer Parkway
Littleton, CO 80127-3732
Tel. 1.720.981.7882
E-mail: seg@segweb.org

Conference Organizer

Etherio
organizer@seg2023.org

For updates and latest information, visit www.seg2023.org.

SEG Awardees for 2023

The SEG Awards ceremony will be Sunday, August 27th, at 7:00 pm. The event will be held at the Hilton London Metropole.

R.A.F. Penrose Gold Medal

RICHARD J. GOLDFARB
(China University of Geosciences Beijing and Colorado School of Mines)

SEG Silver Medal

JONATHAN M.A. HRONSKY
(Western Mining Services)

SEG Waldemar Lindgren Award

JOSEPH M. MAGNALL
(GFZ German Research Centre for Geosciences)

Brian J. Skinner Award

JEFFREY W. HEDENQUIST
(Hedenquist Consulting Inc.)

ANTONIO ARRIBAS
(University of Texas at El Paso)

Honorary Fellow

GERALD G. CARLSON
(Pacific Ridge Exploration)

SEG Distinguished Lecturer

DAVID A. HOLWELL
(University of Leicester)

International Exchange Lecturer

KATHY EHRIG
(BHP)

SEG Thayer Lindsley Lecturer

SALLY GOODMAN
(Newmont)

Regional Vice President Lecturer

SHAO-YONG JIANG
(China University of Geosciences)

INVITED AND KEYNOTE SPEAKERS

Juan Afonso is an associate professor in computational geoscience and geo-data integration at the Faculty of Geo-Information Science and Earth Observation (ITC) at the University of Twente in the Netherlands. He also holds academic positions at Southern University of Science and Technology (China) and Macquarie University (Australia). His research focuses on the thermochemical structure and evolution of the lithosphere, the mechanical and geochemical interactions between tectonic plates and the sublithospheric upper mantle, and their effects on small- and large-scale tectonomagmatic processes and associated mineralization events. Juan is the developer of Multi-Observable Thermochemical Tomography (MTT), a technique that combines multiscale probabilistic inversion of multiple data sets, machine learning, and numerical modeling to address fundamental questions about the nature and evolution of the lithosphere, its mineral prospectivity, and geothermal potential.

Ross Beaty, Chair, Equinox Gold, founded Pan American Silver in 1994 and served as Chair until 2021; he was CEO of the company from its inception until 2004. Over 27 years, the company grew from a start-up to the world's second largest primary silver producer with the world's largest silver reserves and resources, while also becoming a significant gold producer.

Richard Sillitoe graduated from London University in England, where he went on to earn a Ph.D. degree in 1968. After three years with the Geological Survey of Chile and a Shell postdoctoral research fellowship at the Royal School of Mines in London, he has operated for five decades as an independent consultant to mining companies, international agencies, and foreign governments. He has worked on a wide variety of mineral deposits and prospects in 100 countries worldwide but focuses on the epithermal gold and porphyry copper environments. Published research has earned him awards in Europe, Australia, and North and South America, including the Robert M. Dreyer Award (Society for Mining, Metallurgy, and Exploration), the SGA-Newmont Gold Medal (Society for Geology Applied to Mineral Deposits), and the R.A.F. Penrose Gold Medal of the Society of Economic Geologists, of which he was President from 1999 to 2000.

Matthew Steele-MacInnis is a professor of geology at the University of Alberta whose research focuses on geologic processes that involve fluids, especially formation of mineral deposits. Originally from Newfoundland, Canada, he first studied geology at Memorial University of Newfoundland, then did a Ph.D. at Virginia Tech, after which he held a Marie Curie Postdoctoral Fellowship at ETH Zurich. Matthew has been a recipient of the Hisashi Kuno Award from the American Geophysical Union, a CAREER Award from the U.S. National Science Foundation, the Young Scientist Award from the Mineralogical Association of Canada, and the SGA Young Scientist Award from the Society for Geology Applied to Mineral Deposits.

Anna Bidgood is a Fulbright scholar at the Carnegie Institution for Science, where she is conducting phase equilibria experiments on cobalt sulfide minerals. This work is a continuation of her work at the Irish Centre for Research in Applied Geoscience, where her research focused on technology-critical elements in hydrothermal sediment-hosted mineral systems, with a particular focus on the Central African Copperbelt. Anna moved into the field of economic geology after completing her Ph.D. degree in metamorphic geology at the University of Oxford. Her work also takes her to the Arctic, including Greenland, where she has a particular interest in mineral deposits, mineral systems, and the environmental and social considerations of exploration and mining.

INVITED/KEYNOTE SPEAKERS

Mark Bristow is the president and chief executive of Barrick Gold Corporation (NYSE: GOLD, TSX: ABX). Formerly, he was the chief executive of Randgold Resources, the company he built from a small Africa-focused exploration business into one of the industry's most profitable and best-managed gold miners. In January 2019, Barrick and Randgold merged. Bristow restructured and restrategized Barrick, and within months he was the prime mover in the combination of the Nevada assets of Barrick and Newmont, creating the world's single largest gold mining complex, Nevada Gold Mines, majority-owned and operated by Barrick. Bristow leads an organization with mines and projects in 18 countries, focused on long-life, high-margin gold and copper assets and containing six of the world's tier-one gold mines. A geologist by profession, Bristow holds a doctorate in geology. His goal is to make Barrick the world's most valued gold and copper producer, owning the best assets, managed by the best people, and delivering industry leading returns.

Mike Daly is a visiting professor in tectonics and resources in the Earth Sciences Department at Oxford University. After a decade of field mapping and a doctorate in structural geology, Daly joined British Petroleum (BP) as a research exploration geologist. Almost 30 years later, he left BP having been Chief Geologist, Head of Exploration, and an executive vice president on BP's group executive team. Daly then joined Oxford and leads a research group in modern basin analysis, integrating basic geologic principles with leading-edge geophysical and geodynamic tools. He is a principal investigator in the UK's Copper Basin Exploration Science project and sits on the board of CGG, a geoscience technology company. He is a Harvard Business School alumnus and past president of the Geological Society of London.

Quentin Dehaine is a senior researcher at the Geological Survey of Finland (GTK) where he has been leading several national and EU research and innovation projects on ore geology, mineral processing, and traceability and geometallurgy of battery minerals (cobalt, lithium) and critical raw materials (CRMs) such as rare earth elements (REEs). After completing his Ph.D. at the University of Lorraine (Nancy, France), he worked as a postdoc researcher at the Camborne School of Mines, University of Exeter (UK), working on the geometallurgy of cobalt from the Democratic Republic of Congo (DRC). In 2019, Quentin joined the GTK with the objective of developing innovative integrated approaches to supporting mine value chain optimization, responsible sourcing, reducing technical risk, maximizing resource efficiency, and minimizing environmental impacts. Quentin is also the officer for Scandinavia within the International Union of Geological Sciences (IUGS) Initiative on Forensic Geology (IFG).

Laurance Donnelly is a professional Chartered Geologist and head of the technical department and Chief Geologist at AHK International, responsible for global inspection and associated sample preparation. He is also a Chartered Scientist, European Geologist, and elected fellow of the Geological Society of America. He holds a first-class honors degree (mineral exploration and mining geology) and a Ph.D. (fault reactivation and geological hazards) and has 34 years of international experience in mineral exploration, mineral resource evaluation, mining geology, forensic geology, engineering geology, geophysics, and geological hazards. He is the founder of the Forensic Geoscience Group of the Geological Society of London, and served as its first Chair. He is also the founder and current Chair of the International Union of Geological Sciences (IUGS) Initiative on Forensic Geology (IFG).

John Elkington, Founder and Chief Pollinator at Volans, is one of the founders of the global sustainability movement and has worked as a strategic advisor to businesses across the globe for more than 40 years. He has served on over 70 boards and advisory boards and currently chairs an advisory council on sustainability and new markets for Neste (Finland) and an impact valuation advisory council for Novartis (Switzerland). John is also a highly regarded keynote speaker and author; he has addressed over 1,000 conferences around the world and is the author or co-author of 20 books. In September 2021, John won the World Sustainability Award.

Katya Gorbatiouk has over 20 years of experience in international capital markets. As the head of investment funds in primary capital markets for the London Stock Exchange, Katya is responsible for driving the growth and liquidity of the London listed funds market—a conduit of capital into vital sectors of the economy, such as infrastructure, energy transition, and innovation. Katya is also a member of the Group Sustainability working group. Since February 2022 Katya has served on the Best Practice Committee of the Board of the Investor Relations Society. She is chair of its Sustainability Communications subcommittee and a member of its IR Communications and Engagement subcommittee. Her articles on the evolution of corporate communications were published in the recent issues of the IR Society Journal *Informed*. Katya has been a long-standing board member of organizations with high social impact, such as the Global Leaders Institute (Washington, D.C.) and London Music Fund.

Melissa Gregory is a geometallurgist and mineralogist working for Rio Tinto Exploration based at the Bundoora Technical Development Centre in Melbourne, Australia. She has a Ph.D. degree in economic geology plus over 18 years' experience across various deposit types including porphyry and sediment-hosted copper, pegmatite-hosted lithium, orogenic gold, ultramafic-hosted nickel, and unconformity U deposits. Her current role focuses on the geometallurgical characterization of mineralization as part of the exploration process.

Karen Hanghøj is the director of the British Geological Survey. She is a geologist with extensive experience in research and innovation management and the minerals and metals industry. Karen is passionate about understanding the complexity of resource management, about environmental and social sustainability, and about the role of geoscience in finding solutions to societal challenges. Karen holds a Ph.D. in geology from the University of Copenhagen. She worked with research on geologic processes in the lower crust and mantle and their associated mineral deposits before taking senior leadership roles in research and innovation organizations. She is a member of a range of international committees and working groups.

INVITED/KEYNOTE SPEAKERS

Kate Harcourt is a sustainability professional with over 30 years of experience across the mining industry. She works as part of owner's teams with leadership on sustainability issues and for Equator Principles financial institutions and engineering groups. She serves as Non-Executive Director to Condor Gold, Fortuna Silver (transitioning from Roxgold Inc), Orezone Gold Corporation, and Atalaya Mining. Kate works on projects mainly in Europe and Africa, developing policy and environmental management frameworks and navigating local and international ESIA processes, stakeholder engagement and permitting, lender liaison, annual reporting, and construction supervision. She has joined due diligence teams on high-profile new projects and has contributed to documentation on industry good practice. In 2018, she was nominated one of the Top 100 Global Inspirational Women in Mining.

David Holwell is an associate professor in applied geology in the Centre for Sustainable Resource Extraction at the University of Leicester, UK. He completed his Ph.D. research at Cardiff University on the Platreef deposit in the Bushveld Complex, and, after three years working as a consultant exploration geologist on a range of commodities across the globe for SRK Exploration Services, he joined Leicester in 2009. His main research areas focus on magmatic ore deposits, especially Ni-Cu-Co-platinum group element (PGE) sulfides, including PGE deposits in the Bushveld and Skaergaard Complexes, and a range of Ni-Cu sulfide systems across the globe, with a specialization in southern Africa. He has been involved in numerous projects that apply fundamental geoscience to exploration targeting and the understanding of orebodies and the crustal cycling and concentration of critical metals required for the energy transition. He is a Fellow of the SEG and an Associate Editor of *Economic Geology.*

Robert Kaemba is currently an M.Sc. economic geology candidate at the School of Geosciences, University of the Witwatersrand, South Africa, and an exploration geologist who has spent over 13 years working in the Central African Copperbelt at various levels of responsibility. His career has predominantly involved the operation and technical advancement of grassroots to advanced-stage projects of sediment-hosted copper (SHC-type) mineral deposits in Zambia. Robert takes a keen interest in matters surrounding fairness, diversity, equity, and inclusion (DEI) in geoscience. He also maintains an interest in initiatives that can improve stakeholder and local community engagement to foster win-win success in obtaining social license to operate. He is a Fellow of the Society of Economic Geologists and was elected to the SEG Council in 2022.

INVITED/KEYNOTE SPEAKERS

Julian Kettle is the Vice Chairman of Metals and Mining at Wood Mackenzie, having been with the company since 1989. He holds a degree in metallurgy and microstructural engineering and has over 30 years of experience in analysis of the metals and mining industry in relation to base metals, battery raw materials, and bulk commodities. Based in the U.K., he travels extensively and regularly presents to senior industry, government, and financial community stakeholders, leveraging his in-depth knowledge of the key factors influencing the industry and issues that stakeholders face to provide strategic advice to clients. Julian has been involved in market environment assessment, expert witness, raw materials sourcing, contract pricing, portfolio optimization, and strategic advisory relating to brownfield and greenfield investments across the value chain.

Rebecca Major is a corporate and projects partner at international law firm Herbert Smith Freehills LLP and head of the firm's mining and energy practice in Paris. She has significant experience advising the firm's clients on a broad range of legal matters arising from their planned investments and ongoing activities in the mining and energy sectors throughout the world. She has a particular focus on Africa but also works on matters in Europe, Asia, South America, and the Middle East. Rebecca is consistently singled out for her expertise in the energy and mining sectors by the leading independent benchmarking publications (Chambers Global, Legal 500 EMEA, Who's Who Mining). She has been a lawyer for more than 25 years and a partner for over 17 years.

Catherine Mottram completed her undergraduate studies in geoscience at the University of St. Andrews before moving to The Open University to work on her Ph.D. in Himalayan tectonics. She spent time as a Fulbright postdoctoral scholar at the University of Santa Barbara, USA, and as a Killam postdoctoral fellow at Dalhousie University in Canada before moving to the University of Portsmouth in 2017, where she is currently a senior lecturer in structural geology and tectonics. Catherine is a passionate geology educator and science communicator. Catherine's research focuses on directly dating deformation from the ductile to brittle crust using various geochronometers. She has been working on a major project, in collaboration with the Geological Survey of Canada and local exploration companies, focused on understanding structural controls on Au-Cu-Mo mineralization in the Yukon and northern British Columbia, Canada.

Lisa M.P. Munoz is the author of *Women in Science Now: Stories and Strategies for Achieving Equity* (Columbia University Press, 2023). She is the founder and president of SciComm Services, a science communications consulting firm. A former journalist, Munoz was a reporter and managing editor for *Geotimes* magazine and a press officer at the World Meteorological Organization in Geneva, Switzerland. She has more than twenty years of experience crafting science content for scientists and the public alike. Munoz holds an engineering degree from Cornell University with a major in science of the earth systems.

INVITED/KEYNOTE SPEAKERS

Glen Nwaila is the director of the Wits Mining Institute and an adjunct professor at the School of Geosciences, University of the Witwatersrand (Wits). Prior to joining Wits in 2017, he worked in the mining and consulting industries for 12 years, where he led teams of mining professionals and led audits in mineral resources and extractive metallurgy plants. In mining, Glen led the geology functions for various companies. He also served as manager at Deloitte. Glen has been an Erasmus Mundus Scholar at Uppsala University since 2018. He completed his Ph.D. magna cum laude at Würzburg University in Germany, an M.Sc. in chemical engineering at the University of Cape Town, and a geology degree at the University of Johannesburg. Glen is a professional natural scientist with the South African Council for Natural Scientific Professions and is a Fellow of the Geological Society of South Africa.

Sabina Strmic Palinkas is an associate professor in geochemistry and ore geology at UiT the Arctic University of Norway. She also holds an adjunct professor position at the University of Bergen, Norway. After receiving a Ph.D. degree in ore deposit geology from the University of Zagreb, Croatia, in 2009, she worked as a research fellow at the University of Auckland, New Zealand, and as an assistant professor at the University of Zagreb. In fall 2015, she joined UiT, where her main goal is to establish an internationally recognized research program in ore geology and mineral resources. Her area of expertise comprises aqueous and high-temperature geochemistry, geochemical and thermodynamic modeling, and applications of organic geochemistry and stable isotope systematics to ore-forming and environmental processes. Her recent research interest is focused on submarine hydrothermal systems, including recent systems along the Arctic mid-ocean ridges as well as ancient systems preserved along the Scandinavian Caledonides and within the Fennoscandian Shield.

Jim Royall has worked in the mining industry for over 20 years on early-stage and advanced exploration projects for a diverse range of commodities in Europe, Asia, Africa, and South America. This includes 10 years with Rio Tinto and senior management positions for various junior mining companies. Jim has lived in Spain for many years and has extensive experience throughout the Iberian Peninsula, where he has managed a number of important exploration and development projects. Jim has a bachelor of science degree in geology from Oxford Brookes University, is a member of the Australian Institute of Geoscientists (AIG) and a fellow of the Geological Society of London, and served as qualified person for several mineral exploration companies.

Kate Rubingh is a postdoctoral fellow with MDRU (Mineral Deposit Research Unit) at the University of British Columbia, having obtained an M.Sci. in geological sciences from Durham University, an M.Sc. in mineral exploration from Queen's University, and a Ph.D. from Laurentian University. Kate is a field economic geologist with research experience in gold deposits of the Paleoproterozoic and Archean, focusing on the application of volcanic stratigraphy and structural analysis to refine the structural framework at both the deposit and regional scale. Kate's current research in the Golden Triangle area of British Columbia will use new lithogeochemistry and geochronology to refine the paleovolcanic architecture of the Hazelton group stratigraphy and its controls on the distribution of mineral deposits to understand metal endowment in the Golden Triangle area.

INVITED/KEYNOTE SPEAKERS

Charlotte Seabrook has been Exploration Manager at Rupert Resources in Finnish Lapland since 2018 and led the exploration team to the greenfields discovery of the Ikkari gold deposit in 2020. Charlie has nearly 20 years' experience in multicommodity exploration in various terranes including Australia and West and Central Africa. Her experience includes exploring the Central African Copperbelt and Birimian basin of West Africa, where she was integral to several discoveries on the Ivory Coast while working for Newcrest as the district geologist. She completed her Ph.D. at the University of the Witwatersrand, holds an M.Sc. in mineral resources from the University of Wales (Cardiff), and is currently based in Finland.

Libby Sharman is an exploration geologist working as the Technical Lead for BHP Xplor, having spent five years prior with BHP Metals Exploration contributing to global exploration strategy through project generation and prospectivity analysis. She completed her Ph.D. at McGill University and has since gained experience in different facets of the exploration and mining industry, including as a project geologist, researcher, and communications specialist. Libby has experience on varied deposit types across five continents, from project generation to resource drilling for companies of all sizes and scopes. Libby's personal experiences have fostered an interest in improving the mining industry and making it a more inclusive and appealing sector in which to work. She pursues this through various volunteer roles in industry organizations including the Society of Economic Geologists.

Conference Hotel

Hilton London Metropole
225 Edgware Road, London W2 1JU, United Kingdom
Tel. +44 20 7402 4141

Wi-Fi Access
Username: SEG 2023 Conference
Password: #SEG2023

Mother's room: Mezzanine 12

Speaker Ready Room: West Wing Meeting Room 1

Friday:	3:00pm-6:00pm
Saturday-Monday:	7:00am-5:00pm
Tuesday:	7:00am-3:00pm

Registration:

Friday:	8:00am-6:00pm
Saturday-Monday:	7:00am-6:00pm
Tuesday:	7:00am-4:00pm

Exhibits and Posters (Richmond Suite)

Friday:	3:00pm-6:00pm:	Exhibits and Group A Poster Set Up
Saturday:	8:00am-3:00pm:	Exhibits and Group A Poster Set Up
	6:30pm-9:00pm:	Welcome Reception, Exhibits and Group A Posters First on Display
Sunday:	9:00am-10:00am:	Poster and Exhibitor-Only Access
	10:15am-11:00am:	Morning Poster Session
	5:00pm-7:00pm:	Evening Reception, Exhibit and Poster Session
	7:00pm-8:00pm:	Group A Poster Tear Down

**Note all posters from Group A must be removed by 8:00pm. Any remaining posters will be discarded.*

Monday:	8:00am-10:00am:	Group B Poster Set Up, Poster and Exhibitor-Only Access
	10:15am-11:00am:	Group B Posters First on Display, Morning Poster Session
	5:00pm-7:00pm:	Evening Reception, Exhibit and Poster Session
Tuesday:	9:00am-10:00am:	Poster and Exhibitor-Only Access
	10:15am-11:00am:	Morning Poster Session
	2:00pm-4:00pm:	Exhibits and Group B Poster Tear Down

**Note all posters from Group B must be removed by 4:00pm. Any remaining posters will be discarded.*

The information in this program is current through August 4, 2023. Please refer to the conference app or the online schedule of events (www.seg2023.org) for updated information. All sessions held in the Kensington Suite at the conference hotel.

SATURDAY

9:30am-10:00am
Morning Networking Session – Student and Early Career Professionals

10:00am-12:30pm
Career Advice Showcase – Student and Early Career Professionals

12:30pm-2:00pm
Lunch and CV Talk – Student and Early Career Professionals

2:00pm-4:30pm
Round Table Event – Student and Early Career Professionals

4:30pm-5:00pm
Afternoon Networking Session – Student and Early Career Professionals

5:00pm-5:30pm
Global Challenges 1
Supply Chain Issues: Importing Metals and Minerals into Europe/UK in the Next 15 Years
Rebecca Major (Global Challenges Speaker)

5:30pm-6:00pm
Global Challenges 2
Managing and Benefiting from ESG Challenges in the Development of Critical Minerals in the UK
Kate Harcourt (Global Challenges Speaker)

6:00pm-6:30pm
Presidential Address and Brian G. Hoal Tribute
Stuart Simmons, SEG 2023 President

6:30pm-9:00pm
Welcome Reception

SUNDAY

8:30am-8:45am
Welcome and Conference Chair Remarks
Bob Foster, SEG 2023 Conference Chair
Jennifer Craig, SEG Executive Director

8:45am-10:15am
Plenary Session 1
- Does the Energy Transition Start and End with Metals?
- Financing Sustainability
- The Wider Implications for Society at Large

10:15am-11:00am
Poster Viewing Session

11:15am-12:45pm
SESSION 1: Copper – Red Metal for a Green Future I
SESSION 2: Battery Metals I

12:45pm-1:45 pm
Lunch

1:45pm-2:45pm
Career Speed Dating – Student and Early Career Professionals

2:00pm-3:30pm
SESSION 3: Copper – Red Metal for a Green Future II

3:30pm-4:00pm
Break

4:00pm-5:00pm
SESSION 3 (continued): Copper – Red Metal for a Green Future II
SESSION 4: Battery Metals II

5:00pm-7:00pm
Exhibit and Poster Reception

7:00pm-9:00pm
SEG Awards Ceremony (Admiral Ballroom, Ticket Required)

MONDAY

8:30am-10:15am
SESSION 5: Copper – Red Metal for a Green Future III
SESSION 6: Battery Metals III

10:15am-11:00am
Poster Viewing Session

11:00am-12:45pm
SESSION 7: Europe's Metals I
SESSION 8: Transformational Science I

12:45pm-1:45pm
Lunch

2:00pm-3:30pm
SESSION 9: Europe's Metals II

3:30pm-4:00pm
Break

4:00pm-5:00pm
SESSION 9: Europe's Metals II (continued)
SESSION 10: Transformational Science II

5:00pm-7:00pm
Exhibit and Poster Reception

7:00pm-9:00pm
Industry Outlook Dinner (Admiral Ballroom, Ticket Required)
Ross Beaty, Chair, Equinox Gold

TUESDAY

8:30am-10:15am
SESSION 11: Gold – Responsible Discovery and Mining in the 21st Century I
SESSION 12: Advances in Geoscience I

10:15am-11:00am
Poster Viewing Session

11:00am-12:45pm
SESSION 13: Gold – Responsible Discovery and Mining in the 21st Century II
SESSION 14: Advances in Geoscience II

12:45pm-1:45 pm
Lunch

2:00pm-3:30pm
SESSION 15: Gold – Responsible Discovery and Mining in the 21st Century III
SESSION 16: Advances in Geoscience III

3:30pm-4:00pm
Closing Ceremony and Concluding Remarks
Bob Foster, SEG 2023 Conference Chair
Steve Piercey, SEG 2024 President
Anna Nguno, SEG 2024 Conference Chair

4:00pm-6:00pm
Closing Social

Sunday, 7:00pm-9:00pm
Awards Ceremony and Reception
Monday, 7:00pm-9:00pm
Industry Outlook Dinner
Ross Beaty, Chair, Equinox Gold
For additional information on conference events, please visit the conference website at www.seg2023.org.

Exhibit Booths and Posters Map

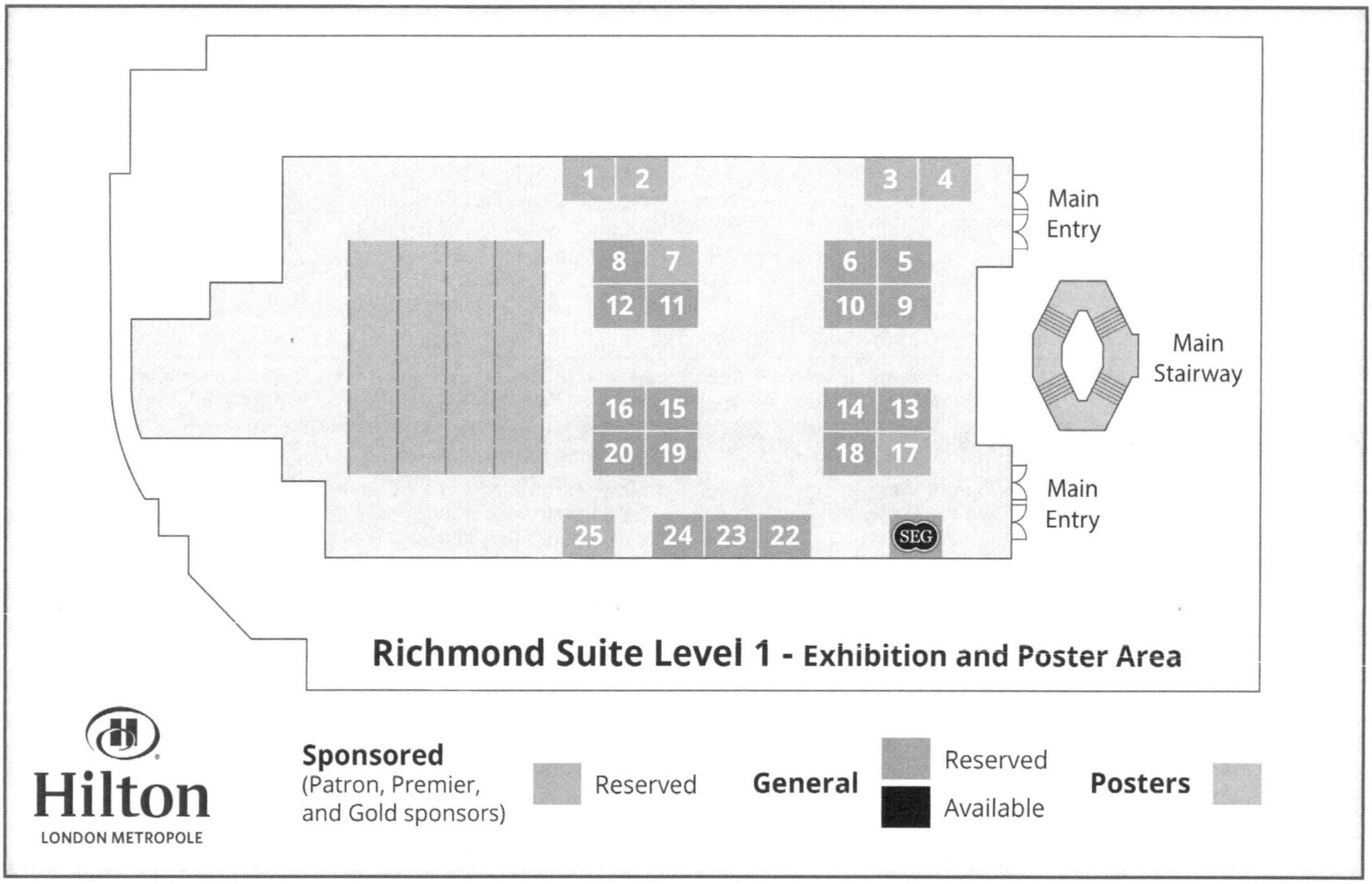

SEG 2023

Resourcing the Green Transition

August 26–29, 2023

Hilton London Metropole, United Kingdom

www.seg2023.org

Hotel Map and Information

www.seg2023.org/venue

Technical Program

DAY 1 - SATURDAY, AUGUST 26, 2023

Global Challenges I

5:00 pm - 5:30 pm **Supply Chain Issues: Importing Metals and Minerals into Europe/UK in the Next 15 Years**
Rebecca Major (Global Challenges Speaker) 1

Global Challenges II

5:30 pm - 6:00 pm **Managing and Benefiting from ESG Challenges in the Development of Critical Minerals in the UK**
Kate Harcourt (Global Challenges Speaker) 2

Presidential Address

6:00 pm - 6:30 pm **Presidential Address and Brian G. Hoal Tribute**
Stuart Simmons (SEG 2023 President) N/A

Welcome Reception and Exhibit and Poster Viewing

6:30 pm - 9:00 pm Welcome Reception

DAY 2 - SUNDAY, AUGUST 27, 2023

Opening Ceremony

8:30 am - 8:45 am **Welcome and Conference Chair Remarks**
Bob Foster, SEG 2023 Conference Chair
Jennifer Craig, SEG Executive Director N/A

Plenary Session 1

8:45 am - 9:15 am **Does the Energy Transition Start and End with Metals?**
Julian Kettle (Plenary Speaker) N/A

9:15 am - 9:45 am **Financing Sustainability**
Katya Gorbatiouk (Plenary Speaker) N/A

9:45 am - 10:15 am **The Green Transition Squeezes As It Embraces**
John Elkington (Plenary Speaker) 3

10:15 am - 11:00 am **Poster Viewing: Saturday and Sunday Posters**

11:00 am - 11:15 am Break

	Oral Session 01 Copper - Red Metal for a Green Future I	Oral Session 02 Battery Metals I
11:15 am - 11:30 am	**Central African Copper Belt** *Michael Daly (Keynote Speaker)* N/A	**Critical Battery Mineral Resources in the United States: Past, Present and Potential** *Jane Hammarstrom* 8

DAY 2 - SUNDAY, AUGUST 27, 2023 (Continued)

11:30 am - 11:45 am	**Central African Copper Belt, continued** *Michael Daly (Keynote Speaker)* N/A	**The Geological Setting of Vanadium Mineralisation in the Neuquén Basin of Central Argentina: An Opportunity for the Clean Energy and Energy Storage Sectors in South America** *Martin Jackson* 9
11:45 am - 12:00 pm	**Copper-Hosting Basins at the Edge of Cratons – the Yeneena Basin in Western Australia as a Case Study** *Weronika Gorczyk* 4	**Canadian Perspective on REE Deposits Hosted by Carbonatite and Alkaline Complexes and Its Implication for Carbonatite and Alkaline Magmatism in Europe** *Anne-Aurélie Sappin* 10
12:00 pm - 12:15 pm	**Numerical Modelling of Basin-Scale Fluid Flow and Cu Transport in the Katangan Basin, Central African Copperbelt** *Meissam Bahlali* 5	**Trace Element Deportment in Sphalerite Across the Bathurst Mining Camp, New Brunswick, Canada** *Foteini Drakou* 11
12:15 pm - 12:30 pm	**Syntectonic Copper Mineralisation in a Back-Thrust Duplex at Pangeni, NW Zambia** *John Wilton* 6	**From Rocks to Batteries: Insights Into Cobalt Mining, Mineralogy, and Sustainability** *Quentin Dehaine (Invited Speaker)* 12
12:30 pm - 12:45 pm	**Experimental Insights Into the Mobility and Extraction of Metals in Sedimentary Basin-Hosted Copper Systems** *Julia Woitischek* 7	**Occurrence and Spatial Distribution of Cobalt in Skarn Deposits: A Case Study from the Middle-Lower Yangtze River Valley Metallogenic Belt (MLYB) in East China** *Taofa Zhou* 13

12:45 pm - 1:45 pm	Lunch	
	Oral Session 03 **Copper - Red Metal for a Green Future II**	**Oral Session 04** **Battery Metals II**
2:00 pm - 2:15 pm	**Global Copper Screening for the Energy Transition - a Holistic Exploration Mindset** *Graeme Nicoll* 14	Lithium Pegmatites in Africa: Understanding Parageneses Kathryn Goodenough 23
2:15 pm - 2:30 pm	**The Role of Crustal Anatexis in Porphyry Copper Formation During Flat-Slab Subduction: Insights from the Laramide Belt, Southwest USA** *Thomas Lamont* 15	Eruption of Extremely Fractionated Magmas: Implications for Magmatic Versus Volatile Processes David R. Lentz 24
2:30 pm - 2:45 pm	**Magma Fertility Evolution Trends: Lessons from the Kerman Porphyry Belt, SE Iran** *Ali Sholeh* 16	Lithium in Manitoba: Production and Exploration Potential Tania Martins 25
2:45 pm - 3:00 pm	**Late Cretaceous Magnetism in the San Matias District, NW Colombia: Insights of a New Ca. 76 - 66 Ma Porphyry Copper Belt in the Northern Andes** *Julian Manco (Student)* 17	**Challenges of Water Use Impact Assessment of Lithium Production from South American Salar Deposits Using Life Cycle Assessment** *Rowan Halkes* 26
3:00 pm - 3:15 pm	**Unraveling the Thermochemical State of the Whole Lithosphere in Space and Time: New Methods and Their Predictive Power for Exploration** *Juan Afonso (Invited Speaker)* 18	**The Jadar Li(B) Deposit (Serbia): Insights Into a Giant Volcano-Sedimentary-Hosted Li-B System** *Francesco Putzolu* 27

DAY 2 - SUNDAY, AUGUST 27, 2023 (Continued)		
3:15 pm - 3:30 pm	Q&A	The Bergby LCT-pegmatite Field, Central Sweden – a New Contender to Produce Swedish Lithium? *Karin Högdahl* 28
3:30 pm - 4:00 pm	Break	
4:00 pm - 4:15 pm	A Rapid Transition to an Electrical Economy Based on Energy from Solar and Wind is Not Resource Viable *Adam Simon* 19	Cornish Lithium: Exploration for Geothermal Lithium Resources in Southwest England *Alistair Salisbury* 29
4:15 pm - 4:30 pm	Copper Mining and Extraction in Cyprus - from Ancient Heritage to a Green Future *Zack Van Coller* 20	Unconventional Magmatic Sulfide Systems: Looking Outside of the Box for the Next New Ni-Cu-Co-PGE Discovery *David Holwell (2023 SEG Distinguished Lecturer)* 30
4:30 pm - 4:45 pm	Unlocking the Full Potential of Metal Leaching in the Oceanic Crust: Assessing Permeability and Fluid Flow *Alannah Brett (Brett-Adams)* 21	Unconventional Magmatic Sulfide Systems, continued *David Holwell (2023 SEG Distinguished Lecturer)* 30
4:45 pm - 5:00 pm	Accelerating Discovery through Cutting-Edge Research *Elizabeth Sharman* 22	ML Modeling of Multiparameter Data for Ni-Cu-PGE Targeting *Wilson Bonner* 31
5:00 pm - 7:00pm	Exhibit and Poster Reception: **Saturday and Sunday Posters**	

DAY 3 - MONDAY, AUGUST 28, 2023		
	Oral Session 05 **Copper - Red Metal for a Green Future III**	**Oral Session 06** **Battery Metals III**
8:30 am - 8:45 am	High Grades in Hypogene Porphyry Copper Deposits *Dick Sillitoe (Keynote Speaker)* 32	Tracing Granite-Hosted Mineralization via Zircon Metal Anomalies *Laurence Robb* 38
8:45 am - 9:00 am	High Grades in Hypogene Porphyry Copper Deposits, continued *Dick Sillitoe (Keynote Speaker)* 32	Quantitative Mineralogy of the Beauvoir Granite and Rare-Metal-Bearing Minerals Characterization *Steven Kahou* 39
9:00 am - 9:15 am	Alteration Geothermometry and Exploration Implications of the Gajah Tidur Porphyry System, One of Two Systems that Formed the Supergiant Grasberg Cu-Au-(Mo) Deposit *Reza Al Furqan (Student)* 33	GREENPEG: New Toolsets to Explore for Pegmatites Hosting Rare Metals and High Purity Quartz to Feed the Energy Transition *Axel Müller* 40
9:15 am - 9:30 am	The Spence Porphyry Copper System Revisited: an Integrated Mineral Systems Study *Edward Bunker* 34	Challenges Associated with the Future Supply of Battery Minerals and Black Mass *Laurance Donnelly (Keynote Speaker)* 41
9:30 am - 9:45 am	Geology, Mineralisation and Hydrothermal Alteration of the Almalyk Porphyry Copper- Gold Deposit, Tien Shan Mountains, Eastern Uzbekistan *Peter Gleeson* 35	Challenges Associated with the Future Supply of Battery Minerals and Black Mass, continued *Laurance Donnelly (Keynote Speaker)* 41

DAY 3 - MONDAY, AUGUST 28, 2023 (Continued)

9:45 am - 10:00 am	**Applications of Multielement Epidote and Chlorite Mineral Chemistry to Locate Mineralized Centers: A Case Study from the Batu Hijau Porphyry Cu-Au District, Indonesia** *David Cooke* 36	**Recovery of Critical Metals from Copper Mine Tailings Utilizing Microanalytical Characterisation for Improved Geometallurgical Outcomes** *Loren Nicholls (Student)* 42
10:00 am - 10:15 am	**A Practical Predictive Targeting Model for Convergent Margin Gold and Copper Deposits** *Jon Hronsky* 37	**Ti-Zr-REE Heavy Mineral Sand Transport Paths and Concentration Processes Revealed by High-Resolution Airborne Radiometric Data** *Anjana Shah* 43
10:15 am - 11:00 am	Poster Viewing Session: **Monday and Tuesday Posters**	
	Oral Session 07 **Europe's Metals I**	**Oral Session 08** **Transformational Science I**
11:00 am - 11:15 am	**Metals and Minerals in Europe - Demand, Exploration, Mining and More** *Karen Hanghøj (Keynote Speaker)* 44	**From Planets to Protons - New Geoanalytical Workflows for Characterizing Rocks of Economic Importance** *Alan Butcher* 49
11:15 am - 11:30 am	**Metals and Minerals in Europe - Demand, Exploration, Mining and More, continued** *Karen Hanghøj (Keynote Speaker)* 44	**Opportunities for Rapid, Field-Based Sulfide Deportment and Trace Element Chemistry Using Laser-Induced Breakdown Spectroscopy (LIBS)** *Cassady Harraden* 50
11:30 am - 11:45 am	**Crystalline Basement of Latvia - An Undercover Opportunity** *Matiss Brants (Student)* 45	**Re-Assessing the Resource Potential of Seafloor Massive Sulfide Deposits** *John Jamieson* 51
11:45 am - 12:00 pm	**An Alternative Formation Environment for Kiruna-Type Iron Deposits: An Example from the Per Geijer IOA Deposit, Northern Sweden** *Amy-Jo Miles* 46	**Machine Learning Algorithms as a Tool for Data-Driven Interpretation of Multiphysics Datasets in Underexplored Areas** *Artem Gorbunov* 52
12:00 pm - 12:15 pm	**Critical Metal Potential of Historic Mining Districts in the Lesser Caucasus and the Eastern Pontides, Central Tethyan Orogenic Belt** *Robert Moritz* 47	**Unlocking the Potential of Predictive Geoscience Through the Adoption of GeoData Science** *Glen Nwaila (Keynote Speaker)* 53
12:15 pm - 12:30 pm	**Ancient Mining, Recent Discoveries and Future Potential for Critical Raw Materials in the Apuseni-Carpathian-Balkan Area of Southeast Europe** *Irena Peytcheva* 48	**Unlocking the Potential of Predictive Geoscience Through the Adoption of GeoData Science, continued** *Glen Nwaila (Keynote Speaker)* 53
12:30 pm - 12:45 pm	**Q&A**	**Q&A**
12:45 pm - 1:45 pm	Lunch	

DAY 3 - MONDAY, AUGUST 28, 2023 (Continued)

	Oral Session 09 Europe's Metals II	Oral Session 10 Transformational Science II
2:00 pm - 2:15 pm	Seafloor Massive Sulfide Deposits on the Arctic Mid-Ocean Ridges: An Insight Into Ore-Forming Processes Along Ultraslow Spreading Ridges *Sabina Strmic Palinkas (Keynote Speaker)* 54	ESG Valuation Model for the Mining Industry *Niklas Grönberg* 62
2:15 pm - 2:30 pm	Seafloor Massive Sulfide Deposits on the Arctic Mid-Ocean Ridges, continued *Sabina Strmic Palinkas (Keynote Speaker)* 54	Gamification for Technology-Enhanced Education in the Mineral Value Chain: Mining Simulation Game for Teaching the Geometallurgy Concept *Glacialle Tiu* 63
2:30 pm - 2:45 pm	New Mineralogical Data on the Mineralizations and Hydrothermal Alteration Zones of the Allumiere-Tolfa Historic Mining District, Central Italy *Francesca Corrado (Student)* 55	Applying Machine Learning Methods to XRF Data to Quantify the Acid-Generating Potential of Waste and Other Intrinsic Rock Properties *Karim Halaseh* 64
2:45 pm - 3:00 pm	Discover the Un-Discovered: New Exploration Concept for the European Kupferschiefer Copper Belt from Rudna (Poland) to Richelsdorf (Germany) *Volker Spieth* 56	PROMT: Philippines Remediation of Mine Tailings *Gawen Jenkin* 65
3:00 pm - 3:15 pm	Mantle Source Enrichment in the Evolution of One of Europe's Largest Rare Element Provinces - the Gardar Province, South Greenland *Adrian Finch* 57	Microbial Impacts on the Circular Economy via Metal Mobility and Carbon Sequestration *Thomas Jones* 66
3:15 pm - 3:30 pm	The Iberian Pyrite Belt, a Premier Mining Jurisdiction *James Royall (Invited Speaker)* 58	Influence of Mineralogy, Texture, and Grain Size on Density and Magnetic Susceptibility Within the Kiruna Mining District, Sweden *Ervin Veress (Student)* 67
3:30 pm - 4:00 pm	Break	
4:00 pm - 4:15 pm	Linking Sn-Li-(W-Cu) Mineralization to Post-Collapse Caldera Evolution Through Cassiterite LA-ICP-MS U-Pb Geochronology *Dino Leopardi (Student)* 59	It's Never Too Early for Geometallurgy – Geometallurgy in Exploration *Melissa Gregory (Invited Speaker)* 68
4:15 pm - 4:30 pm	Integrated Geological, Geochronological and Geochemical Approach to Exploration in the Hod Gold District, Türkiye: Implications for the Origin of the Hod Maden Au-Cu Deposit *Nick Callan* 60	Understanding Your Deposit Better Using Big Data: A Case Study from Kepez North, Balikesir, Turkey *Sinem Oguz* 69
4:30 pm - 4:45 pm	New Investigations on the Geology and Mineralogy of the S'aliderru Bentonite Deposit (Sardinia, Italy) *Sebastiano Coticelli (Student)* 61	The Ikkari Gold Discovery, Finnish Lapland *Charlotte Seabrook (Keynote Speaker)* 70
4:45 pm - 5:00 pm	Q&A	The Ikkari Gold Discovery, Finnish Lapland, continued *Charlotte Seabrook (Keynote Speaker)* 70
5:00 pm - 7:00 pm	Exhibit and Poster Reception: **Monday and Tuesday Posters**	

DAY 4 - TUESDAY, AUGUST 29, 2023

	Oral Session 11 **Gold - Responsible Discovery and Mining in the 21st Century I**	**Oral Session 12** **Advances in Geoscience I**
8:30 am - 8:45 am	**Gold: Responsible Discovery and Mining in the 21st Century** *Mark Bristow (Keynote Speaker)* N/A	**Trace Element Geochemistry of Pyrite and Marcasite as a Vector Towards Mineralization in Irish-Type Zn-Pb Deposits** *Claire Geel* 75
8:45 am - 9:00 am	**Gold: Responsible Discovery and Mining in the 21st Century, continued** *Mark Bristow (Keynote Speaker)* N/A	**Genesis of BIF-Hosted Martite-Goethite Ores: Concept and Application** *Caroline Perring* 76
9:00 am - 9:15 am	**The Jibal Qutman Gold Project, Kingdom of Saudi Arabia. Deposit-Scale Structural Models in an Underexplored Terrane to Drive Discoveries** *Nadia St-Jean* 71	**Hydrothermal Alteration and Volcanic Stratigraphy at the Paleoproterozoic Kay Mine Volcanogenic Massive Sulfide Deposit, Black Canyon City, Arizona, USA** *Aaron Adsit (Student)* 77
9:15 am - 9:30 am	**The Valley Project, a New High-Grade Reduced Intrusion-Related System in the Tintina Gold Province, Yukon Territory** *Andrew Turner* 72	**Random Forest Classification of Epidote Mineral Chemistry from the Aitik Deposit, Northern Sweden** *Michael Baker* 78
9:30 am - 9:45 am	**The Importance of Mineral System Analysis as Inputs to Data-Driven Gold Targeting Projects** *Jean-Philippe Paiement* 73	**Graph Neural Network for Mineral Prospectivity Mapping – a Possible Solution to Maximise Utilization of Geological Relationships** *Felix Sihombing (Student)* 79
9:45 am - 10:00 am	**Porphyry-Like Au-Cu System in Paleoproterozoic Greenstone Belts** *Tramaine James (Student)* 74	**Characterizing Geological Uncertainty with Stochastic Implicit Modeling** *Luiz Gustavo Rasera* 80
10:00 am - 10:15 am	**Q&A**	**Q&A**
10:15 am - 11:00 am	Poster Viewing Session: **Monday and Tuesday Posters**	
	Oral Session 13 **Gold - Responsible Discovery and Mining in the 21st Century II**	**Oral Session 14** **Advances in Geoscience II**
11:00 am - 11:15 am	**Muruntau, Uzbekistan: A Giant, Pluton-Related Auriferous System** *Reimar Seltmann* 81	**The Fluids that Form Magnetite-Apatite Deposits** *Matthew Steele-MacInnis (Keynote Speaker)* 87
11:15 am - 11:30 am	**Hypozonal Orogenic Gold Mineralization Explained by Metamorphic Devolatilization of Komatiites** *Richen Zhong* 82	**The Fluids that Form Magnetite-Apatite Deposits, continued** *Matthew Steele-MacInnis (Keynote Speaker)* 87

DAY 4 - TUESDAY, AUGUST 29, 2023 (Continued)

11:30 am - 11:45 am	**Structural Modelling of the Au Mineral System Using 2D Seismic, Yamarna Greenstone Belt, Eastern Goldfields, Western Australia** *Armelle Kloppenburg* 83	**Fingerprinting Tools to Distinguish Magnetite Deposit Types** *Patrick Krolop* 88
11:45 am - 12:00 pm	**Using Edge Detection Techniques on Aeromagnetic Data to Delineate Structures Control Orogenic Gold Deposits in the Sefwi Granite-Greenstone Belt, Ghana** *Benjamin Boadi* 84	**Tracing the Sources of Rare Earth Elements in IOCGs** *Antony Burnham* 89
12:00 pm - 12:15 pm	**Characterization of Refractory Gold Behaviour in Witwatersrand Mine Tailings: Implications for Gold Recovery** *Steve Chingwaru (Student)* 85	**Mineralogy and Paragenesis of the Boundary Zone Clastic-Dominated Zn Deposit (Macmillan Pass, Yukon)** *Haruna Grema (Student)* 90
12:15 pm - 12:30 pm	**Application of U-Pb Detrital Zircon Geochronology to Delineate Regional Unconformities; Implications for Gold Mineralization in the Red Lake Mining Camp, Ontario, Canada** *Kate Rubingh (Invited Speaker)* 86	**The Paleoproterozoic Amitsoq Graphite Deposit, South Greenland** *Giacomo Fratelli (Student)* 91
12:30 pm - 1:05 pm *(Q&A ends 12:45 pm)*	**SEG's Role in a Global Society: Diversity, Equity and Inclusion** *Panel discussion* N/A	**Q&A**
12:45 pm - 1:45 pm	Lunch	

	Oral Session 15 **Gold - Responsible Discovery and Mining in the 21st Century III**	**Oral Session 16** **Advances in Geoscience III**
2:00 pm - 2:15 pm	**Geological Features Common for Bonanza-Forming Epithermal Gold Deposits** *Keiko Hattori* 92	**Carbonate U-Pb Geochronology: A Novel Tool for Directly Dating Fluid Flow Events in Mineral Deposits** *Catherine Mottram (Keynote Speaker)* 97
2:15 pm - 2:30 pm	**The Electrochemical Flocculation of Colloidal Gold by Semiconductive P-Type Pyrite: A Novel Ore Forming Mechanism for Gold** *Duncan McLeish (Student)* 93	**Carbonate U-Pb Geochronology: A Novel Tool for Directly Dating Fluid Flow Events in Mineral Deposits, continued** *Catherine Mottram (Keynote Speaker)* 97
2:30 pm - 2:45 pm	**LA-ICP-MS Analysis on Sulfide Minerals from the Higashi-Aogashima Knoll Caldera Hydrothermal Field to Understand the Host Phase of Gold** *Tatsuo Nozaki* 94	**New Approaches to Constraining Cu-Co Mineralisation in the Zambian Copperbelt Using Hydrothermal Rutile and Apatite** *Jamie Kelly (Student)* 98
2:45 pm - 3:00 pm	**Gokona: An Archean Porphyry-Epithermal Au Deposit** *Charles Routleff (Student)* 95	**Far Field Geochemistry and C-O Isotope Signatures of the Bingham Porphyry Cu-Au-Mo Deposit in Carbonate Wallrocks** *Michael Kirschbaum (Student)* 99
3:00 pm - 3:15 pm	**Sustainable Futures for Small-Scale Gold Mining Communities: Mapping the Socio-Economic, Environmental, and Geological Baseline for Peixoto De Azevedo, MT, Brazil** *Maria Mesquita* 96	**Mineral Petrography, Trace Element and Pb Isotope Geochemistry of the Black Butte Cu-Co District, Belt Basin, Montana, USA: Constraints on Ore Genesis** *Jeffrey Steadman* 100

DAY 4 - TUESDAY, AUGUST 29, 2023 (Continued)		
3:15 pm - 3:30 pm	**Q&A**	**Kyanite Alteration as an Indicator for Saline Fluids and Metal Transport; Case Studies from the Central African Copperbelt** *Anna Bidgood (Invited Speaker)* 101
3:30 pm - 4:00 pm	**Closing Ceremony and Concluding Remarks** *Bob Foster, SEG 2023 Conference Chair* *Steve Piercey, SEG 2024 President* *Anna Nguno, SEG 2024 Conference Chair* N/A	

Posters

Posters will be on display in two cohorts, with an "A" designation for Saturday and Sunday and "B" for Monday and Tuesday. Poster sessions are posted online.

The presenting author's name is underlined below.

Session A: Saturday and Sunday

Battery Metals - the Drivers of Future Energy Supply

A001 Potential Sources of Rare Earth Elements
102 (REEs) in Loch Borralan, Northwest Highlands, Scotland: Geochemical and Mineralogical Controls on REE Distribution in Carbonatite (*Student*)
Abdulkadir Mohamed Abdulkadir

A002 Mineralogy and Whole-Rock Geochemistry as
103 Exploration Tools for Lithium Pegmatites in Ghana: Examples from the Biriwa-Ewoyaa Pegmatite Fields (*Student*)
Elizabeth Agyekum, Kofi Adomako-Ansah

A003 Structural-Chemical Vectors to LCT-
104 pegmatites in the Varuträsk Area, Sweden
Joel Andersson, Edward P. Lynch, Martiya Sadeghi, Samuel Lundsten, Tobias E. Bauer

A004 Molybdenite from the Palaeoproterozoic REE-
105 line in Bergslagen, Sweden: Textural Relations, Trace Element Chemistry and Re-Os Age Data
Stefan S. Andersson, Erik Jonsson, Thomas Zack, Martiya Sadeghi

A005 The Iberian Pyrite Belt as a Strategic Source
106 of "Green" and Critical Minerals: The Co Enrichment in VMS (*Student*)
Gonzalo Ares, Carmen Conde, Fernando Tornos

A006 Reassessment of Rare-Metal Potential of the
107 Karagaily-Aktas Deposit (Kazakhstan)
Akmaral Baisalova, Yerlan Akbarov, Alla Dolgopolova, Reimar Seltmann

A007 Finding and Quantifying Li Bearing Mineralisa-
108 tions Using the GeoCore X10 Drill Core Scanner
Mikael Bergqvist, Alexander Hansson, Mikael Kördel, Karin Högdahl, Erik Jonsson, Magnus Leijd, Jaroslaw Majka, Pauline Jeanneret

A008 The Yin Deposit, A New REE Discovery in the
109 Gifford Creek Carbonatite Complex, Gascoyne Province, Western Australia
Luke Blais

A009 Investigating Aeration and Filtration as
110 Pre-Treatment Methods for Flowback and Produced Water Prior to Ion-Exchange for Lithium Extraction (*Student*)
Bennett Braun, Karthik R. Shivakumar, Daniel Alessi

A010 Battery Metal Exploration: Integrating Satellite,
111 Airborne, Drone and Field Data with Machine Learning Outputs in Cornwall
Mark Broadley

A011 Scientific Drilling at a REE-Mineralised
112 Carbonatite
Sam Broom-Fendley, Kathryn Goodenough, Michael Marks, John Spear

A012 Mineralogy and Composition of Vanadium-
113 Bearing Shales at the Van Property, Mackenzie Mountains, Northwest Territories
Stefanie M. Brueckner, Turner Green, Merilie A. Reynolds, Daniel Gregory

A013 The Mineralogy and Mineral Associations of
114 Platinum-Group Elements and Base Metal Sulfides in the Down-Dip Critical Zone at Sandsloot, Northern Bushveld Complex (*Student*)
Kate R. Canham, David Holwell, Iain McDonald, Hannah Hughes, Andy Lloyd, Lara Du Preez, Kofi Acheampong

A014 Stream Sediments as an Exploration Tool for
115 Spodumene Pegmatites
Lawrence C. Carter, Julian Menuge, Ben J. Williamson, Teimoor Nazari Dehkordi, Gavyn K. Rollinson

A015 Mapping Critical Minerals in Mineral Systems
116 with the Critical Minerals in Ores (CMiO) Database
George N. Case, Evgeniy Bastrakov, Garth Graham, Albert Hofstra, David Huston, Christopher Lawley, Vladimir Lisitsin, Bronwen Wang

A016 Cobalt Mineralization in the Central African
117 Copperbelt: Preliminary Insights from Petrography and LA-ICP-MS Analyses (*Student*)
Malena Cazorla Martínez, Lingli Zhou, Murray Hitzman

POSTERS

A017 REE Enriched Migmatites and Gneisses of
118 Son Valley Crystalline Area Sonbhadra, Uttar Pradesh: Can It be a Potential REE Deposit in India?
Sakthi S. Chinnasamy, Abhishek Verma

A018 Integration of Multiscale Hyperspectral Data,
119 Geochemistry, and Mineralogy for Assisting Mineral Exploration: The Brownfield Punta Corna Co-Ni Vein-Type Deposit (Italy) (*Student*)
Rita Chirico, Nicola Mondillo, Carsten Laukamp, Jens Najorka, Diego Di Martire

A019 Stratabound Cobalt Mineralization at Kinsevere,
N/A DRC (*Student*)
Robert S. Collar, Herve Ilongo, Takalani Mandiwana, A.E. Williams-Jones

A020 Melt-Fluid and Fluid-Fluid Immiscibility in
120 $Na2SO_4$-SiO_2-H_2O System and Its Implications for the Formation of Rare Earth Deposits
Hao Cui, Richen Zhong, Yuling Xie, Xiaolin Wang, Huan Chen

A021 Magma Mixing in the Huangshandong Ni-Cu
121 Sulfide-Bearing Mafic-Ultramafic Complex, NW China
Yu-Feng Deng, Sheng-Xuan Cao, Feng Yuan, Xie-Yan Song

A022 Textural and Mineralogical Characterization
122 of One Lithium Deposit from the Barroso-Alvão Aplite-Pegmatite Field: Preliminary Study (*Student*)
Filipa Dias, Alexandre Lima, Encarnación Roda-Robles, Tânia Martins

A023 Controls on the High-Grade Gold-Cobalt Min-
123 eralisation at Rajapalot, Finland (*Student*)
Lily Dickson, Taija Torvela, Robert C. Chapman

A024 The Shavazsay Deposit (Uzbekistan) - a Large
124 Caldera-Hosted Volcano-Sedimentary Lithium Deposit
Alla Dolgopolova, Reimar Seltmann, Rustam Mirkamalov, Farid Divaev, Rustam Khalmatov

A025 The Successful Application of Stream Sedi-
125 ment Geochemistry as an Exploration Tool for As-Bi-Co-Cu-Fe-Ni Mineralisation in Volcano-Sedimentary Settings (*Student*)
Adam Eskdale, Sean Johnson, Amy Gough

A026 LREEs, Zr and Th Enrichment of Autometaso-
126 matic Alkaline Dikes in Mamuju, West Sulawesi, Indonesia: Constraints from Petrology and Whole Rock Geochemistry (*Student*)
Fadlin Fadlin, Shaban Godang, Arifudin Idrus, Hartaja M. Wicaksono, Wildan N. Hamzah, Rinaldi Ihkram

A027 Magmatic Glass as a Source of Lithium: A Case
127 Study of the Cerro Galán Ignimbrite of Northwest Argentina
Conor Farrell, Alison Rust, Edward Bunker, Stephen Edwards, Naomi Thompson, Joseph Emmings

A028 Hydrothermal Alteration and the Formation of
128 an Iron-Oxide-Apatite (IOA)-Type Rare Earth Element (REE) Deposit at Kwyjibo, Québec, Canada (*Student*)
Gary Fung, A.E. Williams-Jones, David Martineau

A029 Geochemical and Petrographic Effects of Halo
129 Formation in Host Rocks of Spodumene Pegmatites in Southeast Ireland (*Student*)
Elena J. Geiger, Teimoor Nazari-Dehkordi, Julian Menuge

A030 Mineralogy and Critical Metal Deportment of
130 Fe-Ni-Co Laterite Ores from Sebuku Island, SE Kalimantan, Indonesia
Michele Giorno, Alkiviadis Kontonikas-Charos, Doreen Ebert, Joachim Krause, Robert Möckel, Dirk Sandmann, Mathias Burisch, Ernowo Ernowo, Arifudin Idrus, Yoseph Swamidharma, Yudhi Krisnanto, Max Frenzel

A031 New Lithium Search Spaces - a Mineral Sys-
131 tems Approach
Howard C. Golden, William Heins, Paul Jackson

A032 Lithium Pegmatites in Africa: Understanding
N/A Parageneses
Kathryn Goodenough, Richard Shaw, Anouk Borst, Paul Nex, Judith Kinnaird, Marieke van Lichtervelde

A033 Battery Metals in Diverse Geological Environ-
132 ments of the Lesser Caucasus, Armenia: An Overview for Future Exploration
Samvel Hovakimyan, Robert Moritz, Marianna Harutunyan, Karen Hambaryan, Kima Charatanyan

A034 A Fresh Look at the World-Class "Five-Element" Vein Deposit, Cobalt, ON, Canada: Constraints from Fluid Inclusion and Stable Isotope Systematics, and Mineral Paragenesis (*Student*)
133 *Joshua Jackman, Jacob Hanley, Daniel Kontak, Richard Stern, David Lewis*

A035 Green Transition-Critical Metal and Mineral Resources in the European Nordics
134 *Erik Jonsson, Tuomo Törmänen, Jakob K. Keiding, Terje Bjerkgård, Pasi Eilu*

A036 Project BATTRACE—Using Geo-Based Fingerprinting for Battery Raw Material Traceability
135 *Harri Kaikkonen, Quentin Dehaine, Mari Kivinen*

A037 Market Drivers and Barriers for Battery Raw Material Traceability
136 *Harri Kaikkonen, Mari Kivinen*

A038 Tellurium: Geometallurgy of a Critical Element at the Perseverance VMS Deposit, Quebec, Canada (*Student*)
137 *Filip Kasprowicz, Katharina Pfaff, Thomas Monecke, David Diekrup, Samuel Pierre, Mark Hannington*

A039 "Multi-Step Distillation" Anatexis as a Mechanism for Generating LCT Pegmatites (*Student*)
138 *Lot Koopmans, Tania Martins, Robert Linnen, Richard M. Palin, Catriona Breasley, David Silva, Nicholas J. Gardiner, Lee A. Groat, Laurence J. Robb*

A040 Exploration Model for REE Deposits Associated with Nepheline Syenite Occurrences in Tropical Areas: Case Study of the Eboundja Nepheline Syenite (Cameroon) (*Student*)
139 *Kenneth K. Kwanang, Louise Marie Ngo Bidjeck, Yannick Ngono Onana*

A041 Source to Sink Modelling of Salar-Hosted Lithium Brine Systems
140 *Christopher Lambert, Jean-Jacques Ferguson, Esme Whitehouse, Edward Bunker, Alexandra Scarlet, Jorge Gines*

A042 Eruption of Extremely Fractionated Magmas: Implications for Magmatic Versus Volatile Processes
N/A *David R. Lentz, Reimar Seltmann, Nadia Mohammadi*

A043 Critical Metal Occurrences in Historic Mining Waste in Sweden
141 *Alexander Lewerentz, Johan Camitz, Patrick Casey, Dick Claeson, Roger Hamberg, Erik Jonsson, Daniel Larsson, Stefan Persson, Gunnar Rauséus, Helge Reginiussen, Johan Söderhielm, Björn Wiberg*

A044 Petrogenesis and Ni-Cu Sulfide Mineralization of Early Permian Mafic-Ultramafic Complexes in the Beishan Area, Xinjiang (*Student*)
142 *Hao Lin, Dongyang Zhang*

A045 A Mineral Systems Approach to the Classification of Lithium Deposits
143 *Robert Linnen, Wilson Bonner, Kurt House*

A046 Using Scanning Electron Microscope-Based Automated Mineralogy to Better Understand Magnetic and Conductive Anomalies in Ni-Cu-Co-PGE Exploration Targets in W-Greenland
144 *Stefanie Lode, Björn H. Heincke, Nynke Keulen, Bo Møller Stensgaard*

A047 Geology of the Paleoproterozoic Järkvissle Li-Sn-Ta Pegmatite System, Central Sweden
145 *Edward P. Lynch, Joel B. Andersson, Martiya Sadeghi, Irma Bečelytė, Peter Hedin, Tobias E. Bauer*

A048 Emplacement Mechanisms of Rare Element Pegmatites, Separation Lake Greenstone Belt, Northwest Superior Province (*Student*)
146 *James A. Macdonald, Alfredo Camacho, Garnet J. Ching*

POSTERS

A049 Origin of Alkaline-Carbonatite Rocks and Related Mineral Deposits at the NW Margin of the West African Craton: Inferences from Field and Petro-Geochemistry (*Student*)
147 *Cheikh Elwali Malainine, Muhammad Ouabid, Otmane Raji, Jean-Louis Bodinier*

A050 Lithium in Manitoba: Production and Exploration Potential
148 *Tania Martins*

A051 The REDOX Setting of the Stratabound V Mineralization at the Van Property, Northwest Territories (*Student*)
149 *Chelsey C. Merrick*

A052 Timing and Linkage of the Mesoproterozoic REE-Y-Zr-Nb Mineralized Strange Lake and Flowers River Deposits to the Neoarchean Nekuashu-Aucoin Magmatic Event, Northern Canada
150 *Nadia Mohammadi, Anne-Aurélie Sappin, David Corrigan, Neil Rogers, Nicole Rayner*

A053 Battery Minerals in the Democratic Republic of Congo: Overcoming Challenges to Discover a New World-Class Lithium Deposit in the Kasese-Lulingu Pegmatitic Belts (KLP) (*Student*)
151 *Rub'son H. N'nahano, Huan Li, Mohammed A. Ibrahim*

A054 Application of Soil Geochemistry for Pegmatite Exploration: Spodumene Pegmatites in Ireland and Austria, and High-Purity Quartz-Bearing NYF Pegmatites in Norway
152 *Teimoor Nazari-Dehkordi, Julian Menuge, Lawrence Carter, Ben Williamson, Axel Müller*

A055 Implications of the Magma-Mush Concept for Lithium Enrichment Processes in LCT Pegmatites (*Student*)
153 *Louis R. G. Penfound-Marks, Ben J. Williamson, Julian F. Menuge, Lawrence C. Carter*

A056 The McDermitt Deposit (Oregon, US): Preliminary Results from a Li-Clay Volcano-Sedimentary System
154 *Francesco Putzolu, Robin Armstrong, Rebecca Ball, Paul Meyer, Richard Herrington*

A057 Relationship Between Granitic Intrusions and the Hydrothermal Veins in the Argemela Tin and Lithium Mine (*Student*)
155 *Ricardo Ribeiro, Rui Moura, Alexandre Lima, Charles Gumiaux*

A058 Structural Controls on the Origin of Regolith-Hosted Rare Earth Element Deposits at the Coastal Cordillera, Central-Southern Andes (35-37°S)
156 *Tomás Roquer, Paulina Vergara, Nicolás Bustos, Carlos Marquardt*

A059 Occurrence and Geological Features of LCT Pegmatite in Kalba-Narym Belt, East Kazakhstan (*Student*)
157 *Gleb Smirnov, Azam S. Dehnavi, Reimar Seltmann*

A060 Rare Earth Element Recovery from Apatite (*Student*)
158 *Julia Sordyl, Jarosław Majka, Karin Högdahl, Maciej Manecki*

A061 Geological Analysis of Graphite Deposits in Capim Grosso (BA), Brazil (*Student*)
159 *Thales F. Tandel*

A062 Assessing Vanadium Speciation in Hyper-Enriched Black Shales to Identify the Geochemical Controls Involved: Insights from X-Ray Absorption Near-Edge Spectroscopy (*Student*)
160 *Ozgur Can Tekin, Daniel D. Gregory, Anthony Chappaz, Danielle McGill, Daryll B. Concepcion, Merilie A. Reynolds, Chelsey Merrick, Daisy Zhang, Stefanie M. Brueckner*

A063 The Deeper Platreef at Sandsloot: New Insights Into Controls on Ni-Cu-PGE Mineralisation in the Bushveld Complex (*Student*)
161 *Erin Thompson, David Holwell, Iain McDonald, Marc Reichow, Thomas Blenkinsop, Lara Du Preez, Kofi Acheampong, Andy Lloyd*

A064 Multi-Scale Insights Into the High-Grade Mt Weld REE Deposit from 2D and 3D Active Seismic Survey Data
162 *Greg Turner, Tom Bell, Mohamed Megebry, Ganesh Bhat, Brad Knell, Nigel Miller*

A065 Controlling Factors in Regolith-Hosted Rare Earth Elements Deposits Formation at the Coastal Cordillera, Central-Southern Andes of Chile (*Student*)
163 *Paulina Vergara, Tomás Roquer, Nicolás Bustos, Carlos Marquardt*

A066 Cobalt Distribution and Enrichment in Skarn Iron Deposits: A Case Study of the Zhuchong Skarn Fe-Cu-Co Deposit, Eastern China
164 *Fangyue Wang, Xian Liang*

A067 Characteristics and Origin of the Li-Rich
165 Hydrothermal Altered Granite: A New Type of Clay Li Deposits?
Zhiqiang Wang, Feng Yuan

A068 Ewoyaa Lithium Project—Rediscovery to
166 Resource
Isaac I. Williams, Lennard Kolff van Oosterwijk, Abdul-Razak S. Ballah, Kathryn Goodenough, Richard Shaw, Marieke Van Lichtervelde

Copper - Red Metal for a Green Future

A070 Timing of Cu-Mo Mineralization in the Deep
167 Extension of the Giant Chuquicamata Deposit, Northern Chile (*Student*)
Claudio A. Araya, Antonio Arribas

A071 Geology, Mineralization, Geochemistry, Geo-
168 chronology, and Ore Potential of Cu Deposits, Central African Copperbelt: A Review
Kevin Igor Azeuda Ndonfack, Jilin Zhang, Zhiming Yang

A072 Tectonic Controls on Porphyry Copper For-
169 mation in the Northern Great Basin (Nevada-Utah-Idaho), Western USA (*Student*)
Hero Bain, Thomas Lamont, Dan Bevan, Frances Cooper, Tim Elliott, Adam Gorecki

A073 Alteration Products of Seafloor Massive Sul-
170 fides: A New Source of Copper? (*Student*)
Christian S. Bishop, Bramley J. Murton, Stephen Roberts, Anna Lichtschlag

A074 Application of Alteration and Geochemical
171 Vectors for Blind Porphyry Cu-Au-Ag-Mo Exploration: An Example from Camp Creek, Thorn District, British Columbia, Canada
Farhad Bouzari, Shaun Barker, Robert G. Lee, Cassady Harraden, Gary Thompson, Christina Anstey, Brock Riedell

A075 The Potential Porphyry Cu-Mo Deposit on the
172 Patacual Claim, El Limon-La Cuesta Locations, Municipality of Tecpan De Galeana, Guerrero State, Southern Mexico (*Student*)
Lizbeth G. Carrillo-Marrodan, Jesus Castro-Mora

A076 Magmatic-Hydrothermal Crystal Mush within
173 the Porphyry-Forming Environment—Spatial and Temporal Constraints
Lawrence C. Carter, Ben J. Williamson, Simon Tapster, Gavyn K. Rollinson

POSTERS

A077 The Orion Cu-Au-Ag-Co-Zn Discovery in the
174 West Kimberley Region, Western Australia – A New Mineralisation Style in a Greenfields Terrane
Nicholas Chapman

A078 The Role of Exhumation in Generating High-
175 Grade Copper Deposits: Insights from the Central Andes, Northern Chile
Frances J. Cooper, Byron A. Adams, Simon I. Dahlström, Todd A. Ehlers, Matthijs C. van Soest, Kip V. Hodges, Brian R. Jicha, Bradley S. Singer, Jaime Cortes Yañez

A079 Rapid Cu-Porphyry Indicator Mineral Char-
176 acterization by µXRF: Automated Detrital Indicator Mineral Analysis of British Columbian Cu-Porphyry Exploration Properties (*Student*)
Ben M. Eaton, Shaun L. Barker, Lindsey J. Heagy

A080 Petrology and Geochronology of Pre- to Post-
177 Ore Igneous Rocks from Asarel Porphyry Cu (±Au±Mo) Deposit, Central Sredna Gora Zone, Bulgaria
Stoyan Georgiev, Irena Peytcheva, Elitsa Stefanova, Atanas Hikov, Albrecht von Quadt, Marcel Guillong, Desislav Ivanov

A081 Competency-Controlled Copper Mineralization
178 at the Onganja Mining District, Namibia (*Student*)
Matthew P. Hales, Judith Kinnaird, Paul Nex

A082 Porphyry Cu Fertility of Late Mesozoic–
179 Cenozoic Magmatism Along the East Anatolian Magmatic Belt Based on Zircon Trace Element Compositions
Ali İmer, Rui Wang, Kaan Sayıt, Ersin Koralay

A083 Copper in the Guiana Shield
180 *Salomon B. Kroonenberg*

A084 Improving the Sustainability of Copper Mining
181 Through XRF Sensing, Machine Learning, and Bulk Ore Sorting
Hannah Lang, Karim Halaseh, Giovanni Pamparana

A085 Deciphering Fluid Origins in the Paleozoic Iron
182 Oxide-Cu-Au (IOCG) Deposits, NW China: Constraints from Noble Gases and Halogens
Pei Liang, Yuling Xie, Lingling Xiao

A086 Implementation of a Tectonic and Magmatic
183 GIS-Based Database for Porphyry Copper Deposits Across the Southwest USA (*Student*)
Roseanne Marshall, Thomas Lamont, Frances Cooper, Robert Loucks, Adam Gorecki

A087 Understanding Climate and Weather-Drivers
184 of Tailings Risk in Chile: A Complex Multi-Hazard Challenge
Linda Jakschies, Blake O'Sullivan, John Coll, Stephen Edwards, Katie McFall

A088 Contrasting Tectonomagmatic Conditions for
185 Porphyry Cu and Iron-Oxide-Apatite Deposits in Phanerozoic Arcs
Xuyang Meng, Jingwen Mao, Adam Simon, Chao Duan, Guiqing Xie, Huimin Su, Tong Hou, Ke Shi, Nian Chen

A089 Isotopic Microanalysis of Calcite in the Perm-
186 ian Kupferschiefer System (Saale Subbasin, Eastern Germany) (*Student*)
Mohammed S. Mohammedyasin, Joseph M. Magnall, Sarah A. Gleeson, Hans-Martin Schulz, Maria R. Scicchitano, Michael Wiedenbeck

POSTERS

A090 The Use of Phlogopite as an Indicator of Mag-
N/A matic Cu-Ni Mineralisation in the Curaçá Valley, Brazil (*Student*)
Ellen Oliver, David Holwell, Daryl Blanks, Dongyang Zhang, Fillipe Porto, Taíla Sant'ana

A091 Lithostratigraphy, Mineralogy, and Structural
187 Controls of the Late Ediacaran Tizi N'ousatour Cu-Ag Deposit, Northwestern Anti-Atlas, Morocco (*Student*)
Mehd Ousbih, Mohamed Charroud, Mourad Essalhi, Moha ikenne, Askkour Fatiha, Salih Amarir, Hassan Essaidi

A092 Application of Facies Analysis and Sequence
188 Stratigraphy in the Zambian Copperbelt and Its Implications for the Tectonostratigraphic Evolution Katangan Basin (*Student*)
Martin Purkiss, Bruce K. Levell, Michael C. Daly

A093 Integrated Re–Os, Ar/Ar, and U–Pb Geochro-
189 nology Directly Dates the Timing of Mineralization at the Mina Justa and Marcona Deposits, Peru (*Student*)
Maria Alejandra Rodriguez Mustafa, Adam Simon, Robert Holder, Holly Stein, Andrew Kylander-Clark, Brian Jicha, Daniel Blakemore, Edson Machado

A094 3D Mineralogical Analysis Using X-Ray Atten-
190 uation Computed Tomography (CT) and XRF to Determine Geoenvironmental Reactivity of Future Non-Ore Material (*Student*)
Enrique Saez Salgado, Anita Parbhakar-Fox, Nathan Fox, Regina Baumgartner

A095 Insights into the Origin of Sedimentary-Hosted
191 Copper Mineralization in Kazakhstan
Azam Soltani Dehnavi, Reimar Seltmann, Fereshteh Shabani-Sefiddashti

A096 Hydrothermal Apatite: A Distal Proxy for
192 Potassic Alteration in Porphyry-Cu Systems? (*Student*)
George Stonadge, Andrew Miles, Daniel Smith, Philip Bird

A097 Detrital Zircon Indicators of Metallogenic Mag-
193 ma Fertility: A Potential Tool for Exploration Targeting for Porphyry Copper (*Student*)
Eliana Toro Paz, Frances Cooper, Byron Adams, Thomas Lamont, Anthony Kemp, Robert Loucks, Catherine Belgarde

A098 Mineralogical Characterization of Potential Source, Aquifer and Trap Rocks in the Central African Copperbelt Using the TESCAN TIMA® Automated SEM Mapping System
194 *Veronica G. Trevisan, Jamie J. Wilkinson, Martin Purkiss, Mike Daly*

A099 Alteration Assemblages in the Southern Congolese Copperbelt. A Comparison of Background Stratigraphy with Cu-Vein Hosted Mineralisation and Salt Dissolution Alteration (*Student*)
195 *Helen Twigg, Murray W. Hitzman*

A100 Rhenium - Potential as Proxy for Mineralization: Insights from the Altar Porphyry Cu-(Au-Mo) Deposit (*Student*)
196 *Martin G. Vallejos, Laura Maydagán, David Selby, Kevin B. Heather, Mariano Poodts*

A101 Greenfield Porphyry Copper Prospects in Peru
197 *César E. Vidal*

A102 Mineralization Timing and Characteristics of the Heavy Sulfide Zone at the Grasberg Porphyry Cu-Au Deposit, Papua, Indonesia
198 *Kotaro Yonezu, Shunnosuke Omachi, Thomas Tindell, Adrian Boyce, Euis Yuningsih, Mega Rosana*

Transformational Science, Engineering, and Governance in Economic Geology

A104 A Trace Element and C-O Isotope Study of Sb-Au Vein-Hosting Carbonate-Talc Schists of the Murchison Greenstone Belt, South Africa
199 *Andrea Agangi, Axel Hofmann, Robert Bolhar*

A105 Ambient Noise Seismic Interferometry Applied to Tailings Dam Monitoring in Minas Gerais State, Brazil (*Student*)
200 *Amanda M. Bezerra*

A106 Data Capture from Geoscience Documents Using Transformers
201 *Amit Juneja, Henri Blondelle*

A107 3D Deep Electrical Resistivity Tomography Study of the Calamita Distal Fe-Skarn Deposit, Elba Island (Italy): New Perspectives for Mineral Exploration (*Student*)
202 *Damian Braize, Julien Sfalcin, Matteo Lupi, Kalin Kouzmanov, Andrea Dini, Gianfranco Morelli, Federico Fischanger*

A108 Petrography and Economical Potential of Banded Iron Formations of the Neoproterozoic-Age Buem Formation, Ghana (*Student*)
203 *Ernest Brakohiapa*

A109 A Shift to Natural Gas, Nuclear, and Ocean Mining Could Initiate an Adequate and Resource-Viable Green Transition
204 *Lawrence Cathles, Adam C. Simon*

A110 A Multi-Approach Hyperspectral Analysis for the Mineralogical Characterization of Zn-Cu-Pb Vanadate Ores in the Otavi Mountainland (Namibia) (*Student*)
205 *Francesca Corrado, Anna Sorrentino, Rita Chirico, Matteo Massironi, Sabrina Ferrari, Nicola Mondillo*

A111 How to Empower the Geologist's Expert Knowledge with Machine Learning in Drill Core Logging? (*Student*)
206 *Christian Günther, Filip Siman, Nils Jansson, Foteini Simistira-Liwicki, Marcus Liwicki, Tobias Hermansson, Paul McDonnell*

A112 Exploration and Utilization Prospects of Coastal Placer Deposits
207 *Min Liu*

A113 Determining the Lowest Quantity of Variables Required for Geometallurgical Machine Learning-Based Modelling: A Review (*Student*)
208 *Ethan E. Mack, Bjorn von der Heyden, Margreth Tadie, Tobias Louw*

A114 Deriving Plate Boundary Information from Deep-Time Palaeo-Reconstruction Models and Implications for Regional Mineral Exploration Targeting (*Student*)
209 *Hoang Anh Tu Nguyen, Dène Tarkyth, Bruce Eglington, Jean-Christophe Wrobel-Daveau, Graeme Nicoll, David Evans, Drew Heasman, Samuel Butler*

A115 Unconventional Mineral Resource Potential in a Low-Temperature Geothermal System (*Student*)
210 *Valeria S. Nogales Herrera, Sean A. Fulcher, Karin E. Olson Hoal*

A116 Differentiation Between Endoskarn and Exoskarn Using Portable X-Ray Fluorescence (pXRF) at Antakori Cu-Au-Ag Project, Northern Perú
211 *Luis Paredes Santa Cruz, David Portocarrero*

A117
212
TRIM4Post-Mining: Artificial Intelligence and Virtual Reality for Decision-Making in Remediation for Mining Companies
Luis Pizano, Jörg Bendorf, Andreas Knobloch, Diego Restrepo

A118
213
Quantitative Evidence of a Rift Origin for the Katangan Basin from Subsidence Analysis in the Zambian Copperbelt (*Student*)
Martin Purkiss, Michael C. Daly

A119
214
Maximizing the Value of Geoscience Databases: A Comprehensive Review of Advanced Data Processing Techniques
Mana Rahimi

A120
215
A Novel Approach for Delineation and Mapping of K-Rich Igneous Rocks and their K_2O Content to Support Unconventional Potash Exploration
Otmane Raji, Muhammad Ouabid, Cheikh Elwali Malainine, Youssef Ahechach, Aaron Herve Mbissik, Hicham El Messbahi, Jean-Louis Bodinier

A121
216
Automated Multi-Sensors Core Logging and Analysis for the Assessment of Sn-W Mineralization Pattern and Distribution - Hemerdon Deposit, Devon, UK
Reza Ramdhan, Gavin Hunt, James Shreeve, Briony Shreeve

A122
217
Prospects for Byproduct Critical Mineral Recovery from Mine Waste
Robert R. Seal, Nadine M. Piatak, Sarah Jane O. White, Sarah M. Hayes

A123
218
Environmentally Friendly Low Impact, Low Carbon Footprint, Low Power Electromagnetic Technique for Mineral Exploration Undercover
Gordon Stove

A124
219
Geometallurgy of a Complex Ore: Lappberget Zn-Pb-Ag-(Cu-Au) Deposit, Garpenberg Mine, Sweden
Glacialle Tiu, Christina Wanhainen, Nils Jansson, Yousef Ghorbani

A125
220
Can Gold Mining Provide Critical Metals as By-Products Using Novel Solutions? A Case Study at Björkdal Gold Mine, Sweden (*Student*)
Lara Tritton, Gawen Jenkin, Dan Smith, Helena Moosberg-Bustnes

POSTERS

A126
221
The Green Transition Resource and Supply Chain Model: A Recipe for Global Energy Equity
Edith Newton Wilson

Session B: Monday and Tuesday

Advances in Mineral Deposit Geoscience

B001
222
Geological Settings and Future Ore Potential of the Ordubad Region, Nakhchivan (Naxçıvan), Azerbaijan, Lesser Caucasus
Stephen J. Westhead, Robin N. Armstrong, Anar Valiyev, Rustam Abdullayev, Araz Orujov

B002
223
Trace Element Composition and Cathodoluminescence of Quartz in the Tsunheg Tungsten Deposit in Western Mongolia (*Student*)
Amitan Baatar, Jamsran Erdenebayar, Takeyuki Ogata, Takeyuki Ogata, Yondon Majigsuren

B003
224
An Insight into the Southern Extension of the Egyptian Golden Triangle and Its Potentiality for Mineral Occurrences Related to Shear Activity (*Student*)
Mohamed Badawi

B004
225
Using Chemistry of REE-Bearing Phosphates for ISCG Exploration (*Student*)
Travis D. Batch, Caroline Tiddy, Adrienne Brotodewo, David Giles

B005
226
Assessing the Petrogenesis and Magmatic Sulfide Prospectivity of the Southwestern Laurentia Large Igneous Province (SWLLIP) (*Student*)
Thomas J. Boes, Simon M. Jowitt

B006
227
Enhancing Exploration of Green Energy Transition Metals: Directly Dating Carbonate Vein-Hosted Cu-Au Mineralization (*Student*)
Sarah Bowie, Catherine Mottram, Dawn Kellett, James Darling

B007
228
The Santa Maria Pb-Zn-Cu-(Au, Ag) Deposit, Camaquã Basin, Brazil: Geology, Geochemistry and Geothermometry (*Student*)
Lucas F. Brito, Caetano Juliani

B008
229
Hydrothermal Alteration Zones Estimated by ASTER Satellite Image and ASD Terraspec on the Eastern Zone of Cerro de Pasco, Perú (*Student*)
Evelyn Caiza

B009 The Al:Ti Geochemical Classifier: Distinguishing Between Granites of Similar Characteristics Under Different Degrees of Alteration (*Student*)
230 *Andrés Canales, Diego Guido, Julietta Zamora*

B010 Study on Sedimentary Environments and Implications for Shale Reservoirs in the Permian Longtan Formation, Southeast Sichuan Basin
231 *Yana Chen*

B011 Identification of Ore-Finding Targets Using the Anomaly Components of Ore-Forming Element Group Extracted by SVD and PCA in the Giant Gejiu Sn-Cu Ore Field, Southwestern China
232 *Yongqing Chen, Aoyue Zheng*

B012 Carbonatite Melt as a Travel Agent for Magmatic Sulfide Liquid: Nature vs. Experiments (*Student*)
233 *Maria Cherdantseva, Michael Anenburg, Marco Fiorentini, John Mavrogenes*

B013 Investigating Metamorphosed VMS Mineralization from the Archean Yilgarn Craton, Western Australia: Preliminary Results from the King Zn Deposit (*Student*)
234 *Cendi Dana, Steven Hollis, Darryl Podmore, Megan James, Ian Butler, Florian Fusseis, Adrian Boyce*

B014 São Félix do Xingu's Hydrothermal Gold and Base Metals Mineralization: Using Aerogeophysical Data to Find High Favourable Prospective Areas (*Student*)
235 *Bruna de Freitas, Caetano Juliani*

B015 The Epithermal-Skarn Transition at the La Colorada Deposit, Chalchihuites District, Zacatecas, Mexico (*Student*)
236 *Joshua Ebner, Zhaoshan Chang, Pedro Francisco, Dante Juarez, Patricio Perez, Chris Emerson, Shiqiang Huang, Sidney Hemming, Richard Wendlandt*

B016 The Natural Enrichment of Phosphatic Sediments: Insights from the Moroccan High Atlas Phosphate Series
237 *Radouan El Bamiki, Michel Séranne, Fleurice Parat, Jérémie Aubineau, El Hassane Chellai, Mohamed Marzoqi, Jean-Louis Bodinier*

B017 Geochemical and Mineralogical Characterisation of the Dulaan Khar Uul Pb-Zn-Ag Deposit in Western Mongolia
238 *Jamsran Erdenebayar, Takeyuki Ogata, Batkhuu Baldorj*

B018 Interpreting Machine Learning Neural Network Patterns of an Orogenic Gold Deposit, Yilgarn Craton, Western Australia
239 *David M. First, Yush H. Rajcoomar, Daniel Mogilny, Ilia Sucholutsky, Farzi Yusufali*

B019 UV Fluorescence as an Exploration Tool in Carbonate Replacement Deposits (*Student*)
240 *Spyridon Fragkou, Andrew Steiner, Shaun Barker, Tim Baker, Janja Mihajlovic*

B020 Unravelling the Genesis of the La Colorada Skarn Deposit, Chalchihuites District, Zacatecas, Mexico (*Student*)
241 *Pedro Francisco, Zhaoshan Chang, Joshua Ebner, Dante Juarez, Patricio Perez, Chris Emerson, Shiqiang Huang*

B021 Lagoa Salgada VMS: Geochemistry Applied to Deposit Domaining, Geometallurgy and Exploration
242 *Sergio Gelcich, Vitor Arezes, Luis Lima, Joao Barros, Avelihno Pinheiro*

B022 On the Development of a Microstructural Signature During Alluvial Gold Transport
243 *Juan Gomez-Barreiro, Santos Barrios Sánchez, Juan Morales Sánchez-Migallón, Jose Manuel Compaña Prieto, Kelvin Dos Santos Alves, Vanessa Leonor Escobar Duche, Inés Puente Orench*

B023 244 Applications of Hyperspectral Analysis to VMS and Orogenic Gold Exploration in the Yilgarn Craton, Western Australia
Steven P. Hollis, Cendi D. Dana, Conal Rodgers, Jack Richardson, Darryl Podmore, Megan James, Allan Kneeshaw, Riquan Azri

B024 245 Tourmaline: A Fertility Assessment and Vectoring Tool for Magmatic-Hydrothermal Systems
Wei Hong, Francisco Testa, David R. Cooke, Nathan Fox, Lejun Zhang, Michael Baker, Evan Orovan, Ayesha Ahmed, Pete Hollings, Ivan Belousov, Anthony Harris

B025 246 Melt Inclusions and Chemistry of Detrital Titanomagnetites at Orokolo Bay, Papua New Guinea: Evidence for Genesis by Magmatic Immiscibility (*Student*)
Leeysmon Hulijeli, Yasushi Watanabe, Osamu Nishikawa, Carmela A. Tupaz

B026 247 Paragenesis of Alteration and Silver Mineralization of the Pomperia Veins Hosted in the Tacaza Formation, Southern Peru (*Student*)
Jhoan C. Humpiri Chavez, Milagros Apaza Aquino

B027 248 Stratiform Potash Mineral Systems Prospectivity Analysis in an Andean Setting
Christopher Lambert, Jean-Jacques Ferguson, Esme Whitehouse, Edward Bunker, Alexandra Scarlet, Jorge Gines

B028 249 3D Mineral Prospectivity Modeling for Targeting Concealed Mineralization Within Chating Area, Middle-Lower Yangtze River Metallogenic Belt, China
Xiaohui Li, Feng Yuan, Chen Xue, Yuheng Chen

B030 250 What Lies Beneath? Sub-Seafloor Drilling of the Inactive Ultramafic-Hosted Semenov-1 Seafloor Massive Sulfide Deposit, 13°30'N on the Mid-Atlantic Ridge
Andrew J. Martin, Katie A. McFall, Bramley J. Murton, John W. Jamieson, Sven Petersen, Isobel Yeo, Christopher J. MacLeod, Anna Lichtschlag, Simon Allerton, Christian Bishop, Acer Figueroa, Szu-Ying Lai, Paul Lusty

B031 251 Recognizing Intrusive Phases at the Red Chris Au-Cu Porphyry Deposit Using Integrated Chemical and Geological Drill Core Data Sets
Brian A. McNulty, Shaun Barker, Anthony Harris, Nicholas Fitzpatrick, Alan Wilson

POSTERS

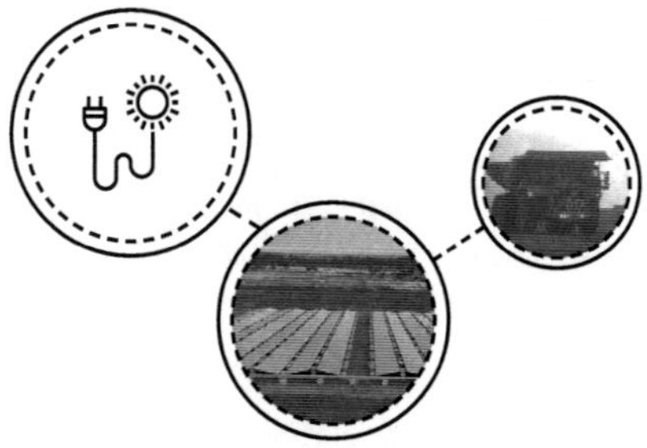

B032 252 Bridging the Gap: Rapid Geochemical and Mineralogical Characterization at the Hand-Specimen Scale
Andrew H. Menzies, Nigel M. Kelly, Alan R. Butcher, Falk Reinhardt, Roald Tagle

B033 253 Cerro Manomó: Carbonatites and BIFs with Rare Earth and Radioactive Mineral Anomalies (*Student*)
Pablo D. Michel, German Uzquiano

B034 254 Mineralogy and Petrology of the Tsagaantolgoi Fe-Pb-Zn Skarn Deposit in South Mongolia (*Student*)
Otgonbayar Munkhtsol, Jamsran Erdenebayar

B035 255 Presence of Vanadium in Epithermal Systems of Patagonia, Argentina (*Student*)
Morena L. Pagola, Sebastián M. Jovic, Diego S. Palma, Daniel Hughes, Diego Guido

B036 256 The West Desert Zinc-Copper-Silver-Indium Deposit, Juab County, Utah, USA
Mark I. Pfau

B037 257 Crosscutting Relations in Advanced Argillic Environment Revealed by Hyperspectral Core Imaging: Antakori Cu-Au-Ag Project Case Study (*Student*)
David G. Portocarrero

B038 A Stratigraphic Framework for the Zn-Pb Ba VMS Deposit, NW BC, Canada, Using New U-Pb Age Dates and Lithogeochemistry; Implications for Mineral Systems in the Golden Triangle (*Student*)
258 *Kyle Powers, Shaun Barker, Kate Rubingh, Brian McNulty*

B039 Role of Hydrothermal Solution in Base Metal Enrichment in Alwar Basin, Rajasthan, India
259 *Sajeev R, Dinesh Pandit, Mallickarjun Joshi, Sanjay Das*

B040 Petrographic Study of Zinc Ore and Subordinate Veins and Veinlets in the North Sucuri Region, Vazante Mine, Minas Gerais - Brazil
260 *Poliana Vidal Salgado, Carolina Penteado Natividade Moreto, Fernando Lucas dos Santos Peixoto de Villanova*

B041 Geological Modelling of the North Sucuri Region, Vazante Mine, Minas Gerais - Brazil
261 *Poliana Vidal Salgado, Carolina Penteado Natividade Moreto, Fernando Lucas dos Santos Peixoto de Villanova*

B042 Geochemistry of Apatite and Biotite in Tin-Mineralized Late Triassic Granites in Bangka Island, Indonesia (*Student*)
262 *Enrico Gilrandy W. Suharjo, Ryohei Takahashi, Andrea Agangi, Syafrizal Syafrizal, Pearlyn Manalo*

B043 Generation of Paired Cu-Au Mineral Systems Within a Suprasubduction Zone Setting in Eastern Proterozoic Australia
263 *Caroline Tiddy, David Giles*

POSTERS

B044 Anglesite-Bearing Polymetallic Hydrothermal-Shear Breccia in the Paja Formation, Eastern Cordillera of Colombia (*Student*)
264 *Nicolás Tique-Ladino, María Alejandra González-Cepeda, Ariel Cadena-Sánchez, Juan Carlos Molano-Mendoza*

B045 Element Fluxes During Alteration in High-Sulfidation Epithermal Systems: A Case Study from the Pueblo Viejo Au-Ag(-Cu) Deposit, Dominican Republic (*Student*)
265 *Ethan R. Tonks, Jamie J. Wilkinson, Robin N. Armstrong, Yannick Buret, Andrew T. Wurst*

B046 IOCG Systems: Are They Just a Problem of Ore Traps?
266 *Fernando Tornos, Tobias Bauer, John M. Hanchar*

B047 Advances in Knowledge: High-Grade Cu-Au-Ag Mineralization Hosted in Tourmaline Breccias at Soledad, Peru (*Student*)
267 *Victor J. Torres Pacheco, David R. Cooke, Lejun Zhang, Julie Hunt, Jerry Blackwell, David L. Kelley*

B048 Dating and Trace-Element Characterisation of Sn-W Skarn-, Greisen- and Vein-Type Occurrences from East-Greenland
268 *Nynke Keulen, Benjamin Heredia, Sebastian N. Malkki, Tonny B. Thomsen, Diogo R. Rosa, David Whitehead*

B049 The Genesis of the Zhijin REE-Phosphorite Deposit in Guizhou Province, China
269 *Yuling Xie, Yunwei Qu, Ziwei Li, Pei Liang*

B050 A New High-Precision SEM-LA-ICP-MS Sample Coordinate Integration Method
270 *Sebastian N. Malkki, Benjamin Heredia, Nynke Keulen, Tonny B. Thomsen*

Europe's Metals - Historic Districts - New Opportunities

B051 Shear-Related Sn-W Mineralisations of the Martinamor Extensional Gneiss Dome (Salamanca, Spain) (*Student*)
271 *Daniel Bermejo-López, Juan Gómez-Barreiro, Lorena Ortega Menor, Santos Barrios, Pedro Castiñeiras García, Elena Crespo Feo*

B052 272 Fluid Evolution of the Ostra Hydrothermal Cu-Zn-Pb-Barite Mineralisation (Eastern Carpathians, Romania) (*Student*)
Balázs B. Boglárka, Árpád Ádámcsik, István Márton, Szabolcs Orbán

B053 273 Re-Assessing the Controls on Magmatic-Hydrothermal W-Sn-Cu-As-Zn-Pb Mineralisation and Zonation Associated with the Cornubian Batholith, SW England (*Student*)
Kiara C. Brooksby, Robin K. Shail, Jens C. Andersen, Jamie J. Wilkinson

B054 274 Discovery, Geology and Timing of the Nautanen North Cu-Au-Ag-(Mo) Deposit, Norrbotten, Sweden
David Drejing-Carroll, Dave Coller, Murray Hitzman, Danny Hnatyshin, Sean Johnson, Franziska D. Wilke, Robert Creaser, Adrian Boyce

B055 275 The Newly-Discovered Ikkari Gold Deposit: Geology and Initial Insights into the Complex Controls on a Major Orogenic Gold Deposit in The Central Lapland Belt, Finland
Craig Hartshorne, Charlotte Seabrook, Kalle-Pekka Kotiaho

B056 276 Developing New Soil-Based Mineralogical and Geochemical Exploration Tools for Sn, W and Li in SW England (*Student*)
Edward Hill, Ben Williamson, Benedikt Steiner

B057 277 Geological Characteristics of Özyurt Gold Deposit: The Discovery of a Shear-Hosted Orogenic Gold Deposit in the Niğde Massif, Central Türkiye
Tuna Kaskatı, Hakan Yalçıntaş, Burak Gürel, Eray Yurtseven

B058 278 Exploration for REEs in Historic Iron Mining Districts - a Focus on Sweden
Aiden J. Lavelle

B059 279 Genetic Study of the Sphalerite from Rudabánya (*Student*)
Kristóf Lipp, János Földessy, Gabriella B. Kiss, Attila Kasó

B060 280 Petrology and Geochemistry of c. 1.8 Ga Granites Linked to Skarn W-F-Mo ± Cu Mineralization, Bergslagen District, Sweden
Edward P. Lynch, Stefan Bergman

B061 281 Geometallurgical Characterisation of Sediment-Hosted Cu-(Ag) Ore from the Spremberg-Graustein-Schleife Kupferschiefer Deposit, Germany (Student)
Zahra Nourizenouz, Bradley M. Guy, Robert Möckel, Doreen Ebert, Max Frenzel

B062 282 Prospecting for Rare Earth Elements in Plio-Pleistocene Sands of the Cenozoic Mondego Basin, Portugal
Michael A. Pointon, Colm S. Pierce, Michael J. Flowerdew

B063 283 Isotope Geochemistry of Native Sulfur, a Tracer of Microbial Activity in Secondary Ore Deposits (*Student*)
Jesús J. Rey-Samper, César Menor-Salván, Fernando Tornos

B064 284 New Insights to the Characterization of the Kiruna Deposit - Contributions from Deeper Exploration
Carlos Santana, Orsolya Katai, Liz Lobo

B065 285 Proximal Hyperspectral Sensing for the Characterization of High-Sulfidation Epithermal Deposits: The Allumiere Quarry Test Site (Central Italy) (*Student*)
Anna Sorrentino, Francesca Corrado, Rita Chirico, Matteo Massironi, Sabrina Ferrari, Stefano Castelli, Barbara Marchesini, Stefano Tavani, Eugenio Carminati, Nicola Mondillo

B066 Trace Element Contents in Sulfides from Porphyry Cu-Au (±Mo) Deposits, Central Srednogorie Zone, Bulgaria
286
Elitsa S. Stefanova, Irena Peytcheva, Stoyan Georgiev, Atanas Hikov, Sylvina Georgieva, Milen Kadiyski, Ventsislav Stoilov, Desislav Ivanov

B067 Piani Resinelli District: Preliminary Data on Ore Precipitation and Multiphase Diagenetic Evolution (*Student*)
287
Luca Summino, Michele Giorno, Luca Barale, Carlo Bertok, Max Frenzel, Marta Gasparrini, Luca Martire

B068 How Fluids Make or Break Critical Metal Deposits: The Ivigtut Cryolite Body, SW Greenland (*Student*)
288
Anna Szreter, Adrian A. Finch, Henrik Friis, Xiao-Ming Liu, Finlay M. Stuart

B069 Sulfide Mineralogy and Cobalt Deportment at the Viscaria Cu-Fe Deposit, Kiruna District, Northern Sweden (*Student*)
289
Koray Tasbicen, Marcello Imaña, William X. Chávez, Jr.

B070 The Origin of Stratabound Scheelite: The Future Resource for Europe?
290
Iñigo Borrajo, Fernando Tornos

Gold - Responsible Discovery and Mining in the 21st Century

B072 Morphotextural and Geochemical Analyses of Gold Particles Related to the Pinalejo Epithermal System (Salamanca, Spain) (*Student*)
291
Kelvin d. Alves, Juan G. Barreiro, Santos B. Sánchez, Lorena O. Menor, José M. Marchena

B073 Microchemical Signature of Placer Gold Nuggets from the Central Iberian Zone (Spain)
292
Santos Barrios Sánchez, Rafael Pablo Lozano Fernández, Juan Gómez Barreiro, Raul Merinero Palomares, Mercedes Suárez Barrios, Kelvin dos Santos Alves

B074 Geological Settings and Mineralisation of Vejnaly Gold Deposit, Lesser Caucasus, East Zangezur, Azerbaijan
293
Stephen J. Westhead, Anar Valiyev, Vusal Amirkhanov, Atabay Nushravanov, Araz Orujov, Aydin Bayramov

B075 Gold Metallogeny During Subduction Initiation and a Source Rock Control on Proto-Arc VMS Endowment
294
Thomas M. Belgrano, James A. Milton, Larryn W. Diamond, Robin C. Wolf, Damon A. Teagle

B076 Structural Control on Fluid Flow Pathways and Vein/Ore-shoot Emplacement at the Au-Ag LSE Pampa Paciencia Deposit, Antofagasta, Chile (*Student*)
295
Andrés Canales, Juan Pablo Radic, Diego Guido

B077 Geochronology of the Tocota Pluton and Au-Cu-Ag Mineralization of the San Francisco de Los Andes Quartz-Tourmaline Breccia, SW San Juan, Argentina (*Student*)
296
Brenda Gaia Carvalho Feitosa, Ticiano José Saraiva dos Santos, Robert Lima Muniz

B078 Mineralogical Characterization of Auriferous Occurrences in Stream Sediments in the Carangola Region, Minas Gerais, Brazil (*Student*)
297
Gabriela O. Carvalho, Alexandre N. Nogueira, Matheus R. Violante, Sofia G. d'Orsi, Gustavo Luiz C. Pires, Silvia R. Medeiros

B079 Nature and Source of Ore-Forming Fluid at Gadag Gold Field, Dharwar Craton, India
298
Sakthi S. Chinnasamy, Debasis Pal, Dnyaneshwar Gaonkar, Manish M. John

B080 Geological and Geochemical Characterization of the Mugomo Gold Prospect, Tete Province, Northern Mozambique (*Student*)
299
Euclesia P. Cossa

B081 Timing of the Orogenic Jouhineva Au-Cu-Co and Huhta Au Mineralization in the Pohjanmaa Belt, Finland (*Student*)
300
Andressa de Araujo Silva, Clifford G. Patten, Jochen Kolb, Simon Hector, Benjamin Walter, Aratz Beranoaguirre, Elisabeth Eiche

B082 Scheelite and Zircon U-Pb Dating Reveals Two Discrete Gold Mineralization Events at the Zaozigou Deposit, West Qinling Orogen, China (*Student*)
301
Shi-Guang Du, Jian-Wei Li

B083 Morphological and Compositional Analysis of Alluvial Gold: Los Ángeles Gold Placer (Cáceres, Spain) (*Student*)
302
Vanessa L. Escobar Duche, Santos Barrios Sánchez, Juan G. Barreiro, Kelvin dos Santos Alves

POSTERS

B084 Geological Setting and Mineralisation Styles of
303 the Zəfər (Zafar) Epithermal Deposit, the Lesser Caucasus, Azerbaijan; a Newly Discovered Deposit in Western Asia
Anar Valiyev, Stephen J. Westhead, Javid Ibrahimov, Aydin Bayramov, Sabuhi Mammadov, Ərzuman Rizvanoğlu

B085 Characteristics of Geology, Mineralization and
304 Alteration of Khan Altai Gold Deposit, Mongolia (*Student*)
Khaliunaa Iderbat, Khashgerel Bat-Erdene, Imants Kavalieris, Mandalbayar Ganbat

B086 Polymetallic Au-Rich Melts in Hydrothermal
305 Systems and Their Roles in Gold Enrichment
Wei Jian

B087 Geology of the Golddigger Intrusion-Related Gold
306 Deposit in British Columbia, Canada (*Student*)
Randall Karcher, Thomas Monecke, James Reynolds, Russell Ashton, Manuele Lazzarotto, Bill Chornobay

B088 Trace Element Compositions of Placer Gold
307 Samples: Implication for Gold Exploration in Northern Cameroon
Yanick Blaise Ketchaya, Taofa Zhou

B089 Using Borehole Data to Discriminate Magmatic
308 and Hydrothermal Processes for Orogenic Gold Exploration (*Student*)
Mosayeb Khademi, David Giles, Caroline Tiddy, Mark Lindsay, Andrej Bona

B090 Carlin-Type Mineralisation at Lookout Mountain
309 in the Eureka District, Nevada, USA
Simon Kocher, Zaid Nadhim, Thomas Monecke, Lesli Wood, Katharina Pfaff, Jens Najorka, Patrick Highsmith, Mark Osterberg, Steve Osterberg

B091 Fluid Evolution and Carbon-Gold Systematics
310 at the Fairview Mine, Barberton Greenstone Belt, South Africa (*Student*)
Rutger La Cock, Bjorn von der Heyden, Long Li, Yifan Du, Matthew Steele-MacInnis

B092 World-Class Gold Deposits and Emerging
311 Exploration Opportunities in the Loulo District, West Africa
Thomas Stapley, Paul Stenhouse, James Lambert-Smith, Andrew Allibone, David Lawrence, Dioumacor Senghor, Germain Crestin

B093 Chromium- and Vanadium-Bearing Phyllosili-
312 cates at the Kerr-Addison Gold Deposit, Virginiatown, ON: Implications for the Sources and Mobilities of Cr and V in Gold Deposits (*Student*)
Derek D.V. Leung, Andrew M. McDonald

B094 Ore-Forming Process of Basalt-Hosted Carlin-
313 Type Gold Deposits, Western Guizhou Province, China
Junhai Li, Jianzhong Liu, Pan Wu, Zhuojun Xie, Zepeng Wang, Shuangju Zhang, Chengfu Yang, Qinping Tan, Jingke Liu, Bingqiang Zhang

B095 Gold Source and Ore-Forming Process of the
314 Linglong Gold Deposit, Jiaodong Gold Province, China
Yayun Liang, Pingye Ma, Xuefei Liu, Lei Shu, Wenhao Xue, Chenxi Zhang, Yao Ma, Chonghui Shen

B096 Structural Controls and Timing of Orogenic Gold
315 Mineralisation in the Baie Verte Peninsula of Newfoundland, Canada (*Student*)
Georgian Manuc, Taija Torvela, Rob Chapman, Nick Roberts, David Copeland, Simon Tapster

B097 Genesis of Gold Placers of the Cordón Baqueda-
316 no, Tierra del Fuego, Austral Andes (54°S)
Carlos Marquardt, Tomás Roquer, Felipe del Valle, José Joaquínara Jara

B098 Compositional Characterization of Placer and
317 Lode Gold Grains from Eastern Interior Alaska, USA
Erin E. Marsh, Robert J. Chapman, Renee Pillers, Heather Lowers, Ryan D. Taylor

B099 Epithermal and Porphyry-Style Gold-Copper
318 Mineralization at Arthur's Seat Project, Jamaica
William Miguel, Stephen Hughes

B100 Uncommon La Virgen High-Sulfidation Epither-
319 mal Deposit, Northern Peru (*Student*)
Raul Montesinos, Zhaoshan Chang, Juan Carlos Castelli

B101 Gold Targeting at the Kalimva Deposit, Kibali
320 Gold District: Hydrothermal Alteration Vectors Interpreted from Multi-Approach Analyses (*Student*)
Yann Mpaka Waku, Bjorn P. von der Heyden, Etienne Mwandale, Gary Hurst, David Lawrence, Trézène Bampata, Fillsmith Luzolo, Prince Cimanuka

B102 High-Sulfidation Epithermal Gold and Porphyry
N/A Copper (Gold) Systems in the Makran Volcanic Arc, Souteast Iran (*Student*)
Pooya S. Naeini, Saeed Alirezai, Mohamad Nabi Noori

B103 Pressure-Temperature-Time Constraints on
321 Synvolcanic Intrusive Activity in the Gold-Rich Doyon-Bousquet-LaRonde Mining District, Abitibi Greenstone Belt, Québec (*Student*)
Kevin Neyedley, Jacob J. Hanley, Patrick Mercier-Langevin, James L. Crowley, Zoltan Zajacz

B104 Structural Controls of Gold Mineralization in
322 Southwest Ghana: The Iduapriem Gold Mine (Pit 5) (*Student*)
Newman T. Odonkor

B105 The Formation of Auriferous Jasperoid Under
323 the Control of Multi-Stage Tectonism at Tavsan, Kutahya, Western Turkey (*Student*)
Furkan Oguz, Burak Mert, Berkin Ugurlu, Erdin Bozkurt, Ahmet Kerim Sener

B106 The Relationship Between the Structural Ori-
324 entation and Copper-Gold Mineralisation in the Alta Floresta Mineral Province, Brazil (*Student*)
Renata A. Paes, Maria J. Mesquita, Thomas Blenkinsop, Guilherme M. Primo

B107 Temporal and Genetic Relationships Between
325 Gold Mineralization, Magmatism, and Deformation: The Neoarchean Windfall Gold Deposit, Northern Abitibi, Canada
Jean-Luc Pilote, Sébastien Castonguay, Patrick Mercier-Langevin, Benoît Dubé, Brandon Choquette, Édouard Côté-Lavoie, Sandra L. Kamo, Natasha Wodicka, Danny Hnatyshin, Robert A. Creaser

B108 A Comparison of Analytical Methods Testing
326 Soils for Gold Exploration in Pembrokeshire, West Wales (*Student*)
Maxwell L. Porter, Iain McDonald, Tristan English

B109 Evaluation of the Potential Intrusion-Related
327 Gold System (IRGS) of the Urcos-Sicuani-Ayaviri Structural Corridor System in Santiago de Pupuja, Puno, Peru (*Student*)
Mirian Puma

POSTERS

B110 Ore Mineralogy and U-Pb and Re-Os Geochro-
328 nology in the Ikkari Gold Camp, Central Lapland Belt, Northern Finland
Jukka-Pekka Ranta, Wäinö Wacklin, Charlotte Seabrook, Holly Stein, Ramiro Lenarduzzi, Matti Kurhila

B111 Recent Advances in Hard Rock Seismic Tech-
329 nology Applications Unlock the Deep
Leigh Rigg, Greg Turner, Nigel Miller

B112 Gold Exploration Through Integrating Machine
330 Learning, Remote Sensing, Aeromagnetic, and Spectrometric Gamma-Ray Data. A Case Study from the Egyptian Eastern Desert (*Student*)
Ali Shebl, Árpád Csámer

B113 Disseminated Orogenic Gold Mineralization in
331 High-Grade Metamorphic Turbidites of the Central Zone of the Damara Belt, Namibia: Controls and Characteristics (*Student*)
Josia Shilunga, Alexander Kisters

B114 Geochemistry of Felsic Rocks in the Yanıklı
332 Prospect, Artvin District, Eastern Pontides, Turkey: Fertility and Implications for Mineral Exploration (*Student*)
Şafak U. Sönmez, Robert Moritz, Serdar Keskin, François Turlin, Alexey Ulianov, Ümit Aydın

B115 333 Geology and Gold Mineralization of the La Debo Permit, Southwestern Côte D'ivoire, West Africa
G. Doma Sorho, Sarah F. Pein, Vagne Kouakou, Sihle Gqoboka, Geoffrey Howarth, Robert Kiefer, Lynnette Greyling

B116 334 Measure Twice - Rochford-Homestake Gold, South Dakota, USA
Holly J. Stein, Zeynep Baran, Gang Yang

B117 335 Gold Fertility of the Inner Lesser Himalayas: Insights from Textural and Mineralogical Investigation of Pyritiferous Mandhali Black Shales, Uttarakhand, India (*Student*)
Abdul Latheef T P, Rajarshi Chakravarti

B118 336 Operating on the Frontier of Discovery: Utilisation of a Multi-Disciplinary Exploration Approach at Hizarliyayla, Northeast Turkey
Berkin Ugurlu, Ahmet Kerim Sener, Izak J. Van Coller

POSTERS

B119 337 Shear-Zone-Hosted Gold Exploration, British Columbia: Unmanned Aerial Vehicle Photogrammetry Mapping and EasyMineXR Application to Inaccessible Steep Terrains
Ronald Uken, Sheila Ulansky, Emre Onsel, Venessa Bennett, Thomas Monecke, Randell Karcher, T. James Reynolds, Manuele Lazzarotto, Russell Ashton, Bill Chornobay

B120 338 Lithological-Structural Setting Mineralisation Styles of the Gilar Epithermal Deposit in the Lök-Qarabağ Volcanic Arc, the Lesser Caucasus, Azerbaijan
Anar Valiyev, Stephen J. Westhead, Aydin Bayramov, Sabuhi Mammadov, Javid Ibrahimov, Shakir Gadimov, Samir Mursalov

B121 339 The Singular Polymetallic Occurence in Zona Basal, Quadrilátero Ferrífero, Brazil: Mineralogical and Geochemical Characterization and Genetic Implications (*Student*)
Felipe Emerson André Alves, Atlas Vasconcelos Correa Neto, Mariana Brando Soares, Reiner Neumann, Gabriel Machado da Silva, Gabriela Pereira da Silva, Ana Clara Varca Pinheiro, Piero Azevedo Berquó de Sampaio, Vitor Diniz Silveira

B122 340 A Study of the Gold-to-Sulphide Ratio in the Cononish Au-Ag Deposit, Tyndrum, Scotland (*Student*)
Amy L. Walsh

B123 341 Calcatreu Project 3D Vein Modelling, Río Negro Province, Argentina
Matías S. Wernert, Darío Vera, Marcelo Paolini, Sebastián M. Jovic, Diego M. Guido

B124 342 Carlin-Type Gold Deposits in Youjiang Basin, Southwest China: Key Characteristics and Genetic Model
Zhuojun Xie, Yong Xia, Qinping Tan, Jianzhong Liu

B125 343 Structural Controls on Carlin-Type Gold Deposits at Huijiabao Trend, Southwestern Guizhou Province, China
Chengfu Yang, Jianzhong Liu, Zepeng Wang, Faen Chen, Wenxing Tai

B126 344 Ore Mineral Textures of the Midas Low-Sulfidation Epithermal Deposit: Implications for Ore-Forming Processes (*Student*)
Lauren R. Zeeck, Thomas Monecke, T. J. Reynolds, Katharina Pfaff, Nigel M. Kelly

B127 345 Identification of Ore-Finding Targets Using the Anomaly Components of Ore-Forming Element Associations Extracted by BEMD and PCA in the Jiaodong Gold Clusters, Eastern China
Aoyue Zheng, *Yongqing Chen*

B128 346 Application of BEMD in Extracting Gravity Anomaly Components Showing Deep Ore-Forming Geo-Dynamic Background of Jiaodong Giant Gold Cluster Region, Eastern China
Aoyue Zheng, *Yongqing Chen*

B129 347 Melt Inclusion Constraints on the Initial Metal Budget of Suspected Causative Magmas in the Au-Rich Doyon-Bousquet-LaRonde Mining District, Abitibi Greenstone Belt, Québec (*Student*)
Kevin Neyedley, *Jacob J. Hanley*, *Patrick Mercier-Langevin*, *Zoltan Zajacz*, *Alexandra Tsay*

SEG 2023 Conference: Resourcing the Green Transition

Supply Chain Issues: Importing Metals and Minerals into Europe/UK in the Next 15 Years

Rebecca Major
Herbert Smith Freehills LLP, New York, NY, USA

The European minerals and metals industry is faced with the trichotomy of climate change, concerns about security of supply, and enhanced requirements for responsible mining. This is a business and strategy question. If your mine/product is not compliant, then you won't be able to sell your product to the whole world. But legislation such as the Critical Minerals Act also brings a number of opportunities to European miners: subsidies, speeding up permitting, European preference, gigafactories, recycling, etc. We will look at the way that the law is shaping metals and mineral business opportunities and risks.

Managing and Benefiting from ESG Challenges in the Development of Critical Minerals in the UK

Kate Harcourt
Independent Environmental and Social Adviser, United Kingdom

The term “ESG” is now widely used but not always clearly understood. A closer examination of the environmental, social, and governance aspects in the context of developing a critical mineral project in the UK shows that there are many challenges but also many opportunities to implement a 21st-century project. The talk will examine the issues of resource use efficiency, the human factors, and the complexity of the governance issues—from legislation, planning and permitting, and lender and investor frameworks to ESG reporting—and suggest some of the approaches to address these.

SEG 2023 Conference: Resourcing the Green Transition

The Green Transition Squeezes as it Embraces

John Elkington
Founder & Chief Pollinator at Volans, Strand, United Kingdom

Expect a 3-D future for mining and minerals. Despite pushback against ESG investment strategies, demonstrating that vested interests feel increasingly squeezed, Decarbonization will shape our economies for decades to come. Already eyes are turning down to the seabed and upwards to space in search of the necessary materials. But the second D, Deglobalization, is very likely to force exploration and production closer to home to ensure security of supply, simultaneously triggering greater interest in circular supply chains and new forms of NIMBYism. The third D, Demographics, will fan the flames, with younger people increasingly feeling betrayed by older generations as the planet comes to the boil - to quote UN Secretary-General António Guterres. The resulting market dynamics will test global and corporate forms of governance to the limit.

Copper-Hosting Basins at the Edge of Cratons – the Yeneena Basin in Western Australia as a Case Study

Weronika Gorczyk[1], Fariba Kohanpour[2, 1], Ian Tyler[1, 2]
1. University of Western Australia, Perth, WA, Australia, 2. CSIRO, Perth, WA, Australia

Basins, orogens, and associated magmatic bodies at the edges of Archean cratons often host significant deposits of strategic (including green) metals (such as Cu, Zn, Pb, Ag, Au, and REEs). Considering that sediment-hosted stratiform copper deposits (SSC) represent one of the most important source of copper in the world, besides porphyry copper deposits, and account approximately for 20-25% of the global production and reserves, understanding the development of ancient basins and their metal potential is of utmost importance. The Yeneena basin—a Neoproterozoic basin deposited at the margin of the Pilbara craton in Western Australia—is a significant example of a mineral-rich (mainly Cu, Au, and U) basin. Until now only a vague understanding of the deposition of metals and processes leading to their enrichment is known, largely due to the scarcity of geological data. In order to obtain a better understanding of the basin, a holistic multi-disciplinary system workflow approach was applied. Novel analytical and numerical techniques were utilized with drill core information and geophysical, geochronological, and spectral data sets to unveil the development history and metal potential of the highly prospective Yeneena basin. With this holistic approach, we were able to link lithospheric-scale architecture and associated mineral systems with the development of basins and sub-basins as well as reactivation of major tectonic boundaries during inversion of the basin.

Numerical Modelling of Basin-Scale Fluid Flow and Cu Transport in the Katangan Basin, Central African Copperbelt

Meissam L. Bahlali, Pablo Salinas, Julia Woitischek, Carl Jacquemyn, Matthew D. Jackson
Imperial College London, London, United Kingdom

Sediment-hosted copper deposits host 23% of the world's copper. The mineral systems approach has been applied to characterise the formation of these deposits, sequentially considering source, transport, trap, and deposition. Here we focus on basin-scale groundwater flow as a mechanism to transport copper from source to trap. Numerical experiments are conducted to investigate controls on copper transport in the Katangan Basin, Central African Copperbelt, using the open-source IC-FERST code, coupling fluid flow, heat and solute transport and employing dynamic mesh optimisation (DMO) to reduce computational cost. DMO allows the mesh to refine or coarsen as necessary during a simulation to capture key aspects of flow. The lower computational cost compared to conventional numerical methods allows for efficient assessment of groundwater flow and metal transport scenarios associated with the mineral systems approach, with application here to 3D models of the basin at various stages of basin evolution.

Results show that density gradients induced by gradients in salinity and temperature play a major role in the initiation of convective groundwater flow. Highly saline, dense brines are created during deposition of salt or by dissolution of salt deposits and form downwards-propagating plumes with complex geometry controlled by the interaction of flow instabilities and geologic heterogeneity. Permeable faults and fractures in basement rocks allow groundwater to percolate through the basement and potentially mobilise copper from extra-basinal source rocks. Otherwise, flow is restricted to the permeable basin-fill deposits, and potential sources are restricted to intra-basinal sedimentary deposits and/or intrusive igneous rocks. The combination of salinity and temperature gradients drives upwelling plumes of groundwater, which can transport copper upwards from deeper, intra- or extra-basinal source rocks, into shallower, organic-rich sedimentary rocks where mineralisation occurs. Development of 3D convection cells may explain why mineralisation is often localised, with deposits potentially corresponding to metal-enriched upwelling plumes.

SEG 2023 Conference: Resourcing the Green Transition

Syntectonic Copper Mineralisation in a Back-Thrust Duplex at Pangeni, NW Zambia

Michael K. Watkeys[1], John Wilton[2], Branko Corner[3], Peter Hildebrand[3], Matthew Hodge[4], Robert Sparrow[5]
1. University of KwaZulu-Natal, Durban, South Africa, 2. BeMetals, Vancouver, BC, Canada, 3. Copper Cross Zambia, Kitwe, Zambia, 4. Remote Exploration Services, Cape Town, South Africa, 5. Sparrow Geoscience Ltd, Shepley, United Kingdom

The Pangeni copper project in the Central African Copperbelt of northwest Zambia is almost entirely concealed beneath Cenozoic sedimentary cover. Nevertheless, by integrating information available from sparse outcrops in drainages over a project-wide area of ~700 km^2 with geophysical data interpretation and drilling, it has been possible to deduce the fundamental controls on copper mineralization, specifically in the Central prospect area.

Structural analysis of data from the limited outcrops reveals that the main deformation involved subhorizontal, NW-SE compression, which formed shallowly to moderately plunging fold axes with axial planes dipping shallowly to moderately northwest and southeast. These structures are also observed in oriented drill core, showing shear zones in which fold asymmetry and kinematic indicators have a dominantly top-to-SE sense of movement. Interpretation of the geophysics data, in particular the first vertical derivative of the reduced-to-pole magnetic survey, shows a pattern indicative of duplex thrust structures involving late Proterozoic Katangan units and their basement.

The copper mineralization in the project area is clearly structurally controlled. At the Central prospect, it generally occurs as chalcopyrite- and subordinate bornite-bearing veins, which are both deformed with and by the shear fabric as well as cutting it. The former two are considered to have formed by fluids migrating along active thrust zones, whereas the latter represent mineralization postdating main thrust movement. These later-stage veins occur as an orthogonal conjugate set developed during oblate strain imposed by vertical loading.

This evidence indicates that the copper mineralization at the Central prospect occurred during development of a top-to-SE thrust duplex, which is a back-thrust in terms of the overall tectonic transport direction of the Upper Proterozoic to Lower Cambrian Lufilian orogeny.

Experimental Insights Into the Mobility and Extraction of Metals in Sedimentary Basin-Hosted Copper Systems

Julia Woitischek[1], Jamie J. Wilkinson[2, 1], Emma Humphreys-Williams[2], Matthew D. Jackson[1]
1. Imperial College London, London, United Kingdom, 2. Natural Hisotry Museum, London, United Kingdom

Understanding the extraction and transport of metals in sedimentary basins is important because basin-hosted copper deposits host more than 23% of the world's discovered copper (Brown, 2007) as well as other metals such as vanadium and cobalt.

Red bed sandstones have been identified as potential metal sources for sedimentary basin-hosted copper deposits because they are a characteristic facies in the mineral system and their iron oxide coatings can adsorb metals. The dissolution of these coatings could lead to the formation of an ore fluid (Parnell et al., 2021). However, previous experiments to understand the extraction and mobility of copper and other trace elements have not considered flow effects, elevated temperatures, or brine chemistry variations.

Hence, we report new experiments to quantify the release of metals from red bed sandstones while in contact with aqua regia and a variety of natural brines. Experiments with aqua regia are used to determine the total leachable metal content of the rocks, whereas brine experiments represent conditions more relevant to metal mobilisation in sedimentary basins. Leaching under aqua regia revealed a significant variability of trace element concentration from the red bed sandstones, with ore-bearing basins displaying higher trace metal concentrations. We found the highest concentrations of copper in samples from the Neuquén Basin (Argentina) and Katangan Basin (Zambia). Preliminary results from the brine-leaching experiments show that sandstones have a highly variable ability to exchange protons and therefore buffer pH.

References:

Brown, A.C., 2007. World-class sediment-hosted stratiform copper deposits: Characteristics, genetic concepts and metallotects. Australian Journal of Earth Sciences 44:3, 317-328.

Parnell, J., Wang, X., Raab, A., Feldmann, J., Brolly, C., Michie, R., Amstrong, J., 2021. Metal Flux from Dissolution of Iron Oxide Grain Coatings in Sandstones. Geofluids, 5513490.

Critical Battery Mineral Resources in the United States: Past, Present and Potential

Jane M. Hammarstrom[1], Laurel Woodruff[1], Graham Lederer[1], Warren Day[2]
1. USGS, Reston, VA, USA, 2. USGS, Denver, CO, USA

Critical mineral commodities essential for current electric vehicle battery technologies (graphite, cobalt, nickel, lithium, manganese, vanadium) were mined in the U.S. intermittently up to the mid-20th century, yet by the early 2000s, except for one lithium deposit, domestic production had ended. No natural graphite has been mined in the U.S. since the 1950s, and manganese mining ceased in 1970. Vanadium production from Colorado Plateau uraniferous sandstones last occurred in 2020. More recently, however, two brine projects produced lithium in 2022, and nickel and cobalt are currently produced at the Eagle mine in Michigan and from historical mine tailings. In addition to about 100 recently active exploration projects for these commodities, there are more than 100 sites in the U.S. with reported past production or identified resources. To guide new data acquisition for domestic critical mineral resources, the USGS Earth Mapping Resources Initiative (Earth MRI) used a mineral system and deposit type approach to identify more than 800 broad focus areas across the country. Of those, 66 focus areas are known to host critical battery mineral commodities. Areas most likely to provide near-term domestic resources for cobalt and nickel include mafic magmatic systems such as the Duluth Complex, the Tamarack deposit, and the Stillwater Complex; IOCG systems in the Idaho cobalt belt; and potentially, low-grade nickel-cobalt laterite deposits in California and Oregon. Projects are actively evaluating lithium in brines and clays in the Great Basin and Li-Cs-Ta-type pegmatites in the eastern U.S.; graphite in Alaska, Alabama, and Montana; manganese deposits in Maine, Minnesota, and Arizona; and vanadium from black shales and phosphorites in marine chemocline systems and sandstone-hosted uranium deposits. Acquisition of new geologic, geochemical, and geophysical surveys of key areas through Earth MRI provides data required to re-evaluate areas explored in the past and identify new exploration targets.

The Geological Setting of Vanadium Mineralisation in the Neuquén Basin of Central Argentina: An Opportunity for the Clean Energy and Energy Storage Sectors in South America

Martin E. Jackson
NewEra Metal Resources Ltd, CEREDIGION, United Kingdom

NewEra Metal Resources Ltd. is Argentina's leading vanadium exploration company, with a strategic project portfolio including former mine sites in a large unexplored vanadium district. Situated in the Neuquén Basin of central Argentina, this vanadium district contains >100 bitumen vein-hosted vanadium occurrences in a region exceeding 500 km in length and 100 km in width, parallel to the Andean trend.

The solid bitumen veins are located within the Malargüe fold-and-thrust belt (FTB) on the eastern side of the Andes in the Neuquén Basin. Dark, organic marine shales of the Late Jurassic- to Early Cretaceous-age Vaca Muerta Formation are the most important hydrocarbon source rocks in the Neuquén Basin. Bitumen is genetically related to oil, and the vanadium-rich veins are hosted by these shales.

In "heavy oils," such as those of South America, asphaltenes contain approximately 40-90% of total vanadium and 25-75% of total nickel. The summed concentration of vanadium and nickel in heavy oil asphaltenes can reach 1 wt.%. The relationship of bitumen to oil explains the high concentration of vanadium in bitumen of the Neuquén Basin.

Research and theoretical modelling of bitumen chemical and physical data show the rock to be very amenable to gasification. A processing flowsheet has been identified whereby synthesis gas ("syngas") and a metal-rich ash is produced. Vanadium (and nickel) is recovered from the ash. Clean, low-carbon hydrogen can be separated from the syngas and is then available for hydrogen fuel cells to produce electricity or to be used as hydrogen in different applications, including the energy, transport, and chemical industries.

While the traditional use of vanadium to strengthen steel continues to lead market demand, it is the metal's use in battery manufacture, namely vanadium redox flow batteries (VRFBs) for energy storage, that is expected to have the biggest impact on future demand.

Canadian Perspective on REE Deposits Hosted by Carbonatite and Alkaline Complexes and Its Implication for Carbonatite and Alkaline Magmatism in Europe

Anne-Aurélie Sappin[1], L. P. Bédard[2], Marc Constantin[3]
1. Geological Survey of Canada, Québec City, QC, Canada, 2. Université du Québec à Chicoutimi, Saguenay, QC, Canada, 3. Université Laval, Québec City, QC, Canada

In 2021, Canada established a list of 31 critical minerals considered essential to the transition of our country and its allies to a digital- and low-carbon economy. Among these minerals, rare earth elements (REEs), which are key components in high-technology industries, have been designated as highly critical because of the risk of supply chain disruptions. Carbonatite and alkaline complexes are the main hosts of REE deposits, and most of the world's REE production is related to these igneous rocks. Although carbonatite and alkaline complexes are relatively unusual on the Earth's surface, representing <1% of all the Earth's igneous rocks, Canada hosts numerous of these uncommon igneous bodies. They range in age from Neoarchean to Mesozoic, albeit many of them are Proterozoic in age, and mainly occur within the Canadian Shield and the Canadian Cordillera. As worldwide, Canadian carbonatite and alkaline complexes are commonly spatially related to major crustal structures in intracratonic anorogenic or post-orogenic extensional settings. The Kapuskasing Structural Zone that hosts the Clay-Howells, Nemegosenda Lake, and Lackner Lake alkaline complexes and Argor carbonatite is a good example of how major crustal-scale weakness zones control the emplacement of these igneous complexes. Canada is currently a small REE producer (Nechalacho deposit—alkaline complex-associated) but is well positioned to become a significant REE producer, as the country has some of the largest known REE geological resources associated with carbonatite and alkaline magmatism (e.g., Wicheeda Lake, Kipawa, Niobec, and Strange Lake deposits). Interestingly, many of these mineralized carbonatite and alkaline complexes share common tectonic history and/or similar formation/emplacement processes with carbonatite to alkaline magmatic events in Europe. Improving our knowledge of REE-bearing carbonatite and alkaline complexes in Canada will therefore help refine petrogenetic and metallogenic models in both Canada and Europe.

Trace-Element Deportment in Sphalerite Across the Bathurst Mining Camp, New Brunswick, Canada

Foteini Drakou[1, 2], Sean H. McClenaghan[1, 2]
1. Trinity College Dublin, Dublin, Ireland, 2. Ireland Research Centre in Applied Geosciences (iCRAG), Dublin, Ireland

The Bathurst Mining Camp (BMC) represents one of the largest volcanogenic massive sulfide (VMS) districts, containing over 600 Mt of massive sulfides from which 180 Mt were developed over 60 years of Zn-Pb-Cu-Ag-Au production. Mineralization is characterised by stratiform massive sulfide deposits occurring along six distinct exhalative horizons hosted by Ordovician volcano-sedimentary rocks deposited in an ensialic back-arc on a Peri-Gondwanan margin. To date, metal deportment studies in the BMC have focussed on the distribution of precious metals and energy critical elements (ECEs) at the deposit scale. This work characterises the trace-element deportment in sphalerite across the entire BMC, comparing exhalative Zn-Pb sulfide horizons of distinct basement crustal blocks. Sphalerite from 45 deposits including past producers such as Caribou, Brunswick No.12, and Heath Steele is assessed for overall trace element signatures as well as ECE potential. The trace element budget of Bathurst sphalerite exhibits significant variability at the camp scale, most notably for Co, Cd, and In. Interestingly, In concentrations in sphalerite exhibit the greatest variability, ranging from 1.00 ppm at the Louvicourt deposit to 1,670 ppm for the Heath Steele HC4 Zone, with samples exhibiting strong variability at the mm scale within Zn-Pb bands. For the most part, Co contents are below detection (LOD <0.1ppm); however, Austin Brook, Louvicourt, and Heath Steele exhibit variable Co signatures with the Heath Steele B zone exhibiting elevated Co concentrations (up to 100 ppm). Low median levels of Sn in sphalerite (1.13 ppm) suggest that bulk Sn signatures (up to 1,900 ppm) are in part controlled by cassiterite present in the ores. Given the capacity of sphalerite to accommodate metals in its mineral structure, identifying the inter-deposit trace element signatures across varied terranes in the BMC may provide insight on metal sources, depositional constraints, and the role of metamorphism in overall metal deportment.

From Rocks to Batteries: Insights Into Cobalt Mining, Mineralogy, and Sustainability

Quentin Dehaine
Geological Survey of Finland, Espoo, Finland

Occurrence and Spatial Distribution of Cobalt in Skarn Deposits: A Case Study from the Middle-Lower Yangtze River Valley Metallogenic Belt (MLYB) in East China

Taofa Zhou[1, 2], Yu Fan[1, 2], Fangyue Wang[1, 2], Shiwei Wang[1, 2]
1. School of Resources and Environmental Engineering, Hefei University of Technology, Hefei, China, 2. Anhui Province Engineering Research Center for Mineral Resources and Mine Environments, Hefei, China

Critical metals, including cobalt, are strategic and key mineral resources that have attracted global attention. Hydrothermal ore deposits, such as skarn and porphyry deposits, contain variable amounts of critical metals and have the potential for usage.The Mesozoic MLYB in east China is part of the circum-Pacific metallogenic belt and hosts a large number of porphyry-skarn Cu-Fe-Au deposits. Critical resources, including Co, Cr, Se, Re, and Te, are produced in these deposits in addition to bulk metals such as Cu, Ag, Fe, and S-rich ores. The subsequent by-product critical metals differ according to the dominant metals in skarn deposits. Most critical metals are hosted in sulfides in these skarn deposits, especially in pyrite in the skarn iron deposits. The average grades of cobalt in the skarn iron deposits are 0.01% to 0.03%, and some iron deposits are estimated to contain more than 10,000 tons of associated cobalt resources. The cobalt content also varies greatly in host minerals, such as pyrite, where Co^{2+} replaces Fe^{2+} in the form of isomorphism. The cobalt content in the same pyrite particle often changes by two to three orders of magnitude, and cobalt can be abnormally enriched in a single pyrite grain or ring rather than the core. Independent cobalt minerals are also developed in the deposit and account for approximately 10 to 15% of the cobalt in the whole deposit. Cobalt can be distributed in the iron orebodies and can also concentrate in the upper portion or upon of the iron bodies in the quartz-pyrite alteration zone. The recovery and utilization of cobalt in skarn deposits still requires further research and technological improvements. However, our pilot study indicates a broad future for this type of cobalt resource.

Funding: This work was supported by the National Natural Science Foundation of China (NSFC) under Grant No. 91962218.

Global Copper Screening for the Energy Transition - a Holistic Exploration Mindset

Graeme R. Nicoll, Paul Helps, Jean-Christophe Wrobel-Daveau
Halliburton, Abingdon, United Kingdom

The Energy Transition is driven by the need to mitigate climate change through progressive global switching to lower-carbon energy sources. Copper and other critical metals will be needed in greater amounts for electrification and providing raw components. Prospecting to expand mineral portfolios increasingly requires looking deeper and beyond known occurrences. New ways of thinking about exploration must be adopted to meet growing material demands.

Working with our mining industry clients, we are taking existing applicable geological subsurface screening methods from the hydrocarbon industry and applying them within a broader mineral system approach to exploration, which often requires turning the terminology on its head. These screening methods rely upon an integrated tectono-stratigraphic framework, compiled from public domain data over the last two decades. We discuss how it can be used to understand and predict likely occurrences of various mineral systems, both igneous and sedimentary.

At a global scale, porphyry copper screening may be possible, mainly from the integration of plate tectonics as a foundation; sedimentary-hosted copper requires integration of more diverse geological data sets (from sources to sinks, tectonics, and gross depositional environments). It requires a basin-scale understanding of the combination of processes (temporal and spatial) that are required for the development of economic sediment-hosted stratiform copper deposits. This includes the presence of basin-scale faults and structures (acting as fluid conduits), favourable metal source lithologies (e.g., continental red beds and volcanic basement), and brine-generating evaporite lithologies that drive fluid circulation and mineralisation.

This highlights that, at a global scale, data can be rapidly integrated using existing geoprocessing workflows, informed by physics-based processes understood at the local scale. The results allow for efficient screening of global locations where the geological conditions were favourable, including subsurface and frontier regions, allowing explorers to then focus attention on more detailed investigations at the local level.

The Role of Crustal Anatexis in Porphyry Copper Formation During Flat-Slab Subduction: Insights from the Laramide Belt, Southwest USA

Thomas Lamont[1], Matthew Loader[2], Frances Cooper[3], Dan Bevin[4], Nick Roberts[5], Tim Elliott[1], Adam Gorecki[6], Simon Tapster[5], Anthony Kemp[4], Robert Loucks[4]
1. University of Bristol, Bristol, United Kingdom, 2. Natural History Museum, London, United Kingdom, 3. University College London, London, United Kingdom, 4. University of Western Australia, Perth, WA, Australia, 5. Geochronology and Tracers Facility, British Geological Survey, Keyworth, United Kingdom, 6. BHP, Tucson, AZ, USA

The prevailing paradigm for the formation of porphyry copper deposits along convergent plate boundaries involves deep-crustal differentiation of metal-bearing juvenile magmas derived from the mantle wedge above a subduction zone. However, many major porphyry districts formed during periods of flat-slab subduction when the involvement of the mantle wedge would have been significantly reduced, if not absent. This leaves it unclear from where the ore-forming magmas were derived. To resolve this paradox, we investigate deep-crustal processes during the formation of the Arizona Laramide porphyry province, which formed between ~80 and 50 Ma during flat-slab subduction of the Farallon Plate beneath North America. We show that (1) Laramide-age granitoids have non-radiogenic Sr-Nd-Pb isotope signatures and (2) Proterozoic basement rocks from the Harcuvar, Harquahala, and Granite Wash Mountains in central Arizona experienced upper amphibolite-granulite facies anatexis at pressure-temperature conditions of 0.75 GPa (equivalent to a depth of ~28 km) and 750° to 780°C. U-Th-Pb monazite geochronology brackets the time of this deep crustal melting to between 73 and 60 Ma, suggesting an elevated geothermal gradient at the time of ~27.5°C/km. Crucially, this thermal climax coincided with (i) the zenith of granitic magmatism and porphyry genesis in the region (73–56 Ma) and (ii) dehydration of the shallowly-subducting Farallon slab (75–60 Ma). To explain the formation of the Laramide porphyry province, we propose that volatiles derived from the subducting slab passed straight into the lower crust, promoting anatexis of Proterozoic crustal sources without the requirement of juvenile mantle wedge-derived magmatism.

Magma Fertility Evolution Trends: Lessons from the Kerman Porphyry Belt, SE Iran

Ali Sholeh[1], Jeremy P. Richards[2]
1. SRK, Almaty, Kazakhstan, 2. Laurentian University, Sudbury, ON, Canada

The Kerman porphyry belt in SE Iran is located in the center of the Mesozoic-Cenozoic Neotethyan orogenic belt and is the most highly explored and well-studied part of the belt. The Iranian part of the Neotethyan orogenic belt hosts several world-class porphyry Cu deposits, which are mostly associated with postcollisional calc-alkaline magmas, including Sungun (21.01 ± 0.15 Ma; 3.446 Gt @ 0.48 Cu) in the Ahar-Arasbaran porphyry belt, Sar Cheshmeh (13.6 ± 0.1 Ma; 3.125 Gt @ 0.41% Cu), Sar Cheshmeh B Anomaly or Sereidun (1.645 Gt @ 0.38% Cu), Meiduk (12.5 ± 0.1 Ma; 792.53 Mt @ 0.67% Cu, 0.01% Mo, 0.08 g/t Au), and several smaller/less explored deposits such as Takht Gonbad (~26 Ma; ~160 Mt @ 0.34% Cu) in the Kerman porphyry belt.

The temporal and geochemical evolution of arc magmatism has been studied at the different magmatic suites in the Kerman porphyry belt. Thirty-six samples from outcrops or drill cores were collected from the various Eocene to Pleistocene porphyry-related and unmineralized magmatic suites along the belt. Whole-rock and zircon geochemical data, together with zircon uranium-lead laser ablation-inductively coupled plasma-mass spectrometry geochronology, were interpreted to compare and contrast characteristics of porphyry-related (fertile) and barren (infertile) suites and consequential fertility trends during temporal evolution of the orogenic belt. Data analysis revealed decreases in zircon Eun/Eu* (Eun/Eu*= Eun/Smn×Gdn1/2), positive Ce anomalies in zircon, and whole-rock Sr/Y and La/Yb ratios, as well as increases in negative Eu anomalies in zircon with age. These findings suggest that in the Kerman porphyry belt, and generally in all fertile arcs, the younger magmatic suites are more evolved, volatile-rich (hydrous), and oxidized and thus are considered more prospective for economic porphyry mineralization. This relative age relationship, together with an understanding of arc erosion level, can serve as a valuable exploration tool.

Late Cretaceous Magnetism in the San Matias District, NW Colombia: Insights of a New Ca. 76 - 66 Ma Porphyry Copper Belt in the Northern Andes

Julian Manco[1,2], Craig J. Hart[1], Hildebrando Leal-Mejia[1], Rob Lee[1,3]
1. MDRU-Mineral Deposit Research Unit, The University of British Columbia (UBC), Vancouver, BC, Canada, 2. Riverside Resources Corp., Vancouver, BC, Canada, 3. Pacific Centre for Isotopic and Geochemical Research (PCIGR), The University of British Columbia (UBC), Vancouver, BC, Canada

Colombia's Western Cordillera (WC) forms the northwestern corner of South America from the Caribbean region to Ecuador, comprising mostly accreted island arc terranes. The region's complex accretionary evolution and the lack of technical data confound the definition of a Cretaceous magmatic arc. Recently recognized porphyry-IOCG mineralization at the San Matias District (SMD) in the northernmost WC extension offers new understanding and insights into this region's magmatic and metallogenic evolution. It also illuminates new exploration opportunities for porphyry copper deposit (PCD) discoveries along the Northern Andes.

Petrography, lithogeochemistry, and zircon trace-element geochemistry of the SMD intrusions indicate low- to medium-K calc-alkaline affinity diorite to tonalite and their division into five magmatic suites. The ca. 74.4-70.0 Ma San Matías-La Jagua Magmatic suite (SMJS) is the most metallogenically significant. The SMJS hosts porphyry-style mineralization (e.g., Montiel E and Costa Azul) and exhibits depleted mantle Sr-Nd values and low radiogenic Pb isotope values similar to Galapagos Islands signatures. Lithogeochemistry (e.g., Sr/Y > 20 (26.2 - 58.8) and Eu*= 0.8 - 1.7) and zircon trace-element geochemistry (e.g., Eu*>0.35, Ce*10 - >225 and Dy/Yb<0.15) are consistent with hornblende fractionation from hydrous oxidized magma and indicate enhanced fertility for PCDs.

With the addition of our new U-Pb ages, two periods of magmatic activity along the WC can be recognized: 1) ca. 93-85 Ma and 2) ca. 76-66 Ma. The latter is coeval with the SMJS ages and considered the fertile period for PCDs. The 76-66 Ma ages extend to the southern WC, where lithogeochemistry and mineral occurrences are consistent with those in the SMD. These findings support the definition of a new Cordillera-scale Andean metallogenic belt with significant and growing copper resources.

Unraveling the Thermochemical State of the Whole Lithosphere in Space and Time: New Methods and Their Predictive Power for Exploration

Juan C. Afonso
ITC, University of Twente, Enschede, Netherlands

It is now widely accepted that the interaction of different processes and systems at lithospheric scale are responsible for the formation of many types of ore deposits. In this context, the thermochemical structure of the whole continental lithosphere holds key information on its origin and evolution that can be used to inform exploration strategies. As such, unravelling the nature of the continental lithosphere, its thermochemical modification through time, and its interactions with the sublithospheric mantle and shallow fluid systems constitutes one of the main goals of modern geoscience. Despite its fundamental importance, imaging the fine-scale thermochemical structure of the lithosphere using indirect (remote) data has been plagued with difficulties.

In recent years, however, "simulation-based" inverse methods that i) integrate multiple geophysical, geodynamic, and geochemical data sets within an internally and thermodynamically consistent platform and ii) utilize state-of-the-art AI and numerical techniques to solve the inverse problem in a fully probabilistic manner have opened new and promising ways to address this "grand challenge." The result of such inversions is a complete thermo-chemical-dynamical model of the subsurface, including the whole lithosphere and sublithospheric upper mantle. These images unravel fertility zones, metasomatized mantle domains, melting regimes, fluid pathways, and thermochemical signatures directly linked to mineral systems. With this information, full thermo-chemical-mechanical inversions of metallogenic processes can be run to aid in the understanding of specific mineral systems and their potential. In this presentation, I will review i) some recent progress, case studies, and future directions on the mapping of the thermochemical structure of the continental lithosphere, ii) their connection with fully dynamic inversion techniques for studying the evolution of specific mineral systems, and iii) their predictive power for the critical minerals and energy sectors.

A Rapid Transition to an Electrical Economy Based on Energy from Solar and Wind is Not Resource Viable

Adam C. Simon[1], Lawrence C. Cathles[2]
1. University of Michigan, Ann Arbor, MI, USA, 2. Cornell University, Ithaca, MI, USA

Transitioning to an energy infrastructure entirely reliant on renewable energy resources requires production of metals from new mineral deposits on an unprecedented scale. For example, to increase the number of U.S. wind turbines by 5%/y over current levels requires 115 kt/y of copper; expanding photovoltaic solar by 5%/y requires 600 kt/y more copper; producing 8 million battery electric vehicles per year requires 730 kt/y more copper. The Bingham Canyon copper mine in Utah produced 144 kt/y of copper in 2022. Thus, to support just these modest steps toward a green economy would require the U.S. put 10 new Bingham Canyon copper mines into production each year, and the number would increase as the wind and photovoltaic base grows. Globally, at least 100 Bingham Canyon mines per year are needed. This seems a practical impossibility considering that permitting and development of a new copper mine takes an average of 16 years. The projected annual global copper shortage will be about 10 million tonnes by 2030. In fact, by 2030 there will be a global shortage of cobalt, graphite, lithium, manganese, nickel, rare earths, and every other metal required for the green energy transition. These projections just consider the energy transition. They do not include the resources needed to bring the underdeveloped world up to developed-world levels of energy use. For equity and feasibility, we must consider zero-carbon strategies that take the pressure off resource supply.

Copper Mining and Extraction in Cyprus - from Ancient Heritage to a Green Future

Zack J. Van Coller[1], Kerim Sener[1], Ruth Bektas[1], Peter van der Borgh[2]
1. Ariana Resources Plc, London, United Kingdom, 2. Venus Minerals Ltd, London, United Kingdom

Cyprus has been associated with copper mining for ~6,000 years. Modern mining commenced in the 1920s, with >74 Mt of ore extracted from ~30 deposits over 50 years. The primary processing method for the sulphide-rich ores associated with the volcanic massive sulphide (VMS) deposits was flotation; oxide material was not amenable and was stockpiled.

Most historic deposits were outcropping and little effort was placed on exploration for "blind" deposits, with activities peaking between 1950 and 1970. The events of 1974 caused the annexation of processing facilities from the mines, leading to accelerated decline in the sector. Operators were forced to adopt new extraction methods, including from low-grade sulphide stockpiles by heap leach, then later (1995-2020) pioneered developments in solvent extraction and electrowinning (SX-EW) technology for processing oxide material.

Most copper mines in Cyprus remain abandoned, while the mine closure and rehabilitation requirements of the past have caused environmental legacies such as acid mine drainage. Despite this, opportunities exist to extract copper and other commodities that were considered uneconomic when first mined. SX-EW technology is capable of being optimised for the recovery of metals from waste at 11 historic mine sites, enabling their holistic rehabilitation to benefit both the environment and local communities. Furthermore, improved energy efficiency and the use of renewable power provide reductions in operating costs. A recent pilot study is considering copper and gold recovery from the Kokkinoyia mine, to reprocess 300,000 tonnes of waste and stockpile material.

Cyprus represents a unique opportunity in Europe for “green” copper. Several sites are largely fully permitted, enabling shorter lead times to commercialisation, with additional environmental, social and governance benefits and strong government and EU support for such initiatives. Further, several deposits contain potentially economic primary resources, and with advancements in exploration technology, the discovery of buried deposits is increasingly probable.

Unlocking the Full Potential of Metal Leaching in the Oceanic Crust: Assessing Permeability and Fluid Flow

Alannah C. Brett (Brett-Adams), Larryn W. Diamond
University of Bern, Bern, Switzerland

The distribution of permeability in the upper oceanic crust controls the extent to which hydrothermal solutions can leach Cu and transport it to form sea-floor volcanogenic massive sulphide (VMS) deposits. The prevailing view in the literature is that the volcanic crust is a fractured aquifer through which seawater circulates dominantly along extensional faults and fracture networks. However, this is at odds with evidence in VMS-rich ophiolites, where hydrothermal alteration is penetrative within the entire rock matrix rather than confined to fracture walls. The relative hydraulic roles of fracture networks versus the rock matrix are unclear, raising the question of how much source rock is actually accessible for metal leaching in VMS systems. To clarify this ambiguity, we have quantified the distribution of permeability in MORB-type lavas in the Semail ophiolite, which are considered to be source rocks for the known VMS deposits. We measured permeabilities of the intensely altered rock matrix (greenschist-facies chl+alb±act spilite) and upscaled the results to outcrop scale. Mapping of hydrothermal veins as proxies of syn-alteration fractures revealed high fracture intensities in km-spaced subvertical fault-damage zones, but low fracture intensities and connectivities in intervening fault-distal blocks. Using *DfnWorks* software, we upscaled the fractures numerically as discrete fracture networks to model fault-distal lava blocks of 0.5 × 0.5 × 0.5 km. Hydraulic simulations using *PFLOTRAN* confirm and constrain the expected high permeability of fault-damage zones, whereas in the fault-distal blocks, the sparse and poorly connected fracture networks are found to have only the same permeability as the rock matrix. These results imply that the entire volume of lavas in the upper oceanic crust is accessible for leaching of VMS metals during hydrothermal circulation, while steep, km-spaced fault zones provide the essential focusing for hydrothermal discharge and localisation of VMS deposits.

Accelerating Discovery Through Cutting-Edge Research

Elizabeth Sharman[1], Sonia Scarselli[2], Nicole Januszczak[3]
1. BHP Xplor, London, United Kingdom, 2. BHP XPlor, Toronto, ON, Canada, 3. BHP Metals Exploration, Toronto, ON, Canada

Resources are fundamental to the way we live now and in the future. With the increased electrification of energy and transportation, the demand for commodities like copper will likely double in the next 30 years. The time to accelerate the exploration of the resources we need into the future is now.

BHP introduces BHP Xplor, a cohort-based accelerator program to support early-stage mineral exploration start-ups to find critical resources, such as copper, to drive the energy transition. BHP Xplor offers candidates in-kind services, mentorship, networking opportunities with industry and investors, and connections.

A key component of the success of this program is the application by the cohort of cutting-edge tools and research. Mineral systems-driven identification of critical work programs provides a foundation to understanding every opportunity. Many key processes in copper mineral systems, including sedimentary copper and porphyry copper, are the subjects of global research initiatives, redefining our understanding of these systems. This research is incorporated into how our cohort thinks about their concepts and supports de-risking of projects through mapping key proxies.

The acceleration of projects is supported by consideration of key above-ground complexities, including land access, social license, and environmental scoping. Program participants are exposed to current thinking and future-looking initiatives in Environmental, Social, and Governance (ESG) issues to further enhance their projects. They are also introduced to key tools to assist them in mapping and tracking those ESG considerations, further de-risking their projects.

We believe that building a better future is a responsibility we all share, and that is why we are committed to accelerating exploration of critical resources needed for the energy transition through this program.

Lithium Pegmatites in Africa: Understanding Parageneses

Kathryn Goodenough[1], Richard Shaw[2], Anouk Borst[3], Paul Nex[4], Judith Kinnaird[4], Marieke van Lichtervelde[5]
1. British Geological Survey, Edinburgh, United Kingdom, 2. British Geological Survey, Nottingham, United Kingdom, 3. KU Leuven, Leuven, Belgium, 4. University of the Witwatersrand, Johannesburg, South Africa, 5. Geosciences Environnement Toulouse, Toulouse, France

Lithium is a key battery raw material and is experiencing significant growth in demand, driving increased exploration for new resources. The continent of Africa has significant potential for lithium pegmatites, with many deposits now being explored and undoubtedly many more yet to be found. Lithium pegmatites occur from the Archaean of the Zimbabwe craton, through the Palaeoproterozoic Birimian belts in West Africa and the Mesoproterozoic Karagwe-Ankole and Kibara belts of Central Africa, to the Neoproterozoic belts in Namibia. Many of these pegmatites share morphological similarities; they are tabular bodies, typically a few tens of metres in thickness, commonly with sharp contacts against their host rocks. However, their detailed mineralogy and paragenesis varies significantly, with major implications for lithium mining and exploration. This talk will compare the paragenesis of lithium pegmatites from different African tectonic belts to illustrate their variability. In Zimbabwe, the Mesoproterozoic Kamativi pegmatite has a complex four-stage paragenesis typical of LCT pegmatites: (1) a coarse-grained magmatic assemblage of quartz, alkali feldspar, and spodumene; (2) widespread albitisation with growth of cassiterite and columbite group minerals; (3) irregular quartz-muscovite greisenisation; and (4) widespread subsolidus alteration. In similar-aged pegmatites in Rwanda, intense kaolinization of spodumene pegmatites (up to 150 m depth) adds further paragenetic variability. In contrast, the Palaeoproterozoic Ewoyaa pegmatites in Ghana have a similar magmatic assemblage of quartz, feldspar, spodumene, and apatite, but show limited albitisation. Localised potassic alteration and greisenisation appears to be structurally controlled. The Mesoproterozoic Manono-Kitolo pegmatites, DRC, similarly lack widespread albitisation. The Archaean Arcadia pegmatite has a different magmatic assemblage, with petalite rather than spodumene, and the lower-temperature alteration has led to the generation of eucryptite. Understanding what controls these parageneses is vital to underpin efficient lithium exploration across the continent.

Eruption of Extremely Fractionated Magmas: Implications for Magmatic Versus Volatile Processes

David R. Lentz[1], Reimar Seltmann[2], Nadia Mohammadi[3]
1. University of New Brunswick, Fredericton, NB, Canada, 2. The Natural History Museum, London, United Kingdom, 3. Geological Survey of Canada, Ottawa, ON, Canada

The resurgence of fluid fractionation as an enrichment mechanism of highly evolved magmas, based on evidence from incompatible element abundances and key element ratios, like Zr/Hf, Y/Ho, and Nb/Ta, has provoked a renewed interest in these fundamental processes and their role in formation of some magmatic-hydrothermal deposits through to pegmatites. Although the Australian Brockman tuff (Niobium tuff) is well known, few would argue about the crystal fractionation mechanism for its enrichment, similarly to other highly evolved peralkalic magma systems, even though significant volatile volumes can also be involved. However, for low-P to high-P, F-, and B-bearing peraluminous systems that are responsible for hydrous high-level (subvolcanic, hypabyssal) Sn-, Ta-, and Li-enriched granites, there are ongoing debates about the rheology of these magmas, their mechanism to facilitate crystal-melt separation to fractionate; most have abundant evidence and some cryptic vectors of high-T alkali alteration, which makes analysis of enrichment processes challenging. In fact, the role of the volatile phases has long been argued in the enhanced evolution of pegmatites. However, extremely evolved aplite and felsite dikes are quenched subvolcanic systems, Argemela (Portugal) can be interpreted as erupted pegmatite melt, and the Macusani tuff (southern Peru) is an ash fall. With the surge in interest in Li, new extremely fractionated volcanic rocks are identified—both explosive and effusive varieties. In the eruptive conduits to subvolcanic and extrusive systems (rhyolites and tuffs), volatile reaction with the quenching melt-glass (gas-melt-glass) cannot be significant, and in fact these melts may not be rheologically challenged, although pyroclastic mechanisms of eruption may enhance their ascent regardless. Depression of liquidi and solidi due to various fluxes (Li, P, F, B) in addition to water, together with enhanced crystal-melt fractionation within an anomalous source and thermal regime, causes extreme fractionation without reaching for fluid fractionation as an additional enrichment mechanism.

Lithium in Manitoba: Production and Exploration Potential

Tania Martins
Manitoba Geological Survey, Winnipeg, MB, Canada

Lithium continues to be in the spotlight due to the continue demand for batteries in tandem with the advent of decarbonization of our economy and the fossil fuel transition. Lithium has other important physical and chemical properties that are suited to a wide range of applications including pharmaceuticals, glass and ceramics, and aerospace technologies. In addition, lithium is considered a critical element by the governments of Canada, the United States, and the European Union.

The Canadian province of Manitoba has a long and rich history of lithium exploration and mining associated with granitic pegmatites. Granitic pegmatites contain the largest known resources of lithium in Manitoba. The most prolific region is the Winnipeg River-Cat Lake pegmatite field, which hosts the world-class Tanco lithium cesium-tantalum deposit, along with numerous other pegmatite occurrences. The Tanco mine was until very recently the sole producer of spodumene concentrate in Canada. The Winnipeg River-Cat Lake pegmatite field is Archean in age and is part of the Superior province. Elsewhere in the Archean Superior province of Manitoba, lithium-bearing pegmatites occur at Red Sucker Lake, Gods Lake, Cross Lake, Red Cross Lake, and McLaughlin Lake.

A second prolific lithium-hosting region in Manitoba is the Wekusko Lake pegmatite field, located just east of the town of Snow Lake. The Snow Lake area in Manitoba is well known for the world-class Flin-Flon-Snow Lake mining camp and its production of gold, copper, zinc, and nickel. The Wekusko Lake pegmatite field is of Paleoproterozoic age (ca. 1800 Ma). The main lithium-bearing pegmatite mineral found in the Wekusko Lake field is spodumene. However, more recently, highly fractionated pegmatite dikes with lepidolite were reported, making this area even more attractive for pegmatite exploration projects.

Challenges of Water Use Impact Assessment of Lithium Production from South American Salar Deposits using Life Cycle Assessment

Rowan Halkes[1, 2], Frances Wall[1], Robert Pell[2], Andrew Hughes[3], Evi Petavratzi[3]
1. Camborne School of Mines - University of Exeter, Penryn, United Kingdom, 2. Minviro, London, United Kingdom, 3. British Geological Survey, Nottingham, United Kingdom

Lithium is a critical raw material for the energy transition due to its use in battery technologies. The primary sources of lithium are spodumene pegmatites and salar brine deposits, with the salar deposits of South America hosting 70% of global lithium resources. Salars are complex groundwater systems where lithium-rich brine is abstracted by pumping. They are also home to valuable ecosystems, reliant on surface water ponds associated with the salar. Due to the region's elevated water scarcity and the salar system's ecological sensitivity, there are sustainability concerns regarding the water use, and associated impacts, of lithium production from salar deposits.

Life cycle analysis (LCA) is a method becoming increasingly prevalent in the assessment of raw materials sustainability. There are several approaches to measuring water impacts or scarcity. LCA utilises a water scarcity footprint methodology called AWARE for water use impact assessment. However, this methodology is widely regarded as deficient for thoroughly capturing and assessing water use impacts of lithium production from salar deposits.

This study investigated the intersection of salar deposit systems, with their complex and often poorly understood geology and hydrogeology, with LCA water use impact assessment. The study identified and explored challenges and opportunities for improvement, broadly categorised as those relating to a) the complexity and uncertainty involved with salar systems hindering impact assessment and allocation, b) generic issues with the data and models that underpin LCA methodology, and c) the limited accuracy and representativeness of current LCA methods when applied to salar systems, specifically assessment of groundwater and the interaction of varying aqueous phases, i.e., brines and water, in the same system.

This work will aid the development of LCA methodology and enable improved assessment of the sustainability of lithium production from salar deposits in South America, and by extension the resourcing of the green transition.

The Jadar Li(B) Deposit (Serbia): Insights Into a Giant Volcano-Sedimentary-Hosted Li-B System

Francesco Putzolu[1], Robin Armstrong[1], Jorge Garcia[2], Nick Hawkes[2], Adrian Boyce[3], Jens Najorka[1], Marie Lefebvre-Desanois[4], Martin Palmer[4], Tobias Salge[1], Isabel Abad[5], Richard Herrington[1]
1. Natural History Museum, London, United Kingdom, 2. Rio Tinto Group, London, United Kingdom, 3. Scottish Universities Environmental Research Centre, Glasgow, United Kingdom, 4. University of Southampton, Southampton, United Kingdom, 5. Universidad de Jaén, Jaén, Spain

Volcano-sedimentary systems (VSS) show enormous potential for economic concentrations of critical metals such as lithium. Lithium VSS are located in block-faulted basins developed within felsic volcanic provinces of crustal affinity. In most conventional VSS (e.g., McDermitt Caldera, US), the alteration of Li-fertile volcanic rocks leads to the formation of an "Li-smectite trap," where hectorite and other related clays represent the main ore repository. The Jadar deposit (Serbia) is the largest lithium deposit in Europe, holding JORC indicated and inferred resources of 143.5 Mt Li at 0.838% Li, and is considered part or the VSS family. However, contrary to conventional VSS, the ore assemblage in Jadar is dominated by the presence of the unique Li-mineral jadarite ($LiNaSiB_3O_7(OH)$), an Na- and Li-borosilicate. In this study, we provide the first geological account on the "jadarite-type" VSS by presenting a combination of petrographic, lithogeochemical, and isotopic data. The mineralization in Jadar is hosted in lacustrine-deposited volcanogenic rocks (i.e., mica-dominated air falls end epiclastic rocks), which experienced post-depositional processes including zeolitization, K-(auto)metasomatism (i.e., K-feldspar alteration), and clay alteration. These processes mirror the closed hydrologic system diagenesis (CHSD) observed in Li clay-type VSS, yet the Jadar deposit shows the following peculiarities: (1) extensive carbonate alteration, leading to the breakdown of early-formed Mg(Li)-smectites and the formation of primary dolomite; (2) a lack of primary quartz and secondary silica; and (3) Li fixation in borosilicates, coeval to zeolitization. These features highlight that the diagenesis at Jadar occurred at higher pH if compared to clay-type VSS. Our data indicate that the hyperalkaline diagenetic conditions at Jadar led to the stripping of silica from Li(B)-fertile volcanogenic material, which was remobilized to form a gel-like material, which was precursor to formation of a jadarite–zeolite assemblage.

The Bergby LCT-pegmatite Field, Central Sweden – A New Contender to Produce Swedish Lithium?

Karin Högdahl[1], Erik Jonsson[2, 1], Magnus Leijd[3], Jarosław Majka[1], Mikael Bergkvist[4], Alexander Hansson[4], Pauline Jeanneret[1], Mikael Kördel[4]
1. Uppsala University, Dept. Earth Sciences, Uppsala, Sweden, 2. Geological Survey of Sweden, Uppsala, Sweden, 3. Bergby Lithium AB, Bollnäs, Sweden, 4. Orexplore AB, Kista, Sweden

The most recently discovered lithium-cesium-tantalum (LCT) granitic pegmatite field in Sweden is located in Bergby, in the eastern central part of the country. Following the find of a small spodumene-bearing boulder in 2007, this pegmatite system was firmly established through core drilling in 2016 by Leading Edge Materials/Woxna, i.e., very nearly 200 years after the discovery of the element Li in petalite from the Swedish island of Utö, some 230 km to the southeast. Recently, several new LCT pegmatite dikes and dike swarms have been located during ongoing exploration by United Lithium. The Li mineralogy varies between the different dikes; some host both spodumene and petalite, whereas others are wholly spodumene-dominated. Lithium is also hosted in Li-mica(s), amblygonite-montebrasite series minerals, cookeite, and Mn-Fe-Li phosphates. Other typical LCT phases like pollucite and columbite-tantalite minerals have also been identified, but both Cs and Ta-Nb contents vary significantly between different dikes.

The presently known LCT-pegmatite field (c. 50 km^2) is located within a folded metasupracrustal package comprising mica schist, 1.89 Ga felsic to mafic metavolcanic rocks, and a <1.85 Ga quartzite in the uppermost part of the stratigraphy. These rocks have recorded at least two fold phases and were metamorphosed in low-PT amphibolite facies, given by the occurrence of andalusite and sillimanite in the schist. Barren granitic pegmatites are ubiquitous in the area, but so far the LCT pegmatites have only been identified in the most competent lithologies, represented by amphibolite (metabasalt) and quartzite. There are also some smaller granitic bodies present, but if they are related to the LCT-pegmatites or not is presently unknown.

In a recently started project (ULiBS), we aim to define the tectonic position of the LCT-pegmatite system, its character and relationships to the metamorphic and structural evolution, barren granitic pegmatites and possible source granitoids.

Cornish Lithium: Exploration for Geothermal Lithium Resources in Southwest England

Alistair Salisbury, Alexander J. Hudson, Adam Matthews, Mike Round, Chris Yeomans
Cornish Lithium Plc, Penryn, United Kingdom

Cornish Lithium aims to develop the UK's domestic lithium supply from novel geothermal lithium resources. These warm geothermal fluids (80°C at approx. 2 km) containing elevated lithium concentrations have been known to ingress Cornish mines since ca. 1864, at which time great expense was dedicated to dewatering mine workings. Now, 250 years on, it is clear there is huge potential to utilise both the lithium and heat contained within these fluids for fuelling the transition towards a low-carbon economy.

Cornish Lithium has access to a wealth of historic data from the extensive tin and copper mining from the 17th century to present. Mining records show that lithium-enriched geothermal fluids are hosted at depth within a network of permeable fractures comprising pervasive ENE-WSW–trending Sn-Cu lode structures as well as kilometre-scale NW-SE–oriented faults, termed cross-courses. We show that historical records, supplemented with modern geological and remote sensing techniques, can be used to target lithium-bearing fracture networks with high accuracy using 3D implicit modelling software. The holistic approach Cornish Lithium is taking to mineral exploration has led to the development of an exploration portfolio based on these 3D models that has intersected target structures at depths of up to 2 km. We will present recent results and from testwork conducted in Summer 2023 from the CRM-geothermal project; these include geological and reservoir assessments in the United Downs area and results from our latest exploration drilling.

To date, Cornish Lithium has proven elevated lithium concentrations and a geothermal gradient of ~45°C/km with fluids encountered throughout the fracture network in the United Downs area with exploration ongoing across a wider area. In March 2022, Cornish Lithium commissioned a pilot plant for direct-lithium extraction, which is the first of its kind in the UK and a major step towards commercial production.

Unconventional Magmatic Sulfide Systems: Looking Outside of the Box for the Next New Ni-Cu-Co-PGE Discovery

David A. Holwell
University of Leicester, Leicester, United Kingdom

Magmatic Ni-Cu-Co-PGE sulfide deposits are the world's major source of battery-grade Ni and PGEs and important contributors to global Cu and Co resources. As such, the discovery and development of these deposits is critical to the energy transition. Magmatic sulfides are hosted by mantle-derived mafic-ultramafic intrusions and can occur at any level in the crust. Established models for their genesis, which thus provide direction for exploration models, are focussed on (1) a craton margin association for many Ni-rich deposits; (2) an intracratonic setting for PGE deposits; (3) the importance of plumes and extensional regimes for both; and (4) moderate to high (>15%) degrees of partial melting of dry, olivine-rich peridotitic source rocks. However, many deposits do not conform to these criteria, and there are a significant number of deposits in orogenic settings formed in more alkaline and hydrous magmas that were sourced from relatively lower degrees of partial melting. The role of metasomatized, hydrous mantle assemblages containing phlogopite is highlighted here, as an important source to magmatic sulfide systems in a wider range of geotectonic settings. A number of global examples of more unconventional deposits are explored to illustrate the variable characteristics of magmatic sulfide deposits formed from hydrous and carbonated pyroxenitic rather than peridotitic mantle sources, emplaced throughout the entire thickness of the continental crust. By identifying greater variability in the components of the Ni-Cu-Co-PGE mineral system, search spaces can be expanded, and the next new discovery could be outside of traditional, established Ni camps.

ML Modeling of Multiparameter Data for Ni-Cu-PGE Targeting

Wilson Bonner, John B. Chapman, Luke M. Fairchild
KoBold Metals, Berkeley, CA, USA

Recent advances in hyperspectral core scanning technology and increased use of machine learning in mineral exploration have resulted in several studies successfully predicting lithologies and alteration assemblages in hydrothermal deposits. Fewer studies have focused on application of these technologies in Ni-Cu-PGE deposits because of the longer spectral wavelengths of key minerals in mafic and ultramafic rocks and lack of alteration vectors. We present an approach combining laser-induced breakdown spectroscopy (LIBS), shortwave to very near infrared (SWIR-VNIR), and portable X-ray fluorescence (pXRF) core scanning with petrophysical measurements on a drill hole through a mineralized, layered gabbroic intrusion to show the same approaches can apply in exploration of critical metals.

The Crystal Lake Gabbro is a layered intrusive complex, approximately 40 km southwest of Thunder Bay, Ontario, that was emplaced late in the 1.1 Ga Midcontinent Rift event. It contains a non-compliant historic resource of 45.6 Mt at 0.334% Cu, 0.183% Ni, 0.12g/t Pt, and 0.59 g/t Pd in the shallowest part of the southeast-dipping intrusion. Mineralization occurs as disseminated chalcopyrite, pentlandite, and PGMs in taxitic gabbro within and at the base of the intrusion. Mineralization is high tenor, with typical (Pt+Pd)/(Cu+Ni) values between 1.8 and 2, peaking at over 4 within short intervals, indicating likely high R-factor for the system. Hence, discovery of a substantial body of mineralization within part of the intrusion represents a very attractive exploration target.

To effectively target such a body during ongoing exploration, we have developed novel analytical routines to integrate chemical, mineralogical, and physical properties data. By bringing multi-parameter drill hole data, both routine and novel, into a machine learning environment, we illustrate strategies to better navigate within multi-phase intrusions and produce testable predictions about locations of larger orebodies.

High Grades in Hypogene Porphyry Copper Deposits

Dick Sillitoe
Independent Consultant, London, United Kingdom

Exploitation of high-grade copper ore accelerates payback of capital expenditure, lowers operating costs, and hence enhances profitability; it also gives rise to lower carbon footprints and less waste per tonne of copper produced. This presentation reviews the geological features responsible for high grades in hypogene porphyry copper deposits, with particular focus on permeability and reactivity of porphyry intrusions and their host rocks. The influence of alteration types and sulphide assemblages on copper grade is assessed. Porphyry copper exploration programs may be designed to take account of features conducive to high grade development.

Alteration Geothermometry and Exploration Implications of the Gajah Tidur Porphyry System, One of Two Systems that Formed the Supergiant Grasberg Cu-Au-(Mo) Deposit

Reza Al Furqan[1], Antonio Arribas[2], Clyde Leys[3], Tetsumaru Itaya[4], Adi Sulaksono[5], Yasushi Watanabe[1]
1. Akita University, Akita, Japan, 2. University of Texas at El Paso, El Paso, TX, USA, 3. Eksplorasi Nusa Jaya (Affiliate of Freeport-McMoran Inc), Jakarta, Indonesia, 4. Japan Geochronology Network, Okayama, Japan, 5. UPN Veteran Yogyakarta, Yogyakarta, Indonesia

Based on geochronological data and field relationships, researchers have recently proposed that the Grasberg Cu-Au-(Mo) deposit comprises two overlapping porphyry Cu systems, the deeper and older (~3.4 Ma) Gajah Tidur (GT) Cu-Mo-(Au) system and the shallower and younger (~3.1 Ma) Main Grasberg (MG) Cu-Au system. Dating of abundant hydrothermal muscovite and subordinate biotite in this study supports the age relationships between the two porphyry systems and indicates that alteration and associated Cu-Mo-(Au) mineralization beneath the now mined-out Grasberg open pit is mainly associated with the GT porphyry system. As well as a ~1-km difference in emplacement depth compared to MG, a different metal signature (Cu-Mo-(Au) vs. Cu-Au), and distinct parent intrusion geometry, Cu mineralization in GT is mainly associated with a lower-temperature alteration assemblage. Sulfur isotope temperatures from sulfate-sulfide pairs show that initial GT magmatic-hydrothermal fluids formed K-feldspar-biotite alteration at temperatures up to 590°C. Cooling of this fluid from ~460° to ~370°C resulted in white mica-dominated alteration associated with anhydrite, chlorite, chalcopyrite, and pyrite. Further cooling (340°–280°C), outward from the core of the hydrothermal system, is characterized by a muscovite-quartz ± pyrophyllite assemblage associated with covellite-pyrite mineralization. While Cu mineralization in the GT porphyry system is largely associated with white mica assemblages (muscovite-anhydrite-chlorite and muscovite-quartz ± pyrophyllite), high-grade Cu and Au mineralization in the MG system is chiefly associated with a K-feldspar-biotite-magnetite assemblage formed at higher temperature than the GT K-feldspar-biotite alteration. The overlapping GT and MG porphyry systems formed the supergiant Grasberg Cu-Au-(Mo) porphyry deposit, within which GT contributed an estimated 30% for Cu, 10% for Au, and 75% for Mo. The existence of two closely spaced, well-mineralized porphyry systems where the older system was emplaced at a greater depth than the younger one is an unusual occurrence that has significant exploration implications.

The Spence Porphyry Copper System Revisited: An Integrated Mineral Systems Study

Edward Bunker, Marianne Parsons, Mark Broadley
CGG, Crawley, United Kingdom

The Spence porphyry copper system in northern Chile was discovered by Rio Algom in 1996 through drilling to bedrock in grids over areas concealed by relatively young geological cover sequences. Despite the serendipity of this discovery, subsequent studies have revealed geological factors that could have facilitated effective exploration targeting for this "blind" discovery.

Spence was emplaced along a major dextral strike-slip fault known as the Antofagasta-Calama Lineament (ACL). Whilst the surface expression of the ACL is poorly defined, evidence for the ACL has been documented both through geophysical methods and in the form of fractures exhibiting mineral concretions and geochemical anomalies expressed within the gravels overlying Spence. Using a mineral systems framework, the ACL has acted as a fundamental geological control on porphyry copper deposit formation in this area, providing both a pathway and a zone of deposition for magmas and fluids, with this relationship being observed from regional to deposit scale.

We utilize the latest in Earth observational and geophysical methods to examine the ACL as a geologic feature and highlight evidence for its relevance in ore-forming processes, using the Spence porphyry copper deposit as a case study. By applying geological constraints such as fault surfaces to our geophysical inversions, it is possible to better delineate geologically reasonable features within the geophysical data. These data are visualized in three-dimensional space to better constrain the geometry of geophysical anomalies and showcase areas of interest. We utilize a range of spectral data sets to examine the surface geology, identify features of interest, and enhance our understanding of the regional- to local-scale geology. By placing this relatively well-studied deposit in a multi-scale holistic framework, we highlight exploration opportunities and derive new insights for the regional controls on mineralization in this mature mineral district.

Geology, Mineralisation, and Hydrothermal Alteration of the Almalyk Porphyry Copper- Gold Deposit, Tien Shan Mountains, Eastern Uzbekistan

Peter Gleeson
SRK Consulting, Southerndown, United Kingdom

The Almalyk porphyry copper-gold deposit in Uzbekistan is one of the longest economically active deposits in the world with JORC compliant resources (inferred, indicated, and measured) of over 19 billion tonnes at 0.23% Cu and 0.28 g/t Au and proven and probable reserves of 8.8 billion tonnes at 0.29% Cu and 0.34 g.t Au (AGMK 2021). It rivals some of the world-class South American and Southeast Asian porphyry deposits. It has been operating as an open pit for over 65 years.. However, despite its size and economic contribution to the Uzbek economy, it is relatively unknown outside of Central Asia. The deposit lies in Eastern Uzbekistan. It is situated geologically in the Valerianov-Beltau-Kurama magmatic arc, part of the Tien Shien Mountains. It has associated epithermal and skarn copper, gold, and base metal deposits of an Andean-type magmatic arc setting. The Almalyk "United Mine" (formed from the individual named deposits of Kalmakyr, Karabulak, Yoshlik, and Balaktyr) is a multiphase porphyry intrusive system emplaced between 337 and 306 Ma, with economic Cu, Au, Mo, and Ag mineralisation associated with quartz monzonite and granodiorite intrusions emplaced in earlier dioritic host rocks. The deposit is a multiphase quartz-sulphide vein stockwork, with a well-established vein paragenesis, and displays classic Lowell and Guilbert (1970) styles of concentric hydrothermal alteration around a central porphyritic intrusive complex. The geometries of the zones of hydrothermal alteration and mineralisation are modified by late-stage faulting to form the four individual deposits (originally a single deposit). Three principal mineralisation styles are present (oxide, transitional, and fresh-sulphide) with only minor supergene enrichment. The main copper sulphide minerals currently being mined are chalcopyrite and bornite. The area surrounding the Almalyk porphyry cluster remains highly prospective for additional porphyry, skarn, and related epithermal-style mineralisation.

Applications of Multielement Epidote and Chlorite Mineral Chemistry to Locate Mineralized Centers: A Case Study from the Batu Hijau Porphyry Cu-Au District, Indonesia

David R. Cooke[1], Matthew J. Cracknell[1], Ivan Belousov[1], Shaun Barker[2], Michael J. Baker[1], Stephen A. Cooke[1], Jamie J. Wilkinson[3], Zhaoshan Chang[5], Jennifer Thompson[1], Noel C. White[1]

1. CODES, University of Tasmania, Hobart, TAS, Australia, 2. Mineral Deposit Research Unit, University of British Columbia, Vancouver, BC, Canada, 3. LODE, Natural History Museum, London, United Kingdom, 4. Department of Earth Science and Engineering, Imperial College London, London, United Kingdom, 5. Colorado School of Mines, Golden, CO, USA

The Batu Hijau porphyry Cu-Au deposit is located in SW Sumbawa, East Sunda arc, Indonesia, and has a mineral resource of 1.64 Gt @ 0.44% Cu and 0.35 g/t Au. Batu Hijau is hosted primarily by a geochemically homogenous protolith of Late Miocene basaltic andesites that were intruded by diorites and mineralised tonalites. There are other porphyry prospects (Sekongkang, Nangka) and intermediate sulfidation epithermal prospects (Bambu) in the district, although Batu Hijau is the only economic ore deposit

A previous study of the Batu Hijau district (Wilkinson et al., 2015) identified systematic compositional variations in chlorite chemistry that allowed for the development of 'proximitor' equations that can be used to predict the distance from the mineralised center, providing new tools for mineral exploration. That study was based on 512 chlorite spot analyses. We have generated an additional 715 chlorite spot analyses from the district, including reanalysis of many samples from the previous study, and have also generated 1,260 epidote analyses. In addition to successfully reproducing the results from Wilkinson et al. (2015) through sample reanalysis, our results illustrate how multielement mineral chemistry using epidote and chlorite data from Batu Hijau can be used in assessments for exploration vectoring (how far and in what direction?) and fertility assessments (potential size of the mineral resource). Our results illustrate the geological controls on mineral chemistry variations in the Batu Hijau district and have significant implications for the formation of alteration and geochemical footprints to porphyry deposits.

Reference

Wilkinson, J.J., Chang, Z., Cooke, D.R., Baker, M.J., Wilkinson, C.C., Inglis, S., Chen, H., and Gemmell, J.B., 2015, The chlorite proximitor: A new tool for detecting porphyry ore deposits: Journal of Geochemical Exploration, v. 152, p. 10–26.

A Practical Predictive Targeting Model for Convergent Margin Gold and Copper Deposits

Jon Hronsky[3], Graham Begg[5]
1. Western Mining Services, West Perth, WA, Australia, 2. Minerals Targeting International, West Perth, WA, Australia, 3. 2. Centre for Exploration Targeting, School of Earth Sciences, University of Western Australia,, Crawley, WA, Australia, 4. 2. Centre for Exploration Targeting, School of Earth Sciences, University of Western Australia, Crawley, WA, Australia, 5. 4. Australian Research Council Centre of Excellence for Core to Crust Fluid Systems (CCFS) and GEMOC, Macquarie University,, Sydney, NSW, Australia

The majority of the world's resources of gold and copper are hosted by convergent margin settings, mostly because these are the primary tectonic environments where fluids are cycled from the mantle into the near-surface. The basic hypothesis of this paper is that, to a first order, we can explain all convergent margin-related Cu and Au deposits by the interplay of only two fundamental processes, operating in a range of geodynamic contexts. In this framework, ore formation is seen as the predictable consequence of certain particular, although relatively rare, segments of the 4D evolution of host terranes, not simply the result of the random coincidence of multiple positive factors.

The two underlying fundamental processes are 1) the Loucks process and 2) the FUME (Fertile Upper Mantle Extraction) process. The Loucks process is most relevant to copper metallogeny, whereas the FUME process is most relevant to gold metallogeny. Critically, both processes require specific anomalous geodynamic processes within the context of the evolution of their host convergent margin settings, thereby providing a potentially robust regional-scale predictive capability.

These two processes commonly coincide because the same geodynamic conditions may favour both at the same location and time. In practice, there is a spectrum between Cu-only (Loucks process-dominant) and Au-only (FUME process-dominant) end members.

This model implies that regional-scale metallogenic fertility has both static and dynamic aspects.

Static fertility is the requirement for a metasomatised upper-mantle source region, which is an inherited property, not dependent on syn-mineralisation geodynamics. It is only relevant to the FUME process and Au metallogeny.

Dynamic fertility is the requirement for a favourable, anomalous geodynamic context that allows the FUME and/or Loucks process to operate. This is transient in both time and space, driven by heterogeneities in regional-scale geodynamics, and is relevant to both Cu and Au metallogeny.

Tracing Granite-Hosted Mineralization via Zircon Metal Anomalies

Nicholas J. Gardiner[1], Chris J. Hawkesworth[2], Laurence J. Robb[3, 6], Jacob A. Mulder[4], Peter A. Cawood[5]
1. University of St Andrews, St Andrews, United Kingdom, 2. University of Bristol, Bristol, United Kingdom, 3. University of Oxford, Oxford, United Kingdom, 4. University of Adelaide, Adelaide, SA, Australia, 5. Monash University, Melbourne, VIC, Australia, 6. University of the Witwatersrand, Johannesburg, South Africa

Granite-hosted magmatic-hydrothermal mineral deposits are major sources of Cu, Sn, Li, and W, originating via mineralizing fluids exsolved from volatile-saturated magmas. We show how trace elements in zircon sampled from the granite-hosted Zaaiplaats Tin Field, Bushveld Complex, South Africa, preserve a record of both incompatible metal enrichment during magma fractionation and that arising from magmatic-hydrothermal mineralization.

Mineralization at Zaaiplaats resulted from closed-system magma differentiation and the eventual appearance of a saline magmatic Sn-rich hydrothermal fluid, which promoted cassiterite precipitation in the central mineralized zone. We sampled across the intrusion, through the magmatic-hydrothermal paragenesis, and, on the basis of mineralogy, subdivided samples into two groups: mineralized and unmineralized.

Unmineralized samples define a fractionation trend on the basis of zircon REE and HFSE arrays. Plots of Sm and Sn versus Gd show consistent trends, where Sn is dominated by Rayleigh Fractionation. In contrast, mineralized samples show enrichment in Sn for the same Gd, attributed to the introduction of an Sn-rich mineralizing fluid resulting from volatile saturation during zircon growth, consistent with the preservation of whole-rock Sn zonation in the Zaaiplaats granites. Zircon Sn/Gd ratios from unmineralized samples define a magma fractionation trend with increasing Y at constant Sn/Gd, whereas those from the mineralized samples are displaced to high Sn/Gd ratios at similar Y, a deviation marking the onset of mineralization processes.

A metal anomaly, Sn/Sn*, is defined that describes the deviation of Sn over that expected through magma fractionation alone (Sn*), and arises from Sn mobilization due to magmatic-hydrothermal mineralization processes. Identification of metal anomalies (e.g., Sn/Sn*, Cu/Cu*) in mineral archives or at the whole-rock level provides an empirical link to the onset of mineralization processes in magmatic-hydrothermal systems and can be coupled with geochemical proxies to yield a better understanding of the conditions leading up to, and subsequent to, volatile saturation.

Quantitative Mineralogy of the Beauvoir Granite and Rare-Metal-Bearing Minerals Characterization

Steven Kahou, Michel Cathelineau, Marie-Christine Boiron
Laboratoire GeoRessources, Université de Lorraine, Vandoeuvre-lès-Nancy, France

The albite-lepidolite-topaz granite of Beauvoir is the last intrusion in a peraluminous granitic complex of Variscan age composed of three successively emplaced units: the hidden granite of La Bosse, the two-micas granite of Colettes, and the Beauvoir granite consisting of three units, B1 to B3. Compared to similar Li-F-rich igneous bodies, the Beauvoir granite is highly enriched in Sn (200-1,400 ppm), Ta (20-400 ppm), and Be (20-300 ppm). The B1 unit is composed of albite and abundant lepidolite laths forming a framework filled with globular quartz and rare crystals of K-feldspar. This work presents a quantitative mineralogical study of the Beauvoir granite (B1 unit) and detailed textural and chemical characterization of the oxides and Li-bearing minerals (i.e., mainly lepidolite and, to a lesser degree, the Li-rich phosphates) from the cores from the PER and EMILI drilling campaigns. The first approcah of this work consisted of estimating the proportion of minerals, and thus determinig their abundance by comparing estimates from in situ thin section mapping from micro-XRF and the calculation of mineral phase proportion from whole-rock geochemistry on drill cores. The main trend is a "mix," for the fresh rocks, between an albite pole (~35%), and a quartz (~25%)-lepidolite (~15-25%) pole. Lepidolite textural study showed a core-to-rim zoning (Rb, Cs), while lithium contents are rather homogeneous between the core and the rim with an average of 28,000 ppm. Amblygonite and montebrasite show high lithium contents (up to 46,000 ppm in montebrasite) but are very heterogeneous at the crystal scale. Oxyde minerals include cassiterite, columbite-tantalite, and microlite, particularly abundant in the upper part of the B1 unit. The results obtained, combined with drilling data, will lead to a mineralogical model block of the Beauvoir quarry and better understandig of the magmatic-hydrothermal evolution of the deposit.

GREENPEG: New Toolsets to Explore for Pegmatites Hosting Rare Metals and High-Purity Quartz to Feed the Energy Transition

Axel Müller[1], Marco Brönner[2], Julian Menuge[3], Ben Williamson[4], Frances Wall[4], Claudia Haase[2], Georgios Tassis[2], Klaus Brauch[5], Claudia Pohl[5], Ana Teodoro[6], Joana Cardoso-Fernandes[6], Alexandre Lima[6], Douglas Santos[6], Encarnación Roda-Robles[7], Kerstin Saalmann[2], Mario Hopfner[8], Kate Smith[4], Idoia Garate-Olave[7], Jon Errandonea-Martin[7], John Harrop[9], Lawrence Carter[4], William Keyser[1], Haoyang Zhou[1], Teimoor Nazari-Dehkordi[3], Wolfgang Reimer[10], Carla Pueyo Lloret[1]
1. Natural History Museum, University of Oslo, Oslo, Norway, 2. Geological Survey of Norway, Trondheim, Norway, 3. University College Dublin, Dublin, Ireland, 4. University of Exeter, Penryn, United Kingdom, 5. terratec Geophysical Services, Heitersheim, Germany, 6. University of Porto, Porto, Portugal, 7. Universidad del País Vasco-UPV/EHU, Leioa, Spain, 8. IFU GmbH Privates Institut für Umweltanalysen, Lichtenau, Germany, 9. Blackstairs Lithium Ltd, Dublin, Ireland, 10. Geokompetenzzentrum Freiberg e.V., Freiberg, Germany

The EU-funded GREENPEG project (2020-2024) aims to develop multi-method exploration toolsets for the discovery of buried, small-scale (<5 million m^3) rare-metal-type pegmatite deposits with focus on lithium and high-purity quartz. These commodities are in increasing demand as they are essential for Li-ion batteries used in electric vehicles and in photovoltaics, respectively. The testing of exploration tool sets is being carried out at province, district, and prospect scale at three GREENPEG case study sites: Wolfsberg in Austria, Leinster in Ireland, and Tysfjord in Norway. These sites represent a wide spectrum of challenges in exploration for European pegmatite deposits, such as variable wall rocks (Leinster), dense vegetation and long winter snow cover (Norway), alpine topography (Wolfsberg), continuous soil cover (Leinster), and tills and glacial sediments hampering conventional soil geochemistry-based exploration (Norway). These characteristics will ensure that the toolsets are robust and flexible enough to be applicable in manifold environments. Toolset development is based on an improved genetic model for pegmatite deposits and utilises the lithogeochemical halo in wall rocks of pegmatite orebodies. Verification of individual tools at the test sites shows that airborne high-resolution radiometry and magnetometry, district-scale geochemical trace-element-in-quartz mapping, and ground-based electrical resistivity measurements are, among others, successful exploration methods to detect buried pegmatites of specific mineralogical types. The next step of tool set development aims to combine methods best suited for different pegmatite types and to optimize the toolset application in different geological, topographic, and climatic settings to increase exploration success and decrease environmental and societal impacts. The developed toolsets will change the focus of available exploration approaches from those applicable to large-volume, high geophysical host-rock contrast targets, such as copper porphyry, VMS, iron ore, and Pb-Zn deposits, to small-volume, low geophysical contrast pegmatite ore deposits.

Challenges Associated with the Future Supply of Battery Minerals and Black Mass

Laurance Donnelly
AHK International Ltd, Prescot, United Kingdom

Civilisation would not exist as we know it without the mineral resources upon which it relies. Minerals were important during Palaeolithic times and underpinned the Bronze Age, Iron Age, Industrial Revolution, and Digital Revolution. As we embark upon the Green Revolution and Battery Revolution, minerals are likely to continue to remain a fundamentally important resource. Globally, society demands a cleaner environment, such as a reduction in the combustion of conventional fossil fuels (i.e., coal, oil, and gas) to reduce rising atmospheric temperatures. This coincides with a projected increase in battery technology, driven in part by the electrification of vehicles in Europe, China, and the USA. However, this "greener" society may ultimately mean more exploration and mining are necessary because batteries contain numerous minerals including lithium, cobalt, nickel, copper, manganese, phosphorous, iron, aluminium, and carbon. From where will these minerals be sourced within the relatively short and optimistic transitional timeframes? Often, these can be located in remote geographical locations and in complex geological settings. Mineral exploration is a high-risk business, with relatively few exploration targets actually becoming a productive mine. What is more, it can take on average 15 to 20 years from exploration to mining and production. The challenges associated with the production of these minerals can be exacerbated by investor confidence variably, economic and market vulnerability, corruption, and increasing illegal and illicit minerals crimes in some countries. Could the emerging circular economy provide part of the solution? End-of-life spent batteries can be shredded and the minerals and metals content recycled via an intermediate product material known as "black mass." But what, specifically, is black mass? Data and information will be presented on the phase characterisation of black mass to demonstrate its textural, chemical, and phase complexity, which might predetermine a hydrometallurgical or pyrometallurgical processing route.

Recovery of Critical Metals from Copper Mine Tailings Utilizing Microanalytical Characterisation for Improved Geometallurgical Outcomes

Loren Nicholls[1], Ashley Norris[2], Anita Parbhakar-Fox[1]
1. Sustainable Minerals Institute, University of Queensland, Brisbane, QLD, Australia, 2. Norris Scientific, Oxford, United Kingdom

Cobalt is a critical metal commonly hosted in cobalt-sulphides and pyrites associated with copper mineralisation. In Australia, cobaltiferous minerals are rarely recovered in copper concentrators, especially if hosted in pyrite, and are often rejected to waste streams. Integration of circular economy principles into mining has encouraged exploration of historic and current mine wastes as supplementary cobalt resources. Understanding cobalt tenor and deportment in mine tailings is essential to modelling these repositories and informing appropriate mineral reprocessing flowsheets.

A case study to determine cobalt endowment in pyrites was undertaken on sulphidic mine tailings from a sedimentary-hosted copper deposit in Queensland, Australia. Field sampling obtained 585 samples from 26 auger holes down to 10m depths. High variability was observed in the facies (16 subgroups), geochemistry and cobalt endowment across the dam. Mineral Liberation Analysis (MLA) and Laser Ablation Inductively-Coupled Plasma Mass Spectrometry (LA-ICP-MS) analyses (1197 spots) were collected from 10 representative samples. The average cobalt grade in pyrite was 485ppm, with individual grains ranging from 1.8ppm to 8% (w/w). Several different pyrite chemistries were identified; including varying combinations of end-member compositions of cobalt-rich, arsenic-rich, and copper-rich pyrite. These pyrites were distributed throughout the tailings profile, however were weakly zoned from arsenic-rich pyrite in upper oxidised tailings (0 to 2m), to cobalt-rich pyrites below the oxidation front (4 to 8m), to copper-rich pyrite at depth (8 to 10m). Cobalt occurred as refractory substitution in the pyrite matrix, zones of enrichment (averaging 791ppm) and micro-inclusions of cobaltite and carrollite (averaging 3,334ppm). Only 45% of pyrites were liberated, with the majority locked in binary associations with non-sulphide gangue. Pyrite p80 (average 46μm) was much finer than the tailings particle size p80 (average 97μm). These geometallurgical observations will help inform the selection of mineral processing technologies if tailings processing is considered a viable option..

Ti-Zr-REE Heavy Mineral Sand Transport Paths and Concentration Processes Revealed by High-Resolution Airborne Radiometric Data

Anjana K. Shah[1], Robert H. Morrow IV[2], William R. Doar III[3]
1. U.S. Geological Survey, Denver, CO, USA, 2. South Carolina Geological Survey, Columbia, SC, USA, 3. South Carolina Geological Survey, Charleston, SC, USA

Shallow concentrations of rare earth elements (REE) can be highlighted by airborne radiometric data when the host minerals contain thorium. Since 2019, over twenty-five new high-resolution airborne magnetic/radiometric data sets have been or are being collected in the U.S. through the U.S. Geological Survey Earth Mapping Resources Initiative (EMRI). These data sets help broaden our understanding of mineralizing systems by providing a fuller geologic context for deposits. Here we examine coastal plain Ti-Zr-REE placer deposits of the southeastern U.S., where the heavy mineral assemblage typically includes Th-bearing monazite. In South Carolina, a pair of EMRI airborne geophysical surveys cover a 212-km-long, 134-km-wide transect from crystalline rock sources in the Piedmont to the Atlantic coastline. The new data show that in areas with Paleogene/Neogene strata closer to the Piedmont, elevated radiometric Th often corresponds to geologic strata of certain ages; ground truth samples show heavy mineral sand sources. The age dependence may be indicative of a favorable depositional environment dependent on changes in sea level. For Quaternary sediments closer to the coast, Th is elevated over highly localized areas within sand units, mostly where 1) the sands undergo concurrent reworking from tidal and marine processes, or 2) sands have undergone multiple episodes of reworking by marine processes, as suggested by the geomorphologic context. Additional information on large-scale sediment transport is provided by radiometric K, associated with immature minerals mica and K-spar. K is elevated over rivers with headlands in the Piedmont and over the youngest Quaternary sediments. Analyses of these anomalies suggest heavy minerals are transported from the rivers to offshore areas, moved 100 km or more by offshore longshore currents, and then redeposited onshore by marine and tidal processes. Such transport has implications for the provenance and composition of Ti-Zr-REE placer deposits from South Carolina to Florida.

Metals and Minerals in Europe – Demand, Exploration, Mining and More

Karen Hanghøj
British Geological Survey, Keyworth, United Kingdom

Raw materials are important for society in general, and for the transition to a green economy in particular. They are key for achieving the goals set out in COP21 and several of the United Nations Sustainable Development Goals, for implementing the European 2030 Agenda for Sustainable Development, and for the European Resource Efficiency Flagship. Metals, minerals, and materials and their sustainable supply and consumption are essential in the move towards a circular economy as laid out in the new EU Circular Economy roadmap.

When a mineral is both essential in use and subject to supply risk, it is considered critical. Emerging energy and mobility technologies create a strong demand for certain raw materials, and some of these are critical raw materials where demand will dramatically exceed current production in the next 10-15 years. Limited access to these materials might negatively impact the transition and reduce the competitiveness of European actors downstream.

Almost all research and other initiatives regarding critical raw materials are aiming at substitution, new materials and technologies, new methods to produce critical raw materials from non-conventional sources, and exploration activities. However, this seems to have not significantly improved the criticality situation, so the challenges perhaps lie elsewhere.

The mismatch between the time a commodity remains critical and the decadal scale of mining development makes it difficult and possibly unattractive for existing mining groups to enter the CRM business. What are we doing about it, and is it helping?

Crystalline Basement of Latvia – an Undercover Opportunity

Matiss Brants
University of Latvia, Riga, Latvia

The mineral-rich, geologically varied, and well-explored Baltic (Fennoscandian) Shield of Scandinavia seemingly dips into the abyss of the Baltic Sea, but in fact, it continues below the sedimentary cover of the Baltic states. In Latvia, this 400- to 2000-m-thick Phanerozoic sedimentary layer has been a critical obstacle for crystalline basement mineral exploration. The main exploratory efforts, including approximately 200 drill holes, were done from the 1960s till the 1980s, after which only reinterpretations and additional analysis of already-extracted material have been performed. But a changing European Union strategy has reignited the interest in searching for valuable minerals.

The main interest lies in the Kurzeme batholith—one of the largest rapakivi-type granite intrusions in the region, with an area of approximately 40,000 km^2, lying 2 km below the surface. The southern part of this layered intrusion has an increased concentration of Ni, Cu, Co, Cr, and platinum group elements.

Additionally, two high-quality Fe-quartzite deposits, Staicele and Garsene, have been of interest due to their high Fe-oxide content of approximately 50%, as well as additional mineralization. Although the depths of 700 and 1000 m from the surface were considered attainable for extraction operations, at the time of exploration, it was not economically viable.

There are also smaller potential ore deposits—for example, Strenci magnetic anomaly—with a notable concentration of elements, like Co, V, Cu, Ni, Sr, and Mn, deemed critical in the European Union, under a thinner, 500-m sedimentary cover and thus more accessible to exploration.

The crystalline basement of Latvia is a previously hidden opportunity for mineral exploration and potentially the extraction of minerals, hopefully utilising the best European practices of sustainable mining.

An Alternative Formation Environment for Kiruna-Type Iron Deposits: An Example from the Per Geijer IOA Deposit, Northern Sweden

Jo Miles[1], Rodney L. Allen[2], Patrick Krolop[1]
1. Luossavaara Kiirunavaara AB, Kiruna, Sweden, 2. Volcanic Resources AB, Enköping, Sweden

The Per Geijer iron oxide-apatite-REE deposit lies ~4 km northeast of the world-class Kiirunavaara orebody, northern Sweden. The >800-m-thick porphyritic Luossavaara Formation separates the Kiirunavaara orebody below from the Per Geijer ores above. Whereas wall-rock brecciation is commonly observed at the contact between the magmatic and/or hydrothermal Kiirunavaara ores and the Luossavaara hanging wall, we suggest the contact between the Luossavaara Formation and the overlying Per Geijer orebody displays a stratigraphic erosive relationship.

In this contribution, core logging observations are combined with lithogeochemistry and in situ mineralogy from thin sections across a 12-m transition zone from the Luossavaara Formation below into magnetite-apatite ore. Alongside QEMSCAN analysis, we aim to provide insight into the environment of formation at the onset of the Per Geijer ores.

The immediate footwall rock to Per Geijer ore comprises a sandy deposit with broken phenocrysts and other angular detritus derived directly from the underlying Luossavaara Formation. QEMSCAN imagery reveals a matrix of albite-quartz-K-feldspar containing angular albite phenocrysts partly overprinted by K-feldspar. This correlates with an Na-rich lithogeochemical anomaly at the base of the orebody that corresponds to the erosive contact. The footwall unit becomes K-rich downhole, reflecting K-feldspar alteration. Above the erosive contact, the magnetite-apatite shows planar lamination parallel to stratigraphy and localised graded bedding of magnetite and apatite clasts, reflecting suspension sedimentation and clastic mass-flow sedimentation, respectively, of magnetite and apatite, in a subaqueous environment.

Reworking of the Luossavaara Formation marks the subsidence and transgression from a subaerial to a deep-water subaqueous environment, thus assigning the Per Geijer ores to the Matojärvi Formation. We interpret the basal ores as hydrothermal magnetite-apatite precipitates or water-settled pyroclastic magnetite-apatite ash. These findings should influence exploration in northern Sweden as we consider magmatic and/or hydrothermal products reached the paleosurface and basin processes influenced the lateral extent of Fe ore.

Critical Metal Potential of Historic Mining Districts in the Lesser Caucasus and the Eastern Pontides, Central Tethyan Orogenic Belt

Robert Moritz[1], Samvel Hovakimyan[2], François Turlin[1], Şafak U. Sönmez[1]
1. University of Geneva, Geneva, Switzerland, 2. Institute of Geological Sciences of the National Academy of Sciences, Yerevan, Armenia

The Eastern Pontides and the Lesser Caucasus belong to a continuous mountain belt extending from northeast Turkey through Georgia, Armenia, and Azerbaijan to northern Iran. This mountain belt is well known for its Mesozoic and Cenozoic metallogeny dominated by base metal volcanogenic massive sulfide, porphyry Cu-(Mo), and precious/base metal deposits. The potential for critical metals and strategic by-products remains largely unknown. The best targets are certainly Re, Te, Li, and REEs, but also Sn, In, and Ge, which are discussed below.

Our best knowledge about various critical metals comes from the Cenozoic South Armenian Block located in the Lesser Caucasus. Eocene and Oligocene porphyry Cu-Mo mines in southernmost Armenia and adjacent Nakhitchevan produce Re as a by-product. The full potential and controls of Re enrichment remain unknown in the entire region. It is only recently that Re enrichment has been reported in Cretaceous porphyry systems of the Eastern Pontides. Some Cenozoic epithermal systems of the South Armenian Block contain abundant telluride minerals, but the Te endowment remains unconstrained. The South Armenian Block also hosts post-collision apatite-magnetite and alkaline magmatic ring complexes, containing nepheline syenite, where the potential of REEs and other critical metals has not been investigated. Another target is Li, which remains to be evaluated in post-collision Cenozoic clay-altered rhyolite, in tourmaline-bearing felsic shoshonitic intrusions, and in thermal waters.

There is only fragmentary knowledge about critical metals in older environments, such as Precambrian, Variscan, and Jurassic-Cretaceous settings. Recently, we recognized that Carboniferous intrusions of the Eastern Pontides have geochemical compositions that are comparable to Sn-bearing Variscan intrusions of western Europe. High In concentrations have also been reported in sulfides from a Jurassic-Early Cretaceous(?) porphyry-epithermal setting in southernmost Georgia. Finally, the Eastern Pontides host abundant volcanogenic massive sulfide deposits, where the potential of by-products (Ge, In) must be tested.

Ancient Mining, Recent Discoveries, and Future Potential for Critical Raw Materials in the Apuseni-Carpathian-Balkan Area of Southeast Europe

Irena Peytcheva[1], István Márton[2], Vasilios Melfos[3], Panagiotis Voudouris[4], Stoyan Georgiev[1]
1. Geological Institute, Bulgarian Academy of Sciences, Sofia, Bulgaria, 2. Dundee Precious Metals Inc., Sofia, Bulgaria, 3. Department of Mineralogy, Petrology and Economic Geology, Aristotle University of Thessaloniki, Thessaloniki, Greece, 4. National and Kapodistrian University of Athens, Athens, Greece

The Apuseni-Carpathian-Balkan (ACB) area of southeast Europe is one of the world's oldest mining districts and played a major role in the history of European civilizations until present day. The rising demand on metals and the increasing risk of supply with mineral raw materials are expected to revive the exploration and mining interest and change the vision for the ACB region. Three main magmatic and metallogenic belts with specific deposit types make the area a premier Cu-Au-(Pb-Zn-Ag) province. The Late Cretaceous (~94–67 Ma) Apuseni-Banat-Timok-Srednogorie belt formed on the European margin during closure of the Neotethys Ocean. The continental margin magmatic activity produced world-class porphyry Cu-Au(-Mo) deposits, high-sulphidation epithermal, as well as VMS-like Cu-Au-Ag(-Pb-Zn), IOCG(?), sediment-hosted and carbonate-replacement deposits. The Cenozoic Serbo-Macedonian-Rhodope belt (<55 Ma) is directly connected to post-collisional episodes of back-arc extension. The magmatism is sourced in the subcontinental mantle lithosphere and lower crust that were enriched during previous subduction. Diverse deposit types include intrusion-related Au-Ag-(W), mesothermal Pb-Zn-(Ag-Au) vein and carbonate replacement, porphyry Cu-Au-(Mo), high- and intermediate-sulphidation epithermal Cu-Pb-Au-Ag-Te, and sedimentary rock-hosted low-sulphidation epithermal Au(-Ag) deposits. The Miocene (13.61–7.24 Ma) Apuseni Mountains district is linked to magmatism generated during the late post-collisional deformation stages and north-eastward incursion of the Adriatic-Pannonian micro-continent that caused extensional strike-slip motions. The Au-Cu-(Te) mineralization is hosted in porphyry and epithermal deposits.

Recent discoveries in the ACB area result from exploration activities including field work, novel geochemical methods, geophysics, and geomodelling. The national legislations are formally not restrictive, but the social acceptance for mining might be low due to environmental and political issues. Critical raw materials have potential as main or by-products but require further studies and assessment.

Acknowledgments: The study is supported by Horizon Europe AGEMERA project (Agile Exploration and Geo-modelling for European Critical Raw Materials), grant no. 101058178.

From Planets to Protons - New Geoanalytical Workflows for Characterizing Rocks of Economic Importance

Alan R. Butcher
Geological Survey of Finland, Espoo, Finland

The dream of some geologists is to be able to image and analyse our planet at various scales, with different techniques, and in different dimensions. Although we are still far from reaching this ultimate goal, new geoanalytical workflows developed over the past 10 years or so have allowed us to get fairly close to imaging the rocks that make up the Earth down to the atomic scale, starting with space-borne instruments, progressing to land-based field-based technologies, and then ultimately using laboratory-based techniques that can unlock the secrets of petrogenesis that are so important for the complete understanding of ore-forming processes. This so-called bridging-of-scales-of-observation approach is a powerful tool for the development of geometallurgical practises, whereby mineralogical, textural, and geochemical information can be integrated to solve, predict, or troubleshoot challenges in exploration, mining, mineral processing, product traceability, and rehabilitation of ore deposits and associated mines. Such workflows are practiced routinely at the Geological Survey of Finland, and this talk will highlight some cases where this approach has led not only to the discovery of new deposits, but also enabled a fresh look at the formation of orebodies once they have been found and delineated, as well as an improved understanding of sustainable mineral extraction.

Opportunities for Rapid, Field-Based Sulfide Deportment and Trace Element Chemistry Using Laser-Induced Breakdown Spectroscopy (LIBS)

Cassady Harraden[1], Shaun Barker[1], Andrew Somers[2]
1. Mineral Deposit Research Unit, University of British Columbia, Vancouver, BC, Canada, 2. SciAps International, Woburn, MA, USA

The grain size, mineral associations, and trace element chemistry of sulfide minerals are key controls in the successful extraction, liberation, and recovery of economic metals. Traditionally, sulfide deportment and trace element chemistry are determined by microanalytical techniques such as scanning electron microscopy (SEM) or laser ablation-inductively coupled plasma-mass spectrometry (LA-ICP-MS). While providing high precision, these methods are completed on lab-prepared materials and as a result, sulfide deportment information is often insufficient to capture sulfide variability prior to designing the processing flowsheets. This lack of initial data can then result in costly modifications to already established milling and recovery circuits.

Laser-induced breakdown spectroscopy (LIBS) is an atomic emission method that can be used to rapidly assess mineral chemistry. Field-portable LIBS systems allow for low-cost, rapid analysis and require no sample preparation beyond a clean analysis surface. This method is capable of detecting most elements in the periodic table, including elements with low atomic numbers (e.g., Li, Na, and Mg), which are difficult to distinguish using most other portable field instruments. The rastering capabilities of some LIBS systems can produce geochemical maps reflecting intra-mineral chemical variability as well as textural and spatial relationships of minerals in the analysis area. Sulfide grain size, intra-mineral chemistry, and mineral grain boundary associations (e.g., chalcopyrite in contact with bornite) ultimately have implications for the successful, economic extraction of raw materials and treatment of tailings in the mineral industry. The application of field-based portable LIBS provides an opportunity for rapid and cost-effective sulfide deportment and trace element chemistry on most geological material available throughout the exploration and mining cycle. The ability to collect fit-for-purpose, high-density deportment data through a project's life provides better statistical sampling of ore and gangue minerals, which can be used to build more robust geometallurgical models and mine plans.

Re-Assessing the Resource Potential of Sea-floor Massive Sulfide Deposits

John W. Jamieson[1], Mark D. Hannington[2], Sven Petersen[3], Thomas Monecke[4]
1. Memorial University of Newfoundland, St. John's, NL, Canada, 2. University of Ottawa, Ottawa, ON, Canada, 3. GEOMAR Helmholtz Centre for Ocean Research Kiel, Kiel, Germany, 4. Colorado School of Mines, Golden, CO, USA

Metal-rich deposits that form on the modern ocean floor, including sea-floor massive sulfide (SMS) deposits, manganese nodules, and cobalt-rich crusts, are considered by many to be a future resource to satisfy the growing metal demand for the green economy transition. An understanding of the resource potential of the sea floor, including the distribution of deposits and deposit sizes and grades, is required to assess whether marine minerals are indeed a viable future source for metals. However, compared to traditional land-based exploration techniques, the logistical and technological challenges associated with deep-sea mineral exploration, combined with the uncertainty with respect to the environmental impacts of sea-floor mining, remain an obstacle to fully understanding the potential for deep-sea mining to significantly impact global metal supply. Resource uncertainty also limits our ability to develop effective policy to regulate this emerging industry.

In 2011, we produced the first ever resource assessment for hydrothermally derived SMS deposits, in which a modest ~600 Mt of massive sulfide deposition was predicted to occur along the global neovolcanic zones of ocean floor. In this presentation, we will discuss how our understanding of the resource potential has evolved after more than a decade of subsequent exploration off-axis. We will examine the rates of discovery, the development of new geological models for deposit genesis, and emerging exploration technology and approaches. We will provide a critical assessment of our understanding of how hydrothermal deposits are sampled and how deposit grades and tonnages are reported. The results of this re-assessment of the global resource potential of SMS deposits will provide insights into the potential for these deposits to contribute to society's pursuit of new mineral resources for the green economy.

Machine Learning Algorithms as a Tool for Data-Driven Interpretation of Multiphysics Data Sets in Underexplored Areas

Artem Gorbunov[1], Joseph Lebbie[2], Ahmed Swaray[2], Tim Archer[1]
1. Reid Geophysics Limited, Eastbourne, United Kingdom, 2. National Minerals Agency, Freetown, Sierra Leone

Machine learning (ML) and artificial intelligence (AI) algorithms have become very popular in various mining industry sectors in recent years. The increasing popularity of the algorithms can be explained by their extreme efficiency in extracting valuable and objective information from enormous data volumes with minimal human input. Some of the algorithms have become an essential part while studying areas with dense data coverage. Various cases have shown that interpretation of geological-geophysical data sets can benefit from the application of ML techniques for tasks such as delineation of lithological units (Kuhn et al, 2018; Melo et al, 2021), anomaly detection (Zhong et al, 2019; Triskova, 2017), prospectivity analysis (Yeomans et al, 2020; Granek, 2016), etc. In this paper, we demonstrate the usefulness of supervised and unsupervised machine learning algorithms as a tool for inferring complementary geological information while studying poorly explored areas. As input data for the algorithms demonstrated, we use results of a high-resolution nationwide airborne geophysical survey flown in Sierra Leone in 2019 (Gorbunov et al., 2023), terrain measurements, and satellite imagery. We show that convolution neural networks allow extracting valuable information on structural features from acquired geophysical data. Moreover, we present results of self-organizing map clustering as a pseudo-lithological map.

The huge amount of data collected required efficient (but also sparse) interpretation techniques. Using high-quality geophysical and open-source remote sensing data sets has allowed us to confirm, challenge, and complement previous results of geological-geophysical interpretation of acquired data and to chalk out previously unknown prospective areas for further ground verification and detailed evaluation. The results of applying ML algorithms to geological-geophysical data sets have confirmed their appropriateness and value due to the fully data-driven approach.

Unlocking the Potential of Predictive Geoscience Through the Adoption of GeoData Science

Glen T. Nwaila[1], Steven E. Zhang[2], Julie E. Bourdeau[3], Yousef Ghorbani[4]
1. University of Alaska, Anchorage, AK, USA, 2. Geological Survey of Canada, Ottawa, ON, Canada, 3. Department of Civil, Environmental and Natural Resources Engineering, Luleå University of Technology, Luleå, Sweden, 4. School of Chemistry, University of Lincoln, Joseph Banks Laboratories, Lincoln, United Kingdom

The mineral value chain is an evolutionary and complex system that is currently subject to changing internal or external drivers or evolutionary pressure. One major driver is the increasing demand in quantity and velocity of critical raw materials (CRMs). The mineral value chain is evolving to adapt to the evolutionary pressure through several axes of integration. Two key axes are emerging—across (e.g., predictive geometallurgy and material fingerprinting) and transdisciplinary (e.g., GeoData Science) techniques within the mineral value chain.

Innovation and integration along these axes can exhibit an evolutionary characteristic, which gradually adapts primarily traditional processes, or a transformative characteristic, which creates entirely new processes that expand the niche or scope of traditional domains. We demonstrate transformative integrations of transdisciplinary methods using geochemical exploration, geologic modeling, and mine/metallurgical waste valorization. We provide examples of challenges and rewards in our journey and identify potential solutions.

A major challenge in GeoData Science is the hidden bifurcation of traditional notions of data quality and modern purposes. This challenge encompasses data engineering and specification, scientific philosophies, and the distinction between evolutionary versus transformative integration of transdisciplinary techniques. However, we demonstrate that these challenges can be overcome with the aid of legacy data and discipline knowledge, scientific philosophies, and creative thinking. We share examples of the deployment of predictive geosciences from geochemical exploration (Trans-Hudson orogen and Churchill province, Canada), mineral resources estimation, and predictive geometallurgy (Witwatersrand Basin and Bushveld Igneous Complex, South Africa). Using predictive geoscience, those working in the field can make significant improvements in the sustainable management of CRMs, leading to an efficient and sustainable future.

Seafloor Massive Sulfide Deposits on the Arctic Mid-Ocean Ridges: an Insight Into Ore-Forming Processes Along Ultraslow Spreading Ridges

Sabina Strmic Palinkas[1, 2], Rolf B. Pedersen[2], Fredrik Sahlström[1, 3], Marie Wold[2], Eszter Sendula[1]
1. UiT The Arctic University of Norway, Tromsø, Norway, 2. University of Bergen, Centre for Deep Sea Research, Bergen, Norway, 3. Pontifical Catholic University of Peru, Lima, Peru

The Arctic Mid-Ocean Ridges (AMOR), an oceanic ridge system located north of the Arctic circle (66°N), consists of several slow and ultraslow spreading ridges associated with abundant hydrothermal activity. The ultraslow Mohns ridge segment of the AMOR hosts numerous seafloor massive sulfide (SMS) deposits and represents a natural laboratory to study recent hydrothermal ore-forming processes that drive their formation.

One of the main characteristics of the SMS deposits found along the Mohns ridge is their diversity in terms of mineral assemblages and metal contents. This study brings new lithology, mineral chemistry, stable isotope, and fluid inclusion data obtained from the selected types of SMS deposits.

The Loki's Castle active SMS deposit (73°34′N, 8°9′E; 2,400 m) is located near the summit of a 30-km-long and 800-m-high axial volcanic ridge (AVR). The deposit is characterized by a basalt-hosted sediment-influenced type of the Zn-Cu-Pb mineralization enriched in Au, Ag, and Tl. The Mohns Treasure inactive SMS deposit (73°N, 7°E; 2,600 m depth) is hosted by basalts. The mineralization can be subdivided to the Cu-, Zn-, and Au-rich types. The Fåvne active SMS deposit (72°45′N, 3°50′E; 3,000 m) occurs at the floor of the rift valley northwest of an AVR. The hydrothermal activity is strongly controlled by a normal NE-SW–trending fault that underlies the western portion of the field. The deposit is characterized as a basalt-hosted ultramafic rock-influenced type of the Zn-Cu-Co mineralization. The Gnitahei extinct SMS deposit is hosted by the footwall of the normal fault that underlies The Fåvne deposit. The mineralization is characterized by an enrichment in Au.

While their future role as a source of metals, including those from the critical raw materials (CRMs) list, remains uncertain, recent SMS deposits provide unique insights into the hydrothermal processes that may support both deep-sea and onshore exploration.

New Mineralogical Data on the Mineralizations and Hydrothermal Alteration Zones of the Allumiere-Tolfa Historic Mining District, Central Italy

Francesca Corrado[1], Barbara Marchesini[2], Giuseppina Balassone[1], Eugenio Carminati[2], Stefano Tavani[1], Nicola Mondillo[1, 3]
1. Department of Earth, Environment and Resources Science, University of Naples “Federico II”, Naples, Italy, 2. Department of Earth, University of Rome "La Sapienza", Rome, Italy, 3. Natural History Museum, London, United Kingdom

The Tolfa-Allumiere mining district is located approximately 60 kilometers northwest of Rome (Italy) and has been characterized by a long-lasting history of exploitation for base metals (Pb, Fe), alunite, and kaolinite mineralizations. These deposits are genetically associated with the emplacement in an arc setting of a Plio-Pleistocene volcanic dome consisting of latites and rhyolites, which intruded sedimentary carbonate and siliciclastic country rocks. The intrusive event first produced contact metasomatism of the limestones and lately triggered hydrothermal activity, which caused widespread epithermal alteration of the volcanic rocks.

This study presents the preliminary results of mineralogical and geochemical investigations aimed at better understanding the development of the hydrothermal activity and the genesis of the mineralizations in the Tolfa-Allumiere area. The analysis performed on the altered volcanic rocks collected in old pits and outcrops allowed identification of three main alteration facies: residual silica, advanced argillic, and argillic facies. The residual silica facies is commonly sulfide-bearing, containing pyrite, As-pyrite, and cinnabar, with traces of sphalerite, galena, and chalcopyrite. In the advanced argillic facies, alunite occurs as aggregates of planar crystals in veins and cavities, is commonly zoned to natroalunite, and is intimately associated with svanbergite. Smectite marks the zones interested by argillic alteration. Regarding the metasomatized host rock, the carbonates show recrystallization with presence of sulfides (pyrite and cinnabar), Sr-sulfates, thorutite, thorianite, and fluorite. The epithermal assemblage in the volcanics and the metasomatized sedimentary rocks are affected by supergene alteration and gossan formation, which overprint the hydrothermal assemblage.

In conclusion, the preliminary results are in agreement with the occurrence of a magmatic-related hydrothermal system in the Tolfa-Allumiere area. More detailed investigation is needed to determine if mineralized zones of greater economic importance occur in the surroundings of the investigated sites as proposed in typical genetical models of precious metals epithermal systems.

Discover the Un-Discovered: New Exploration Concept for the European Kupferschiefer Copper Belt from Rudna (Poland) to Richelsdorf (Germany)

Volker Spieth[1]
1. Mineralogical Institute University Stuttgart Germany, TUCSON, AZ, USA, 2. VS.GLOBALMETAL LLC, TUCSON, AZ, USA

An assessment by the USGS of the European Kupferschiefer Copper Belt estimates in the permissable tracts a discovery potential of 31 deposits with 50% probability, including 14 deposits with 90% probability. Current Polish and German exploration shows positive results confirming the literature and expert-based USGS predictions.

Geological, mineralogical, and geochemical observations have shown in Spremberg, Sangerhausen, and Rudna that metal concentrations are associated with Zechstein hot hydrothermal volcanic brine mud slurries that surfaced onto a shallow marine Weissliegend silica extrudite paleosurface. Rich copper-silver-gold-polymetallic mineralization is present. The deep-seated origin of the Zechstein-Kupferschiefer-Weissliegend mud brine and mud volcanic system is manifested by exotic element chemistry (PGE, Co, Ni, Cr, V, Li) and exotic minerals such as talc, serpentine, and clinochlore.

The dynamic origin of the European Copper Belt is the continent-sized area of the Permian-aged, W-E–directional failed rifting in the Saxo-Thuringian Zone bordered in the north by the Rheno-Hercynian Zone and in the north-east by the Variscan Front. This zone shows the typical multi-layered metal deposits that were mined in Richelsdorf and Mansfeld, have been discovered in Spremberg, and are in production as world-class deposits in the Lubin-Rudna district.

The sulfide-metal-rich mineralized Weissliegend sand extrudites cover km-sized large areas and can be used as exploration targets. They can be located by geophysical methods and are known from previous geological survey assessment programs. The targets are flat lying and can easily be drilled. Successful drill holes can be expanded horizontally at low cost with oil- and gas-field drilling technology. The exploration concept components can be demonstrated. The successful exploration discovering the deposit at Spremberg is the proof. The predictions of the USGS from 2010 could become reality in the European Kupferschiefer Copper Belt.

Mantle Source Enrichment in the Evolution of One of Europe's Largest Rare Element Provinces - the Gardar Province, South Greenland

Charles D. Beard[2], Anouk M. Borst[3], Adrian A. Finch[1], William Hutchison[1], Robert Webster[1]
1. University of St Andrews, St Andrews Fife, United Kingdom, 2. University of Cambridge, Cambridge, United Kingdom, 3. University of Leuven, Leuven, Belgium

One of the big questions in economic geology is why some regions and periods of Earth history gave rise to significant ore deposits and others did not. For mantle-derived magmatic ore deposits, such as the carbonatite or alkaline igneous rocks targeted for high-field strength elements (HFSEs) and rare earth elements (REEs), the composition and temperature of the mantle control the abundance of metals and ligands within resultant magmas, and thus the potential for the formation of the ore. Yet the processes that form enriched mantle sources are poorly constrained.

The mid-Proterozoic (1.32–1.14 Ga) Gardar Province in South Greenland contains several globally significant rare earth, niobium, vanadium, and titanium projects, formed during breakup of the Columbia/Nuna supercontinent. Gardar intrusions have multiple surface ($\delta^{34}S$, $\Delta^{33,36}S$) and Archaean ($\Delta^{33,36}S$, εHf, εNd) crustal signatures inherited from their mantle source. These enriched signatures vary across the province, showing temporal and spatial heterogeneity.

We interpret the metallogeny of the Gardar rift to be controlled by mantle source composition, which in turn reflects the geodynamic history and the recycling of continental sediments through subduction along the margin of the North Atlantic craton, presumably during the assembly of Columbia/Nuna. Our data provide evidence for fertilisation of the mantle by subduction as a precursor for these HFSE and REE deposits—such a model may be widely applicable for other Proterozoic critical metal deposits associated with the breakup of Columbia/Nuna.

The Iberian Pyrite Belt, a Premier Mining Jurisdiction

James Royall
Pan Global Resources, Huelva, Spain

The Iberian Pyrite Belt (IPB) is one of the world's premier mining jurisdictions and one of Europe's most important mining centres. Located in southwest Europe, the 250-km-long metallogenic belt extends from near Seville in Spain to south of Lisbon in Portugal.

The IPB hosts over 80 Carboniferous age volcanogenic-hosted massive sulphide (VMS) deposits associated with intermediate to acid volcanic sequences and associated sedimentary packages that were strongly affected during the Variscan Orogeny and by a lesser amount during the Alpine Orogeny. The IPB hosts some of the largest VMS multimetal sulphide deposits in the world, such as Rio Tinto, Tharsis, and Aguas Teñidas, and also some with very high grades such as Las Cruces, Neves Corvo, and Magdalena.

The IPB has a long and important mining history that dates back to the Chalcolithic period nearly 2,000 years BC, with important expansion periods during the Bronze Age, the Roman Period, later during the Industrial Revolution, and in the modern era of the 20th century.

Currently there are seven large operating mines in the belt, operated by major international mining companies, and the area is being very actively explored with several additional advance projects in both Spain and Portugal. In both countries there is a well-established mining code and supportive administration. The region has excellent road and rail infrastructure, seaports, and also a smelter.

Pan Global Resources Inc (TSX PGZ) has an exciting exploration project in the southeast of the belt that is adjacent to the Aznalcollar Los Frailes Project and only a short distance from First Quantum Minerals' Cobre Las Cruces mine. In 2019 the Company completed its first drill holes at the La Romana target resulting in the discovery of a near-surface copper-rich VMS deposit with significant tin and associated silver. The company has drilled over 130 holes at the La Romana deposit and continues to test the extensions as part of a property-wide exploration program, with multiple drill rigs targeting additional VMS mineralization within a more than 5700-hectare contiguous land package.

Linking Sn-Li-(W-Cu) Mineralization to Post-Collapse Caldera Evolution Through Cassiterite LA-ICP-MS U-Pb Geochronology

Dino Leopardi[1], Axel Gerdes[2], Richard Albert[2], Jens Gutzmer[1], Bernd Lehmann[3], Mathias Burisch[1, 4]
1. Helmholtz-Zentrum Dresden-Rossendorf, Helmholtz Institute Freiberg for Resource Technology, Freiberg, Germany, 2. FIERCE, Department of Geosciences, Goethe University Frankfurt, Frankfurt, Germany, 3. Mineral Resources, Technical University of Clausthal, Clausthal, Germany, 4. Department of Geology and Geological Engineering, Colorado School of Mines, Golden, CO, USA

The Sadisdorf Sn-Li-(W-Cu) prospect, eastern Erzgebirge/ Krušné hory, is one of several Sn-W vein- and greisen-style deposits that are associated to the Altenberg-Teplice Caldera (ATC) system. Cassiterite mineralization is hosted by greisen alteration and stockworks intimately related to the composite Sadisdorf leucogranitic porphyry as well as a NE-SW–striking vein system. The latter appears spatially—and possibly also genetically—unrelated to the Sadisdorf leucogranitic porphyry. Here, we applied LA-ICP-MS U-Pb geochronology to 16 cassiterite samples from the Sadisdorf prospect in order to constrain the timing of ore formation in the regional context of the evolution of the ATC.

All cassiterite ages range between 315.1±3.3 and 311±4.0 Ma and overlap within uncertainty across the Sadisdorf prospect—irrespective of the style of mineralization. The calculated weighted mean average ages for greisen/stockwork and vein-hosted mineralization are 313.07±0.56 and 313.2±1.6 Ma, respectively. This suggests that cassiterite mineralization at Sadisdorf is related to one single magmatic-hydrothermal event, albeit possibly associated to two spatially separated magmatic centers.

The cassiterite ages also coincide with the intrusion age of several microgranitic and rhyolitic dikes (314-313 Ma) that were emplaced late during the collapse of the ATC. This relates Sn mineralization to the late stages of the caldera evolution and suggests that Sn mineralization in the eastern Erzgebirge occurred 5-12 m.y. later than previously assumed. The occurrence of fertile intrusions during the late stages of a caldera evolution is documented elsewhere (e.g., Mt. Aetna, USA and Mt. Pleasant, Canada) and provides useful criteria for regional exploration targeting.

SEG 2023 Conference: Resourcing the Green Transition

Integrated Geological, Geochronological and Geochemical Approach to Exploration in the Hod Gold District, Türkiye: Implications for the Origin of the Hod Maden Au-Cu Deposit

Nick J. Callan[1], Timothy Baker[2], Orkut Baysal[3], Thomas Herbort[2], Ahmet Tukac[3]
1. NJC Geological and Exploration Services, Taunton, United Kingdom, 2. Eldorado Gold Corp, VANCOUVER, BC, Canada, 3. Tüprag Metal Madencilik, Ankara, Turkey

Since discovery of the Hod Maden "hybrid VMS"- type Au-Cu deposit (2.8 Moz Au) in 2015 by Lidya Madencilik/Marianna Resources, Eldorado Gold/Tüprag Metal Madencilik has been conducting detailed exploration within the surrounding district. Driving this initiative has been a comprehensive program of geological mapping complemented by U-Pb dating and geochemical fertility studies, which together have provided a robust geological framework for ongoing exploration. Key findings include (1) that highly prospective Late Cretaceous-age strata, and more specifically the distinctive volcano-sedimentary/felsic volcanic interval ("Hod Maden Complex", HMC), which constitute the immediate host to the Hod Maden deposit, extend for at least 7 km north along strike from the Hod Maden deposit; (2) locally hydrothermally altered, concordant felsic flow domes within the HMC at the Kurak, Hizarli, and Derinkoy prospects overlap in age (86-84 Ma) with felsic domes spatially associated with Au-Cu mineralization at Hod Maden (85-84 Ma); and (3) indicators of submarine hydrothermal activity contemporaneous with rhyolitic volcanism include stratiform and/or stratabound zones of hydrothermal alteration, exhalative-type chemical sedimentary horizons, banded disseminations of base-metal sulphides within fine-grained volcaniclastic units, and discordant quartz-sulphide veins associated with hydrothermally altered zones. Trace element geochemistry of rhyolites in the district are consistent with fertile VMS environments, whereas fertility studies on zircon suggest the Cretaceous suite is unproductive for porphyry copper systems. Geochronology also highlights a younger (83-80 Ma) intrusive suite of intermediate to felsic composition that intruded the thick andesite-basalt units in the stratigraphic footwall and hanging wall to the HMC (e.g., Kırmızıkaya and Hızarlıyayla). These younger intrusions are overprinted by IS-type base metal-rich quartz vein mineralization within propylitic and advanced-argillic alteration zones. These relationships suggest a possible intrusion-related or deep epithermal overprint on pre-existing volcanogenic mineralization within the HMC following Late Cretaceous thickening of the volcanic pile.

New Investigations on the Geology and Mineralogy of the S'aliderru Bentonite Deposit (Sardinia, Italy)

Sebastiano Coticelli[1], Piergiulio Cappelletti[1], Concetta Rispoli[1], Giuseppina Balassone[1], Fabio Granitzio[2], Nicola Mondillo[1]
1. Dipartimento di Scienze della Terra, dell'Ambiente e delle Risorse, Università degli Studi di Napoli Federico II, Napoli, Italy, 2. Clariant Prodotti (Italia) Spa, Milano, Italy

Bentonites are a group of rocks formed essentially by smectites, mainly belonging to the montmorillonite-beidellite series. Bentonites derive from the alteration of volcanic and/or sedimentary rocks, with a prevalent siliceous-vitreous component, by aqueous fluids that operate at temperatures between 90° and 120°C. During the alteration, chemical elements are remobilized and the composition and structure of the minerals are modified. The resulting mineral assemblage depends on the nature of the protolith and the composition of the altering fluids (Cuadros et. al., 1999). Bentonites have a wide range of properties (CEC, thixotropy, swelling, viscosity, water absorption) that make them very important industrial minerals, which are periodically tested also by the EU for possible inclusion in the list of critical raw materials. The aim of the present study is to investigate the genesis of the S'Aliderru deposit, located in north-western Sardinia (Italy), which represents one of the largest bentonite deposits of the Mediterranean area. The bentonite formed at the expense of arc-related Cenozoic tuff and ignimbrite products. From fieldwork, distinct bentonite horizons were identified, belonging to multiple pyroclastic flows that were progressively emplaced in a pull-apart basin hosted in the Mesozoic carbonate basement. The occurrence of travertine bodies and conglomerate and sand lenses hosted in the pyroclastic sequence suggests that the deposition and alteration took place in a marine-transitional environment. Mineralogical and petrographic analyses on several samples collected in the S'Aliderru deposit allowed the classification of the bentonites into four classes and discrimination of the featuring minerals of the other lithologies. Petrographic analyses confirmed that the alteration of the volcanic glass into clay minerals occurred in a marine environment, but the presence of barite as well as other phases suggests that low-temperature hydrothermal fluids also contributed to determining several peculiar features of this deposit.

ESG Valuation Model for the Mining Industry

Lisa Ek, Niklas Grönberg
Nordic Strategy Partners, Luleå, Sweden

If the mining industry is to claim we are taking sustainability and responsible mining seriously, we need to prove that decision-making at all levels is affected by these factors. Our project aim is to develop a new decision model for the mining industry that incorporates ESG [environmental, social, governance] factors in a comparable way to today's more easily calculated directs costs and potential revenues—to put monetary value to soft factors. With a new model, the industry can not only show that ESG factors have a real impact on investment and project decisions, but also that by considering these factors, new types of financially relevant short- and long-term risks and opportunities will be included, and hence improve financial transparency and performance.

A pre-study in collaboration with stakeholders from the Swedish mining industry, such as Boliden, LKAB and Lunding Mining, showed that that today's decision-making is mainly governed by net present value calculations [NPV], which in most cases are short term and lack ESG factor impact. When ESG factors are considered, it is as a stand-alone framework. The analysis indicates that the current method, with inadequate models, makes mining companies miss opportunities that can have significant positive impact.

Together with the Swedish mining industry, we have developed a hypothesis of an ESG valuation model with focus on a risk-driven, top-down approach to attach financial values to ESG factors. Model design is based on a global best-practise analysis, with input from mining industry actors as well as investors, lenders, other financial stakeholders, and industries outside of mining. At SEG 2023, we would like to present the model and engage in discussions on how to improve it further and implement it outside of Sweden.

Gamification for Technology-Enhanced Education in the Mineral Value Chain: Mining Simulation Game for Teaching the Geometallurgy Concept

Lilian Schleret[1], Glacialle Tiu[1], Christina Wanhainen[1], Nils Jansson[1], Jan Jan Rosenkranz[1], Yousef Ghorbani[1, 2]
1. Luleå University of Technology, Luleå, Sweden, 2. University of Lincoln, Lincoln, United Kingdom

The similarity between gaming and learning has led to a growing interest among educators in gamifying the learning process to make it more appealing and engaging. The incorporation of game elements in non-gaming situations to motivate learners is what gamification in education involves. In preparing young adults for future roles in the mineral industry, digital teaching and learning concepts, coupled with virtual technologies like gamification, can be useful in helping students grasp complex concepts. As part of the "Innovative and hybrid education in mining value chains - Nordic context" work package within the EIT Raw Materials project *Emeraldinho*, this study aims to improve student learning in the area of geometallurgy by developing an educational mining simulation game. The game allows the students to immerse themselves in decision-making aspects in the different stages of the mining value chain: exploration geology, mining, mineral processing, and mine rehabilitation. The game utilizes a three-dimensional synthetic exploration and geometallurgical database of an iron ore deposit. It is the student's task to find, mine, and process the ore in the best possible way given the engineering, financial, and environmental constraints in the game. The student is also provided all the information required to establish a mineralogy-based geometallurgical program that can be utilized to create predictive models necessary for successful mine planning. This hands-on approach enables students to comprehend concepts beyond theory and see how changes in one area can impact other sections of the mining value chain. The game's geometallurgical approach promotes a holistic and responsible mindset and encourages sustainable mining and beneficiation methods to minimize the mineral industry's environmental impact.

Applying Machine Learning Methods to XRF Data to Quantify the Acid-Generating Potential of Waste and Other Intrinsic Rock Properties

Karim Halaseh, Hannah Lang
MineSense Technologies Ltd, Vancouver, BC, Canada

MineSense Technologies equips shovels and loaders with XRF sensors to provide real-time bucket measurements of material as it is being excavated. The primary purpose of the measurements is to predict the grade of material in each bucket. These grade predictions are integrated with the mine's fleet management system to provide synchronous ore sorting to optimize recovery and reduce the quantity of waste processed by the mill — adding both economic and environmental value. Further investigation of these XRF spectra using key mine data sets and the application of techniques from machine learning has led to the discovery of relationships between extracted features of the spectra and other intrinsic rock properties. A noteworthy example is the classification of NAG and PAG waste, which can be used to sort trucks for optimal waste management with significant downstream environmental implications. Another is the classification of material by copper mineral speciation, which is consequential for mill processing performance. These associations are identified by engineering problem-specific feature sets, leveraging physical and geological knowledge as a starting basis to extract informative predictor variables from the XRF spectrum that are both directly and indirectly related to elemental composition. Care must be taken to account for confounding factors such as the spatiotemporal correlation of these properties, and these models can be validated by measuring the relationships between predictions and relevant mill processing KPIs downstream. Here we discuss the methodology used in developing and validating a rock property prediction model with XRF data, as well as the broader implications for sustainable mining.

PROMT: Philippines Remediation of Mine Tailings

Gawen R. Jenkin[1], Carlo A. Arcilla[2], Daniel J. Smith[1], Jessie O. Samaniego[2], PROMT consortium
1. Centre for Sustainable Resource Extraction, University of Leicester, Leicester, United Kingdom, 2. Philippine Nuclear Research Institute, Manila, Philippines

Tailings storage facilities at both operational and closed mines pose environmental hazards; failure could cause contaminated materials to be released, affecting people and ecosystems. On the other hand, tailings are significant resources of unrecovered metals, especially in older facilities that were produced using less efficient mineral processing. The fact that the material is already finely ground makes reprocessing of tailings appealing, but challenges here include the danger of damaging the structural integrity of the tailings by re-excavation, handling costs, and the potential for release of contaminants.

PROMT involves over 40 UK and Philippine researchers working together with mining companies in the Philippines to develop new sustainable technologies to manage Cu-Au mine tailings, to recover metals, and to make soils to support plant growth and allow the land to be reused. PROMT combines three innovative science areas:

1. Demonstrating how tailings storage facilities can be monitored in real time to allow reactive management to environmental changes, including in situ leaching. We are employing emerging technology in geophysical tomography and remote sensing to monitor and understand tailings behaviour in 4-D.

2. Investigating novel environmentally benign solvents to dissolve metals in situ from tailings, in tandem with electrokinetics to infiltrate solvents into low-permeability material. This will allow more metals to be recovered with economic value and will decontaminate hazardous components.

3. Understanding how plants and microbes colonise mine wastes and how this is affected by the use of solvents, and identifying the best ways to promote biological growth. This will not only rehabilitate the land and allow it to be reused for agriculture or wildlife, but will also minimise environmental hazards by improving the stability of the tailings and decreasing toxicity.

Microbial Impacts on the Circular Economy via Metal Mobility and Carbon Sequestration

Thomas R. Jones, Anita Parbhakar-Fox, Gordon Southam
University of Queensland, Brisbane, QLD, Australia

Climate change, caused by greenhouse gas emissions, has led to an increasing demand for companies to 1) reduce their carbon footprint and 2) explore green energy technologies. Microorganisms can accelerate the nucleation and growth of secondary carbonates, a process that can sequester carbon dioxide from the atmosphere. This can be particularly useful in mafic and ultra-mafic materials, such as kimberlite residue, produced as a by-product of diamond mining. This transition to green energy is different from traditional methods of energy generation and as a result requires very different ingredients—known today as critical metals. These metals are more difficult to mine and are failing to meet an ever-growing demand. Microorganisms can play a significant role in the extraction of critical metals from low-grade ore and tailing materials that are not amenable to conventional processing methods. Bioleaching, which uses microbes to extract metals by producing sulfuric acid, is increasingly being used for metal extraction.

To track the effectiveness of these microbial interventions, various analytical tools such as X-ray fluorescence mapping (XFM), mineral liberation analysis (MLA), and high-resolution SEM have been used in conjunction with geochemical analysis to track metal mobility and secondary carbon precipitation across time. Bioleaching columns of both the Princess Creek and Mt. Garnet Tailings have shown the accelerated release of Pb, Ni, Cu, and Li relative to abiotic controls, while pilot study experiments with mined kimberlite material from the Venetia diamond mine reveal that bio amendments accelerate the increase in precipitation of secondary carbonate materials—to a CO_2e offset of around 20% per year when scaled.

Overall, the application of biotechnology in the mining life cycle can contribute to the green transition by aiding in the extraction of critical minerals, remediation of mining waste, and carbon sequestration via bio-mineralisation.

Influence of Mineralogy, Texture, and Grain Size on Density and Magnetic Susceptibility Within the Kiruna Mining District, Sweden

Ervin Veress, Oskar Rydman, Mathis Warlo, Joel Andersson, Tobias E. Bauer
Luleå University of Technology, Luleå, Sweden

The shift to fossil-free energy systems will lead to increased demand for minerals used in clean-energy technologies. In addition, the decelerating rate of new shallow ore deposit discoveries further amplifies the challenging circumstances for mineral exploration that must focus on deeper exploration targets. As improving our understanding of the deeper underground becomes more important, reliance on geophysical exploration techniques increases. In greenfield exploration or areas without dense drill hole coverage, the geological understanding of the subsurface is limited and the interpretation of geophysical responses requires a good understanding of the physical and chemical properties of the rocks. Petrophysical characterization of the lithological units serves as a link in integrating geophysical models and understanding their geological implications.

The Kiruna mining district is situated in northern Sweden and is recognized as the type locality of the Kiruna-type iron oxide-apatite (IOA) deposits, hosting the Kiirunavaara orebody. The stratigraphic sequence constitutes well-preserved Rhyacian greenstones followed by Orosirian volcanic, volcanoclastic, and sedimentary rocks, affected by low-grade metamorphosis. Based on 32 outcrop samples, we investigated the signatures of density and anisotropic magnetic susceptibility in accordance with lithology, secondary geological processes, location, and porosity within the Kiruna mining district. The bulk (modal mineralogy), grain, and texture control of the samples were studied using whole-rock lithogeochemistry, micro-X-ray fluorescence, and X-ray computed tomography. Crossplots of density and magnetic susceptibility values result in clusters corresponding to specific lithological units, while alteration zones can be effectively recognized if the spatial location of the samples is considered. The resulting characterization of the local lithologies supports the analysis of geophysical measurements in combination with geological information and facilitates the integration of the local geological framework with geophysical surveys. Furthermore, integrated characterization of an active mining district serves as a base for further development of exploration campaigns in similar geological settings.

It's Never Too Early for Geometallurgy – Geometallurgy in Exploration

Melissa Gregory
Rio Tinto Exploration, Melbourne, VIC, Australia

Most geologists would agree that geometallurgy is the integration of geological and metallurgical data to build a model that predicts the mineral processing response of all domains in an orebody. Metallurgical testwork data are key to such models, but there is still much that can be done in terms of geometallurgy well before the first metallurgical test samples are selected. From the first exploration drill hole into significant mineralization, we can start to think about the geometallurgical characteristics. As more drill holes are completed, geometallurgical domains can be interpreted. To do this we need to understand the geology, geochemistry, and modal mineralogy of the mineralization and interpret the mineralogy data in terms of mineral processing options.

At the exploration stage, for base (± precious) metal deposits, we are looking at whether the mineralogy would be suitable for metal recovery via flotation-based mineral processing or leaching. From a flotation viewpoint, we can use mineralogy data together with the known flotation response of the various sulfides (and other minerals) to assess the applicability to our mineralization. We can also predict likely trace metal credits and penalties. The grade alone cannot tell us if a metal is likely to be a penalty or a credit; we need to look at what minerals the trace metals are hosted by and their association with the major ore minerals. From a leaching viewpoint, we can use the presence or absence of leachable ore minerals and acid-consuming gangue minerals to assess whether this option would be economic.

Understanding mineralogical variability across an orebody and its potential implications on metallurgy at the exploration stage makes for a rapid transition from exploration to the studies phase. At this stage metallurgical testwork can be combined with steadily increasing geological, geochemical, and mineralogical knowledge to build a truly predictive geometallurgical model.

Understanding Your Deposit Better Using Big Data: A Case Study from Kepez North (Kizilcukur), Balikesir, Turkey (Türkiye)

Mikail Mert Gumus[1], Sinem Oguz[1], Ahmet Kerim Sener[2], Ruth Bektas[2], Berkin Ugurlu[1]
1. Galata Mineral Madencilik San. ve Tic. A.Ş., Ankara, Turkey, 2. Ariana Resources plc, London, United Kingdom

There is an imperative towards development of multi-sensor core logging and machine learning algorithms, which generate large multi-variable databases of drilling data, improving the efficiency, accuracy, and detail captured from drill core. This paper summarises an approach to characterise drill core more efficiently and consistently using multi-sensor core scanning within a project-specific workflow to increase the quality of data acquisition and complement core characterisation practices.

Compared to traditional techniques of core logging, digital data collection allows geologists to create repeatable, accurate, and consistent results, providing objective data rather than a set of otherwise subjective interpretations. Such data is obtained either continuously or on a point-by-point basis by a semi-automated multi-sensor analyser directly from the samples in core trays using X-ray fluorescence, magnetic susceptibility, and mineral spectrometry, among other techniques, without sample preparation or significant operator intervention required. Within the context of this study, unsupervised methods such as principal component analysis and K-means clustering algorithms were applied to multi-element results for grouping and defining lithological units, alteration, and mineral composition at Kepez North, where three lithological groups are identified from the lithogeochemistry-derived variables.

Defined lithological units are compared with the results of pXRF geochemical analysis, mineral assemblages from spectrophotometry, and magnetic susceptibility, allowing for multivariate statistical analysis methods. Geochemical studies show gold mineralisation is associated with arsenic and antimony, whilst mineralogy studies show an intensity of smectite and mica group minerals in the host rocks to the mineralisation. Magnetic susceptibility data show high magnetic values and anomalies in ophiolitic rocks. All data is reviewed in 3D, providing further support to the modelling of the deposit. Acquiring multi-variable data, which may not commonly be collected until later in a programme, provides robust understanding of the mineralisation from an early stage, allowing better targeting, thus saving time and money.

The Ikkari Gold Discovery, Finnish Lapland

Charlotte Seabrook, Kalle Kotiaho
Rupert Resources Ltd, Toronto, ON, Canada

The discovery of the 4 Moz Ikkari gold deposit in Finnish Lapland in 2020 was the culmination of a systematic and focused regional grassroots exploration program that took advantage of local conditions to overcome challenges that had effectively stymied previous exploration.

Thick cover sequences composed of transported glacial till that effectively masks bedrock signatures and extremely poor outcrop coupled with difficult terrain and restrictive environmental legislation presented a working environment that required a focused and effective approach.

Benefiting from good-quality regional data collected by the GTK, an initial assessment of prospectivity was compiled using existing geophysics to outline structural targets, which were refined using cheap and quick UAV (drone) magnetics. This provided a framework to design a field program that would effectively test the potential "fertility"of the identified structures. Following several geochemical orientation programs, the base of till (BoT) sampling method was selected as the only method capable of providing a high-confidence and meaningful sample. Early trialling and in-house refinement of the BoT technique yielded quality results across targeted structures which led to board support and additional funding to continue the program. Learnings from lithological and geochemical observations re-focussed sampling locations and motivated infill programs. At Ikkari, a single point anomaly of 0.7 ppm Au was drilled in Spring 2020 into sericite-altered felsic sediments and highly strained ultramafics, which contained mineralized, disseminated pyrite throughout, yielding 54 m @ 1.5 g/t Au from base of till subcrop surface (18 m deep).

Ongoing target prioritization allowed for a disciplined approach to drill targeting. Initial drill testing was purposefully limited and economic potential assessed rapidly so that weaker targets did not attract additional meters. Persistence was key, and flexible, reactive drilling schedules facilitated rapid follow up on new information and ideas, supported by a motivated and ambitious local team and adaptive contractors.

The Jibal Qutman Gold Project, Kingdom of Saudi Arabia. Deposit-Scale Structural Models in an Underexplored Terrane to Drive Discoveries

Nadia St-Jean
Gold and Minerals Co., Bisha, Saudi Arabia

The Arabian-Nubian shield is one of the largest Neoproterozoic cratons on earth and covers significant areas of Eastern Africa and the Middle East. It comprises accreted intra-oceanic arc volcanic terranes, multiple events of batholith intrusions, and late metasedimentary molasse basins. The shield extends from Egypt in the NW to Yemen in the SE, with a significant portion of the overall area located in Saudi Arabia. Despite being host to several Tier 1 deposits (Sukari Gold mine, Jibal Sayid VMS deposit, Mansourah/Masarrah Gold mine), the shield is largely underexplored. In particular, the tectonic and structural history remains poorly understood. In recent years, the political and legislative changes implemented by the Kingdom of Saudi Arabia has resulted in the promotion of and renewed interest in mineral exploration for all commodities (Au, Cu, Ag, Zn, Lithium, REEs). The Jibal Qutman gold project lies in the Asir Terrane of the Arabian Shield, along the regional-scale Nabitah-Tathlith fault zone. The JQ project has recently undergone renewed exploration work as it advances toward production. The deposit is hosted within tholeiitic-intermediate volcanic and metasedimentary rocks and is a classic example of greenstone-hosted gold. Mineralization style is present as disseminated pyrite and shear-hosted quartz-carbonate veins. The structural paragenesis of the deposits shows an early F_1 fold event resulting in isoclinal NW- to WNW-oriented folds associated with a pervasive regional foliation. The early fold event is crosscut by brittle-ductile N-S–oriented thrust faults hosting thick, mineralized quartz-carbonate veins. Both structural events are interpreted to be associated with mineralization and predate the regional-scale Najd strike-slip event, which has also been historically linked to gold mineralization. Whereas previous exploration focussed on N-S–trending anomalies, this additional deposit-scale structural control aids in future target generation and may support the discovery of blind deposits at depth.

The Valley Project, a New High-Grade Reduced Intrusion-Related System in the Tintina Gold Province, Yukon Territory

Nicolas Piette-Lauziere, Sergio Gamonal, Andrew Turner
Snowline Gold Corp, Vancouver, BC, Canada

The Valley target is a reduced intrusion-related gold system (RIRGS) hosted in a Cretaceous granodiorite intrusion in the Selwyn Basin, eastern Yukon Territory. Although the mineralization is similar to classic RIRGS found within the Tintina Gold Province between Alaska and Yukon, such as Fort Knox and Dublin Gulch, Valley hosts unusually high gold grades for this deposit type. The key characteristics of this system are presented after two years of exploration.

Valley mineralization was discovered following up a 100-ppb Au stream-sediment anomaly located downstream of vein-mineralized outcrops. NW-oriented faults control the emplacement of an elongated biotite + hornblende granodiorite stock composed of three intrusive facies: a fine-grained, medium-grained, and porphyritic granodiorite. Potassic to sericite + chlorite alteration is restricted to vein selvages and fault zones. Hornfelsed country rock is characterized by strong biotite and silica alteration obliterating primary sedimentary textures of sandstones, mudstones, and siltstones of the Ordovician Road River Group. All intrusive facies host early pyrrhotite + pyrite + chlorite veins that are crosscut by steeply dipping quartz-carbonate sheeted vein arrays that parallel the regional fault system. The sheeted quartz vein system hosts multiple gangue and ore assemblages that mostly carry visible gold as free grains contiguous to Bi-Te-Pb-Ag sulphide alloys. Drill results from 2022 demonstrate continuous, high-grade intersections up to 2.5 g/t Au for 319 m from surface indicating bulk tonnage potential associated in the granodioritic intrusion and its quartz sheeted vein system.

These results indicate that the Valley system is a new discovery of RIRGS type that significantly enhances the metallogenetic potential of eastern Yukon. At the district scale, similar mapped intrusions form a WNW-oriented corridor, indicating potential for further discoveries in this underexplored district.

The Importance of Mineral System Analysis as Inputs to Data-Driven Gold Targeting Projects

Jean-Philippe Paiement
Mira Geoscience Ltd., Montreal, QC, Canada

Mineral exploration targeting is a key step in ground investigation area reduction that allows for lesser ground disturbance when exploring for gold. In recent years, data-driven mineral targeting has become a trend in the mining industry, which involves using advanced data analytics and machine learning techniques to identify and locate gold deposits with greater accuracy and efficiency. However, this approach requires careful interpretation and geological modelling that should be guided by mineral system analysis (MSA). MSA is an approach to better understand the formation and distribution of mineral deposits within a particular region or mineral belt by identifying the geological processes and events that led to the formation of thedeposits and the geological settings in which they occur.

QAQC of predictive model results is of primary importance. Overreliance on output can be problematic, which is why it's essential to ensure that model results make geological sense and are reliable. By understanding the mineralizing system, geoscientists can better judge the target areas resulting from data-driven targeting. One issue that can arise with data-driven mineral targeting is the introduction of cognitive bias in model training. This bias can be mitigated through MSA, which allows the geoscientist to build an inclusive data store and compile all existing exploration ideas and models at all scales. Bad or incomplete data is frequent in geosciences, which can be mitigated through the MSA review. Additionally, an idea of the predictive importance of different exploration features can be used for later interpretation of predictive model results and the interpretation of feature importance in the light of known mineral system components.

MSA can provide a means of designing, documenting, and QAQC'ing data-driven predictions. It provides context for the review of data-driven models and acts as a base case and lens for understanding data-driven results.

Porphyry-Like Au-Cu System in Paleoproterozoic Greenstone Belts

Tramaine James[1], Stéphane Perrouty[1], Daniel Kontak[1], Michael Tedeschi[1], Phillip Lypaczewski[1, 3], Robert A. Creaser[2]
1. Laurentian University, Sudbury, ON, Canada, 2. Department of Earth and Atmospheric Science, University of Alberta, Edmonton, AB, Canada, 3. College of the North Atlantic, St. John's, NL, Canada

The Toroparu Au-Cu(-Ag) system, located within the Mazaruni greenstone belt of the Guiana Shield, Guyana, South America, is a magmatic-hydrothermal system overprinted by an orogenic system. The ore is hosted in a cogenetic bimodal calc-alkaline igneous suite of intermediate volcanic, volcaniclastic, and intrusive host rocks metamorphosed to greenschist facies assemblages and lies on a regional NW-trending, sub-vertical, high-strained zone with a protracted deformation history.

The main mineralization stage has an assemblage of gold, chalcopyrite, bornite, pyrite II, covellite, molybdenite, and magnetite hosted within NW-trending extensional quartz carbonate ± chlorite (QCBCHL) veins, veinlets, and fractures associated with dextral strike-slip movement. Ore-stage molybdenite yielded an Re-Os age of ~2160 Ma, which overlaps with a ~2160 Ma U-Pb zircon age for the host intrusion that is a fine- to medium-grained porphyritic tonalite-granodiorite. Hyperspectral imaging in shortwave infrared (SWIR, 1,000–2,500 nm) indicates that metamorphic chlorite (background) within the project area has absorption bands between 2,255 and 2,252 nm (Mg# 42–47). In contrast, proximal Au-Cu(-Ag) mineralization zones have chlorite that is Mg rich (Mg#>90, <2,245 nm); white mica has absorption bands between 2,205 and 2,215 nm. Fluid inclusion studies indicate the QZCBCHL hosts secondary two-phase aqueous types (L_{H2O}-V_{H2O}) with Th = 100° to 160°C and moderate to low salinities (22.0 to 0 wt. % equiv. NaCl).

Most of the base and precious metals at Toroparu are syn-magmatic and represent an early mineralization episode in the Mazaruni greenstone belt. The Las Cristinas Au-Cu deposit in Venezuela, the Gaoua Au-Cu deposit in Burkina Faso, and the Aitik Au-Cu deposit in Sweden are hosted in rock of similar ages and geological settings. These major ore systems, which differ from the typical orogenic gold of greenstone belts, emphasize the need for new exploration approaches to lead toward new magmatic-hydrothermal Au-Cu deposit discoveries elsewhere in Precambrian greenstone belts.

Trace Element Geochemistry of Pyrite and Marcasite as a Vector Towards Mineralization in Irish-Type Zn-Pb Deposits

Claire Geel[1], Aileen Doran[1], Maurice Brodbeck[2], Murray Hitzman[1]
1. iCRAG SFI Research Centre in Applied Geosciences, Dublin, Ireland, 2. Trinity College Dublin, Dublin, Ireland

Irish-type Zn-Pb deposits were formed via the replacement of carbonates during the mixing of metal-bearing hydrothermal fluids and saline brines with bacteriogenically reduced sulfur resulting in the precipitation of polymetallic sulfide assemblages. Both pyrite and marcasite are associated with the Zn-Pb orebodies and occur in both proximal and distal positions relative to the economic portions of the deposits. The iron-sulfides incorporate trace elements through solid-solution within the crystal lattice and by nano- and/or micro-inclusions. Therefore, the geochemistry of pyrite and marcasite can record changes in fluid chemistry and the evolution of the hydrothermal system. In this study, multidimensional, multimethod statistical classification of trace element geochemistry of pyrite and marcasite is applied to new and historic data from the Rathdowney Trend, which includes the Lisheen and Galmoy deposits and the Rapla prospect. Laser ablation-inductively coupled plasma-mass spectrometry (LA-ICP-MS) is used to collect trace elements from the different pyrite and marcasite paragenetic phases utilizing drill cores positioned proximal and distal to orebodies and interpreted with whole-rock geochemistry of the same samples. Trace element (LA-ICP-MS) maps accompanied by petrographic imaging show that morphological and textural variation (e.g., framboidal, spongey texture, growth banding, and plumose textures) correlate with a distinct geochemical signature and therefore the paragenetic phase. Trace elements obtained from LA-ICP-MS spot analyses show hydrothermally sourced pyrite and marcasite are enriched in As and Tl in comparison to diagenetic pyrite, which is enriched in Mn, Mo, and Cu, with correlation observed between Co and Ni; As and Tl; and Mn and Mo. On a regional scale, pyrite and marcasite distal to orebodies show greater enrichment in Mo, Cu, and Sb. Thus trace elements of iron-sulfides can be used to vector towards zones of potentially economic Zn-Pb concentrations by informing proximity to hydrothermal sources.

Genesis of BIF-Hosted Martite-Goethite Ores: Concept and Application

Caroline S. Perring
BHP, Perth, WA, Australia

A new genetic model for supergene martite-goethite ore formation in the Hamersley Province of Western Australia provides a process understanding of how the ores formed that underpins successful extraction and allows us to maximise the value of our existing data sets. The model was developed using exploration drilling data, 2D geophysical data sets, 3D modelling, and ore petrography to understand controls on ore development from camp to grain scale.

Iron is leached from the BIF host rock, in the vadose zone, by meteoric-derived fluids that are reduced and acidic. Gravity drives the ore fluid downdip within the BIF and flow is focussed along bedding planes, which represent zones of enhanced permeability. The mineralisation takes place in three stages: i) replacement, ii) leaching, and iii) cementation. Further complexity is introduced by the fact that supergene ore formation can envelop a much older hypogene style of mineralisation and that both types of mineralisation are lateritised near surface.

There are several ways in which an ore-genetic understanding informs processes across the mining value chain. The ore-forming process modifies the mineral composition, texture, and porosity of the rock, which in turn influences its hardness and the material handling properties and lump to fines ratio of the mined product; it is thus a critical input to mine planning and scheduling. Increasing reliance on mathematical models and data analytics within geoscience requires an ore-genetic understanding to inform the selection of appropriate training data and/or application of appropriate domains. This is particularly important in the Hamersley context, where there are multiple ore-forming pathways (hypogene/supergene/weathering). Finally, some geoscience data sets are time consuming and costly to acquire. It is desirable to be able apply key learnings from operating mines to genetically similar deposits at an early stage of definition; here the concept of deposit analogues comes into play.

Hydrothermal Alteration and Volcanic Stratigraphy at the Paleoproterozoic Kay Mine Volcanogenic Massive Sulfide Deposit, Black Canyon City, Arizona, USA

Aaron J. Adsit[1], Thomas Monecke[1], Mark D. Hannington[2], Miranda R. Lehman[1], David Smith[3]
1. Center to Advance the Science of Exploration to Reclamation in Mining, Department of Geology and Geological Engineering, Colorado School of Mines, Golden, CO, USA, 2. Department of Earth and Environmental Sciences, University of Ottawa, Ottawa, ON, Canada, 3. Arizona Metals Corp., Toronto, ON, Canada

The Kay volcanogenic massive sulfide deposit in Black Canyon City, Arizona, is hosted by bimodal volcaniclastic rocks assigned to the 1.79 to 1.76 Ga Townsend Butte Formation of the Black Canyon Creek Group, which forms part of the Proterozoic Yavapai Supergroup. Detailed core logging of the strongly deformed volcanic host rock succession shows that the stratigraphic footwall to the massive sulfides is dominated by fine-grained intermediate to mafic volcaniclastic rocks. The favorable interval is marked by the occurrence of carbonaceous mudstone that occurs immediately below coherent to brecciated rhyolite or rhyodacite. Mudstone matrix monomict rhyolite or rhyodacite breccia interpreted to be peperite is widespread along the contact. Textural evidence suggests that the massive sulfides at the Kay deposit formed primarily through subseafloor infiltration and replacement of the volcaniclastic host rocks. The sulfides are commonly intergrown with nodular dolomite and ankerite. The massive sulfides are underlain by a zone of intense black chlorite alteration hosting chalcopyrite stringers. Sericite is the most common alteration product in the hanging wall that is dominated by mafic-intermediate volcaniclastic rocks, and basalt that forms massive or pillowed flows. Geochemical and mineralogical gradients around the massive sulfides are established through core scanning using a Minalyze continuous X-ray fluorescence core scanner and a short-wave infrared core scanner by Hyspex. Recent software advancements allow for co-registration of the geochemical and mineralogical data sets produced by both instruments. Unraveling alteration vectors through core scanning will assist in future target generation and potential expansion of the historic resources and reserves estimate of 5.8 million tonnes grading 2.2% Cu, 3.03% Zn, 54.9 g/t Ag, and 2.8 g/t Au at a cutoff grade of 2% Cu equivalent.

Random Forest Classification of Epidote Mineral Chemistry from the Aitik Deposit, Northern Sweden

Michael J. Baker[1], Matthew Cracknell[1], David R. Cooke[1], David Drejing-Carroll[2], Evan Orovan[1]
1. CODES / University of Tasmania, Hobart, TAS, Australia, 2. Boliden, Boliden, Sweden

The ability of propylitic epidote mineral chemistry to record the subtle geochemical signatures of mineralised magmatic-hydrothermal systems has been established by several studies in recent years and is becoming a novel tool for copper and gold exploration in terranes that can supplement traditional geochemical exploration tools (e.g., assay geochemistry, geophysical surveys). Due to their low natural abundance, the tell-tale signatures of trace element anomalism in propylitic epidote from magmatic-hydrothermal systems are often difficult to quantify, let alone define, using simple statistical and graphical interpretation. We have developed a workflow based on results of random forest classification of epidote mineral chemistry analyses from the Aitik Cu-Au-Ag-Mo deposit, northern Sweden, which has potential to help recognise domains of porphyry, IOCG, and metamorphic-dominant mineral assemblages.

The random forest classification model for epidote trace element data can be used as a predictive modelling tool in greenfields and brownfields terranes containing epidote-bearing propylitic alteration assemblages by providing an indication of ore deposit type. This advance can assist mineral explorers by allowing early implementation of predictive ore deposit models when prospecting for ore deposits. Furthermore, the ability of the classifier to identify epidote of metamorphic origin can allow researchers and explorers to effectively screen prospective deformed and metamorphosed samples that are affected by hydrothermal overprints.
This work formed part of the AMIRA P1202 research program, which is part of the Centre for Ore Deposit and Earth Sciences (CODES) at the University of Tasmania.

Graph Neural Network for Mineral Prospectivity Mapping – A Possible Solution to Maximise Utilization of Geological Relationships

Felix M. Sihombing[1, 3], Richard M. Palin[1], Hannah S. Hughes[2], Laurence Robb[1]
1. The University of Oxford, Oxford, United Kingdom, 2. Camborne School of Mines, University of Exeter, Cornwall, United Kingdom, 3. Universitas Indonesia, Depok, Indonesia

Numerous attempts at mineral prospectivity mapping (MPM) have been made with different machine learning models employing various exploration data as inputs, which have met with different degrees of success. In particular, information about geological relationships is still not well utilised as a data input type, largely due to the limitations of the models used. This omission reduces the impact of geological input for the training process, which in turn affects the degree of success of such models. This situation also limits the ability to incorporate insight from geologists as domain experts in order to improve model performance. In this study, we address this issue by presenting a simple case study that demonstrates the use of neighbouring relationships between geological units to generate mineral prospectivity maps. We represent this relationship by converting geological map data into a graph format that consists of nodes (objects) and edges (relations). The training process uses graph neural network (GNN) algorithms, which are a group of deep neural network algorithms that can operate in graphical space. The approach was applied to generate a prospectivity map for three metal commodities (tin, lead, and copper) in the United Kingdom, and we compared this performance with several machine learning approaches (support vector machine, random forest, gradient boost, and deep neural network). Our study shows that utilising GNN in MPM is beneficial since it can highlight certain lithological boundaries that may potentially be related to mineral occurrences. Despite the model not having explicitly received information about lithological boundaries as a data input, the model can infer such geological relationships implicitly from the provided graph. In addition, GNN also shows higher accuracy and precision scores in certain scenarios when compared to other approaches.

Characterizing Geological Uncertainty with Stochastic Implicit Modeling

Luiz Gustavo Rasera[1], Jeff Boisvert[2], Shaun Barker[1]
1. University of British Columbia, Vancouver, BC, Canada, 2. University of Alberta, Edmonton, AB, Canada

Geological models are inherently uncertain due to limited access to subsurface information. Implicit modeling techniques enable rapid construction and assimilation of new data into geological models. Nevertheless, most techniques rarely account for the uncertainty associated with model predictions, and even fewer assess the reliability of uncertainty estimates. To address these issues, we propose a stochastic framework to characterize geological uncertainty in implicit modeling. First, our method creates an initial implicit model based on available drill hole information. Multiple alternative models are then generated by integrating additional synthetic data into the original model using geostatistical simulation. The ensemble of alternative scenarios is subsequently used to characterize geological uncertainty. Possible frameworks for calibrating model parameters and validating uncertainty estimates are discussed. The approach enables propagation of geological uncertainty into mineral resource estimation and provides additional information for decision-making in exploration and resource management. We demonstrate the approach with illustrative examples.

Muruntau, Uzbekistan: A Giant, Pluton-Related Auriferous System

Reimar Seltmann[1], Alla Dolgopolova[2], Rustam Mirkamalov[3], Farid Divaev[3], Rustam Khalmatov[4], Yulia Mun[5], Nigora Isokova[6]
1. Natural History Museum London, Centre for Russian and Central EurAsian Mineral Studies (CERCAMS), London, United Kingdom, 2. Natural History Museum London, Centre for Russian and Central EurAsian Mineral Studies (CERCAMS), London, United Kingdom, 3. Institute of Mineral Resources / Centre for Advanced Technologies (CAT), Tashkent, Uzbekistan, 4. University of Geological Sciences / Centre for Advanced Technologies (CAT), Tashkent, Uzbekistan, 5. The Arctic University of Norway, Tromsø / Centre for Advanced Technologies (CAT), Tashkent, Uzbekistan, 6. National University of Uzbekistan / Centre for Advanced Technologies (CAT), Tashkent, Uzbekistan

Muruntau is situated in the Central Kyzylkum of Southern Tian Shan, Uzbekistan, with gold production of ~60 tonnes per year. The main geological characteristics of the giant Muruntau auriferous system are revisited to improve genetic and exploration models. Field-based research in the Muruntau district, its deep drill holes, as well as complex geoscientific data have been integrated with regional geology and geophysics.

Deposit formation occurred through a multi-stage process involving sedimentation, regional metamorphism including thrusting, magmatism with formation of hornfels aureoles, and several stages of hydrothermal activity. Late-orogenic granitic magmatism (294-288 Ma) occurs roughly within the same time interval as the main hydrothermal gold precipitation (~288 Ma) expressed by intense, high-temperature fluid-wall-rock interaction resulting in the formation of gold-bearing cone-like stockworks with veins, veinlets, and metasomatites.

Muruntau and other gold deposits in the Kyzylkum are hosted by a distinctive metasedimentary package (the "variegated Besapan," BS3, a Lower Palaeozoic siliciclastic continental margin assemblage) that is dominated by psammopelites and contains major carbonaceous units. Ore-stage veining, mineralization and metasomatism are focused in the hornfelsed package, the more ductile carbonaceous pelites constituting local "screens" partitioning strain and fracture. The auriferous systems underlie a Devonian-Carboniferous (D-C1) carbonate-dominated package which acted as a low-permeability seal/cap on these systems.

Gold ores occur mainly in south-dipping zones enveloped by lower-grade material. High-grade ore zones comprise sheeted quartz±feldspar vein networks and enveloping biotite-orthoclase-quartz–rich metasomatised metasediments with ore-stage arsenopyrite, pyrite, and pyrrhotite.

Main stage gold mineralization is accompanied by biotite±amphibole stable, potassic alteration that is thermally compatible with host-rock thermal metamorphic assemblages. The latter overprint deformation fabrics resulting in massive textures in psammites and psammopelites and outlining a broad thermal aureole related to block uplift serving as fluid conduit that is intruded by underlying granitoid plutons.

Hypozonal Orogenic Gold Mineralization Explained by Metamorphic Devolatilization of Komatiites

Richen Zhong[1], Chang Yu[1], Andrew G. Tomkins[2], Hao Cui[1], Yanjing Chen[3]
1. University of Science and Technology Beijing, Beijing, China, 2. Monash University, Clayton, VIC, Australia, 3. Peking University, Beijing, China

The widely accepted metamorphic devolatilization model, or supracrustal metamorphic model, for orogenic gold deposits emphasizes that gold and auriferous fluid are originated from metamorphic devolatilization of pelites or altered basaltic rocks when crossing the greenschist-amphibolite facies boundary. However, this model is thought to be invalid when applied to hypozonal deposits that are hosted in amphibolite facies terranes in greenstone belts. Except for metasedimentary and metabasaltic rocks, komatiites are important components in Precambrian greenstone belts and have potential to release metamorphic fluid during metamorphism. In this study, we simulated the devolatilization processes of gold- and sulfur-bearing komatiite to evaluate its feasibility as the source for ore-forming fluids of hypozonal gold deposits.

The situation results of komatiite show that there are two pulses of fluid and gold liberation. First, ~4 wt % of water and ~1 to 1.4 ppb of Au (relative to rock mass) can be released when crossing the greenschist-amphibolite boundary at ~550°C due to breakdown of antigorite. After entering amphibolite facies, a sharp rise in gold liberation occurs at ~700°C as a response to dehydration of talc and chlorite, which liberates ~0.5 wt % of water and ~2 to 5.5 ppb of Au. Although the second step of devolatilization releases less water than the first, more gold can be scavenged because of the higher solubility of gold in high-temperature metamorphic fluids. This simulation result indicates that devolatilization of komatiite at amphibolite facies may account for the formation of hypozonal gold deposits in greenstone belts.

Structural Modelling of the Au Mineral System Using 2D Seismic, Yamarna Greenstone Belt, Eastern Goldfields, Western Australia

Armelle Kloppenburg[1], Jamie Robinson[2]
1. 4DGeo | Applied Structural Geology, The Hague, Netherlands, 2. Gold Road Resources, Perth, WA, Australia

Developing an understanding of the 3D architecture and structural evolution is a key part of building terrane-scale geological framework to support exploration targeting within the Yamarna Terrane. Three bespoke 2D seismic lines (approx. 25 km long and 7 km deep) have been used to constrain the internal greenstone belt architecture and recognise interconnected fault pathways that have focussed mineralising fluids. This belt-scale seismic data was integrated with a nearby government crustal-scale seismic line (approx. 45 km deep; L154-N1) as well as geological and geophysical maps to provide context during interpretation in 3D structural modelling space.

Most of the Yamarna gold deposits are located around structures that define the greenstone belt's architecture today; the thin-skinned fold-and-thrust belt. However, structures in the belt are not continuous along length, and this complexity appears favourable for gold mineralization.

A scenario is proposed in which deep-seated inherited extensional faults and related transfer faults played a significant role. Reactivation and inversion of these extensional faults occurred during subsequent crustal shortening, influencing the geometry and linking to the newly formed structures at upper crustal levels. A rotated, non-ideal angle of shortening has resulted in the complex but systematic pattern seen today.

This combination of "tramlined" and newly formed structures contributed to focussing of mineralising fluids by forming connected fluid pathways and structural traps. At Yamarna, understanding how these fault zones interconnected in 3D and to depth at critical moments in time informs ongoing exploration.

Using Edge Detection Techniques on Aeromagnetic Data to Delineate Structures that Control Orogenic Gold Deposits in the Sefwi Granite-Greenstone Belt, Ghana

Pv S. Raju, Benjamin K. Boadi
CSIR-NGRI, HYD, TN, India

The orogenic gold deposits in the Sefwi granite-greenstone belt are associated with quartz veins and occur along shear zones. Edge detection techniques (i.e., analytic signal, vertical and tilt derivatives) were employed to delineate structural lineaments related with shear zones using aeromagnetic data. The enhancing techniques aided the delineation of NE-striking D_2 and D_3 deformation lineaments associated with the Kenyase-Yamfo, Hwidiem, Ketesso, Bibiani, and Afema shear zones in the study area. These shear zones are within the Sefwi belt and at the contact zones between the belt and the adjacent metamorphosed volcano-sedimentary rocks of the Sunyani-Comoé and Kumasi-Afema basins. Low-displacement N- to NNE-striking D_4 faults, which are oblique to the D_2 and D_3 shear zones, were also delineated. The local bend and offset along the D_2 and D_3 shear zones created by the cross-cutting N-to NNE-striking D_4 faults serve as favourable trap zones to focus mineralised fluids migrating along these shear zones. Fry analysis of known orogenic gold occurrences in the study areas shows that the NE-trending structures, with azimuths of 40°-50°, are predominant in controlling most mineralisation and distribution of deposits; for example, regions of the Kenyase-Yamfo, Hwidiem, and Ketesso shear zones in the N-S to NNE-SSW direction with azimuths 0°-10° and 10°-20°. Integrating results from each edge detection filter effectively help to diffuse and separate different anomalies from shallow and closely spaced sources. The filters (first vertical derivative, tilt angle derivative, TAHG derivative, and STDR derivative) were used to separate and delineate closely spaced structural lineaments associated with the Kenyase-Yamfo, Hwidiem, Ketesso, Bibiani, and Afema shear zones. According to the Fry analysis, most orogenic gold occurrences are associated with NE-trending D_2 and D_3 structures and confirmed by ground checks (Newmont Ghana Gold Limited). The secondary structural trend in the N to NNE direction (D_4 deformation) influences orogenic gold mineralisation.

Characterization of Refractory Gold Behaviour in Witwatersrand Mine Tailings: Implications for Gold Recovery

Steve J. Chingwaru, Bjorn Von der Heyden, Margreth Tadie
Stellenbosch University, Stellenbosch, South Africa

The Witwatersrand Basin is the largest gold province in the world, having produced more than 6 billion tons of mine tailings over the past century of mining. As the basin has matured in terms of gold production, traditional cyanidation techniques are now used to extract gold from the tailings concurrently with primary gold ores. Despite these efforts, the tailings remain refractory, and an estimated 1,325-1,855 tons of gold are still present in tertiary tailings. The complete mineralogical composition of this remaining gold is not yet fully understood. This study uses separation, mineralogical, and chemical analyses to characterise the refractory behaviour of Witwatersrand tailings material gold obtained from spatially varying goldfields. Results indicate that the heavy mineral concentrate has a higher gold concentration of 0.71-10.12 ppm, while the light (<2.95 $g.cm^{-3}$) and slimes (<10 μm) fractions have 0.05-0.22 ppm and 0.17-0.76 ppm concentrations of gold, respectively. Detailed in situ analyses and leach tests of the different mineral fractions suggest that the gold is encapsulated by tailings mineralogy in the form of "invisible" or "solid-solution" gold, mainly in pyrite and arsenian pyrites. These minerals have gold concentrations ranging up to 2,730 ppm and, from mass balance, may account for up to 420 tons of gold locked as refractory gold in the existing surficial tailings heaps. Additionally, the tailings sulphides host a significant concentration of deleterious and critical heavy metals such as As, Ni, and Co. Identifying the invisible gold represents an under-utilized gold resource, as it accounts for up to 31% of the 50-70% of gold unrecovered during secondary processing. The primary Witwatersrand ores are known for their native liberated gold, which explains the refractory behaviour of remaining re-dump gold during traditional extraction. The study's results have implications for mining, economics, environment, optimization of flowsheets, and the origin of the Witwatersrand.

Application of U-Pb Detrital Zircon Geochronology to Delineate Regional Unconformities; Implications for Gold Mineralization in the Red Lake Mining Camp, Ontario, Canada

Kate E. Rubingh
University of British Columbia, Vancouver, BC, Canada

The Archean Red Lake greenstone belt is host to several gold deposits including the world-class Campbell-Red Lake mine hosted in the Mesoarchean mafic-ultramafic volcanic rocks of the Balmer assemblage (2.99–2.96 Ga), which lies proximal to a folded regional angular unconformity, and the Madsen mine, which lies proximal to another regional unconformity. This spatial relationship between gold deposits and regional unconformities is characteristic of Archean greenstone gold mineralization and is recognized in the Red Lake mining camp for other gold deposits. Therefore, identifying their spatial distribution is a key exploration target. Several of these unconformities are delineated by sedimentary rock packages, and U-Pb detrital zircon geochronology is one of the main tools to unravel the stratigraphic position of these rocks and establish a structural and stratigraphic framework of the mineralization. In the Red Lake camp, two main stages of mineralization have been previously identified. The main gold mineralization event formed before 2712 Ma at the Campbell-Red Lake mine and is coeval with the emplacement of a series of stocks, the main penetrative deformational event, and the deposition of a package of polymictic conglomerates. At the Madsen mine, the main mineralization event is at least 10 m.y. younger, interpreted to have formed between 2700 and 2698 Ma, and is itself associated with a younger plutonic event and a coeval package of conglomerates and finer clastic sediments. New U-Pb detrital zircon geochronology suggests that younger Madsen-aged conglomerates are present along a broader structural corridor and proximal to the Campbell-Red Lake mine, which has implications for connectivity of large-scale regional unconformities across the Red Lake camp from the Madsen mine to the Campbell-Red Lake mine.

The Fluids that Form Magnetite-Apatite Deposits

Matthew Steele-MacInnis
University of Alberta, Edmonton, AB, Canada

The origins of magnetite-apatite deposits (also known as iron oxide-apatite, or IOA, deposits) are extremely contentious. Even basic questions, like what type of fluids can produce this type of mineralization, are still vigorously debated. Fundamentally, the problem is that we have been missing a known geologic fluid capable of mobilizing and depositing massive amounts of ferric iron and calcium phosphate, especially without simultaneously depositing voluminous quartz and other minerals. Here, I will describe recent progress in characterizing the fluids that formed these enigmatic rocks by analysis of fluid inclusions in minerals. The results reveal some surprises—especially, that a key commonality between magnetite-apatite deposits worldwide and in diverse geologic settings is a central role of hitherto unexpected liquids: salt melts dominated by sulfate, carbonate, and chloride components, and with appreciable concentrations of iron and titanium oxides and calc-silicate components besides. Melt inclusions representative of this novel category of ore-forming fluids are found in magnetite-apatite rocks from every continent and from at least Mesoproterozoic through to Quaternary times—in other words, such melt inclusions are a fundamental feature of magnetite-apatite deposits worldwide. Moreover, analysis of these inclusions shows that they represent samples of hot, chemically aggressive, low-viscosity ionic liquids that are capable of mobilizing high concentrations of a variety of metals and other components that are only sparingly soluble in hydrothermal aqueous fluids. This discovery paints a new picture of what magnetite-apatite deposits represent and helps explain many of the previously puzzling characteristics of these rocks.

Fingerprinting Tools to Distinguish Magnetite Deposit Types

Patrick Krolop[1], Christin Schulz[2], Sabine Gilbricht[2], Thomas Seifert[2]
1. Luossavaara Kiirunavaara AB, Kiruna, Sweden, 2. TU Bergakademie Freiberg, Freiberg, Germany

Magnetite is the prominent ore mineral in several deposit types including iron oxide-apatite (IOA), iron-rich ultramafic pegmatites (IRUPs), magnetite bands in layered intrusions, magnetite-rich carbonatites, magnetite skarns, and banded iron formations (BIFs). Mineral chemical data has commonly been used to utilize discriminating tools (e.g., provenance studies, pathfinder for mineral exploration, and deposit-formation scenarios). In this contribution, we provide both modified existing discrimination diagrams and new tools for separating deposit types by magnetite chemistry. A set of comprehensive EPMA and LA-ICP-MS data of magnetite from several deposits of different genetic types were obtained. Suitable fingerprinting elements are Ti, V, Ni, Cr, Co, Al, Mg, Mn, Si, Ca, Ga, Sn, and Zn. These elements display systematic variation and mostly detectable concentrations in the respective magnetite-bearing samples investigated in this study. Especially Ni vs. Ti+V seems to allow discrimination for all deposit types, with distinct fields when LA-ICP-MS data is used. Since some of the elements (e.g., Cr, Ga, Zn, and Sn) are commonly below the detection limit of the electron microprobe, a thorough textural analysis needs to be conducted to compensate the larger beam diameter of the laser, lowering contamination of the signal by exsolutions and inclusions in some magnetite types.

Tracing the Sources of Rare Earth Elements in IOCGs

Antony D. Burnham[1, 3], Jonathan Cloutier[2, 3], Vickie C. Bennett[1], Ross Chandler[1], John A. Mavrogenes[1]
1. Research School of Earth Sciences, Australian National University, Canberra, ACT, Australia, 2. Centre for Ore Deposit and Earth Sciences, University of Tasmania, Hobart, TAS, Australia, 3. Geoscience Australia, Symonston, ACT, Australia

Developing reliable mineral potential and exploration models depends on understanding the formation histories and source characteristics of different ore types. Iron oxide-copper-gold deposits have multiple models proposed for their formation, and their link to magmatism as a source of heat and metals is contentious. Fortunately, many IOCGs host abundant rare earth element (REE) minerals that are amenable to radiometric dating and Nd isotope microanalysis. We examined the REE-rich SWAN IOCG in the Mount Isa Inlier, Australia, and assessed its potential link to local magmatism. Drilling shows that SWAN and the neighbouring Mt Elliott IOCG deposit sit in the contact with a mafic intrusion. Using laser ablation ICP-MS, we measured εNd of REE-bearing minerals (allanite, apatite, and titanite) within ore samples ranging from highly mineralized to distal disseminated mineralization; these values were compared to the bulk-rock εNd as measured by ID-TIMS.

The SWAN deposit contains abundant allanite with a TIMS ^{207}Pb-^{206}Pb age of 1519.6 ± 1.1 Ma. The εNd_{1520} of the mafic intrusion is 3.2 (± 0.4; 1σ), in contrast to the ore samples, which consistently had εNd_{1520} = –4.7 (± 0.1; 95% CI) regardless of mineral type or location within the deposit. Most of the local granites have values of εNd_{1520} ~ –3, whereas the local siliciclastics have typically εNd_{1520} ~ – 8. Thus, a juvenile mafic source cannot have contributed chemically, though may have provided a driving heat source. The oxygen isotope composition ($\delta^{18}O$) of allanite at SWAN is +7.6 (± 0.5; 1σ)‰, indicating that the host marble (typical $\delta^{18}O$ > +20) has not contributed significantly to the oxygen in allanite. Hence, the REEs in the SWAN IOCG deposit were possibly sourced from a combination of the local granites and clastic metasediments, with brines as the transport vector.

Mineralogy and Paragenesis of the Boundary Zone Clastic-Dominated Zn Deposit (Macmillan Pass, Yukon)

Haruna M. Grema[1, 2], Joseph M. Magnall[1], Sarah A. Gleeson[1, 2], Jack E. Milton[3], Vitor R. Barrote[4], Marta Sośnicka[1], Hans-Martin Schulz[1]
1. GFZ German Research Centre for Geosciences, Potsdam, Germany, 2. Institute of Geological Sciences, Freie Universität Berlin, Berlin, Germany, 3. Fireweed Metals, British Columbia, BC, Canada, 4. Paul Scherrer Institute, Villigen, Switzerland

In the Selwyn Basin (Yukon, Canada), clastic-dominant (CD-type) Zn-mineralized rocks are located across three broad stratigraphic intervals representing metallogenesis during the Cambrian, Ordovician-Silurian, and Late Devonian. The Boundary Zone prospect in the Macmillan Pass district is the most recent Zn-Pb discovery in the Selwyn Basin. The mineralized rocks at Boundary Zone are hosted by the Mid-Late Devonian Portrait Lake Formation and Late Ordovician-Early Silurian Duo Lake Formation, which provides a unique opportunity to investigate metallogenetic processes at different stratigraphic levels. This study provides a petrographic investigation of the mineralogy and paragenesis of the mineralized rocks, with sphalerite trace element chemistry (LA-ICP-MS) and fluid inclusion microthermometry.

Hydrothermal sulfides in the Duo Lake Formation comprise stratiform spherical grey sphalerite (Sp-I) locally overgrown by metallic pink coarse-grained euhedral sphalerite (Sp-II). Late pale-yellow sphalerite (Sp-III) occurs in veins associated with cavity-filling quartz and apatite with pyrobitumen. In the Portrait Lake Formation, very fine grained laminated sphalerite (Sp-I) is hosted in mudstone of the Fuller Lake Member in an assemblage with barite and barium feldspar. A second stage of sphalerite (Sp-II) forms in veins and breccias, associated with pyrite, galena, chalcopyrite, siderite, and Ba-mica. Similar to the occurrence in the Duo Lake Formation, straw yellow coarse-grained sphalerite (Sp-III) occurs, and is associated with very coarse grained quartz, pyrobitumen, and siderite. Preliminary fluid inclusion homogenization temperatures in Portrait Lake Sp-II sphalerite are between 194° and 225°C, within the range of estimated Ga, Ge, In, Mn, and Fe (GGIMFis) sphalerite geothermometry constraints (163°–279°C). Textural and trace element variations suggest variable fluid conditions, with the moderate to high temperature of the hydrothermal fluids consistent with the temperature range for the nearby Tom and Jason deposits of the Macmillan Pass district. Further work is being undertaken to constrain the fluid characteristics, metal zonation, and deposit age(s).

The Paleoproterozoic Amitsoq Graphite Deposit, South Greenland

Giacomo Fratelli[1], Alberto Vitale Brovarone[1, 3, 4], Nynke Keulen[2], Kristoffer Szilas[2], Eva E. Stüeken[6], Thomas F. Kokfelt[2], Tonny B. Thomsen[2], Stefan Bernstein[7]

1. Dipartimento di Scienze Biologiche, Geologiche e Ambientali (BiGeA), Alma Mater Studiorum Università di Bologna, Bologna, Italy, 2. Geological Survey of Denmark and Greenland, Øster Voldgade 10, DK-1350, Copenhagen, Denmark, 3. Institut de Minéralogie, de Physique des Matériaux et de Cosmochimie (IMPMC), Sorbonne Université, Muséum National d'Histoire Naturelle, UMR CNRS 7590, IRD UR206, 75005, Paris, France, 4. Institute of Geosciences and Earth Resources, National Research Council of Italy, Pisa, Italy, 5. Department of Geosciences and Natural Resource Management, University of Copenhagen, Øster Voldgade 10, 1350, Copenhagen, Denmark, 6. University of St Andrews, School of Earth & Environmental Sciences, Queen's Terrace, Bute Building, KY16 9TS, St Andrews, United Kingdom, 7. Greenroc Mining plc, 60 Gracechurch Street, London, United Kingdom

The Amitsoq graphite deposit of South Greenland's Nanortalik region is one of the richest graphite deposits worldwide, with a JORC-resource estimate of more than 23 Mt at 20.41% graphitic carbon. The Amitsoq deposit was mined in the early 20th century, yet the origin of the graphite is presently unknown. The ore is hosted by two main bodies, the UGL and LGL, of which the LGL is the thickest (up to 20 m) and richest (typically 20-24% graphite). The orebodies display a simple planar geometry, dipping 30° westward measured from the surface outcrops on the eastern flanks of the Amitsoq Island. A remarkable feature of the LGL and UGL is their consistent grades across the orebodies. The ore is a graphitic-sulphidic schist in which graphite is present as elongate millimetric flakes, often piled in stacks, and as cryptocrystalline agglomerates. Graphite flakes are interleaved with biotite and pyrrhotite and are variably folded. The deposit likely formed in the context of the ca. 1.8 Ga Ketilidian orogeny, during the Paleoproterozoic, which is a geological period in which the vast majority of the worlds richest graphite deposits formed. The amphibolite facies psammite hosting the ore is thought to represent part of the fore-arc of the northward-dipping Ketilidian subduction zone and was formed by the erosion of the nearby Julianehåb batholith. Shortly after the orogeny, the area was intruded by several granitoid plutons. The combination of these factors brought the area to experience HT-LP metamorphism with local anatexis. The texture of the ore indicates that graphite is either pre- or syn-kinematic with respect to the metamorphic events that occurred in the area. $\delta^{13}C$ analysis of graphite yields an average of -33‰ vs. PDB, indicating a biotic carbon source for the carbonic fluids involved in graphite precipitation.

Geological Features Common for Bonanza-Forming Epithermal Gold Deposits

Keiko Hattori
University of Ottawa, Ottawa, ON, Canada

High-grade (bonanza) epithermal quartz-adularia-gold veins with illite and carbonates occur in volcanic arcs around the world, from Pliocene to at least Jurassic age. Characteristic features include fine-grained colloform laminations due to colloidal silica deposition from boiling, gas-rich but low-salinity solutions. The deposits form at <500-m depth at 180°–250°C. The relationship of mineralization to magmatism has not been well established.

The plate configurations and lithospheric structures plus temporally associated magmatism with young deposits show striking similarity in the geological development leading to the mineralization. Studied deposits include the Hishikari deposit in southern Japan (>440 t Au, ~1 Ma age, extension related to lithospheric rotation, primary host Cretaceous Shimanto accretionary sediments, contemporaneous with basaltic-andesitic volcanism of Lower Hishikari Andesite); the Sado deposit of western Japan (>77 t Au, 14 Ma, back-arc extension during Japan Sea formation, underlain by organic-rich Tertiary basin sediments, contemporaneous with mafic dike swarms); the McLaughlin deposit, California (109 t Au, 1 Ma, pull-apart extension along a San Andreas Fault splay, underlain by Franciscan accretionary prism sediments, contemporaneous with basaltic-andesite Clear Lake volcanism); the Hauraki gold fields, NZ (372 t Au, 6-8 Ma, rotational extension, underlain by Cretaceous accretionary sediments, contemporaneous with andesite volcanism); and the Ada Tepe deposit, Eastern Rhodopes, Bulgaria (~30 t Au, 35 Ma, underlain by Cretaceous accretionary sediments formed during Africa-Europe convergence, contemporaneous with a range of volcanism).

These deposits are hosted or underlain by rootless, immature sediments of accretionary prism that underwent sub-greenschist metamorphism. After the uplift of the marine sediments, subaerial magmatism derived from the asthenopheric mantle overlies the sediments. The volcanic rocks are contemporaneous with the mineralization. Plate configurations reveal lithospheric-scale extension that allowed asthenospheric mantle upwelling for the volcanism and associated hydrothermal activity (with asthenospheric N isotope signature for high-grade gold ore at Hishikari).

The Electrochemical Flocculation of Colloidal Gold by Semiconductive P-Type Pyrite: A Novel Ore-Forming Mechanism for Gold

Duncan F. McLeish[1, 2], Anthony E. Williams-Jones[1], James R. Clark[1], Travis Murphy[2]
1. McGill University, Montreal, QC, Canada, 2. Newcrest Mining Ltd., Vancouver, BC, Canada

A growing body of evidence suggests that bonanza-grade hydrothermal gold deposits are formed by the physical transport of gold as a colloid (i.e., a suspension of ≤10 nm, negatively-charged nanoparticles in an electrolyte solution) rather than direct deposition from dissolved, aqueous Au complexes. While there is general consensus that colloidal suspensions can explain how gold may be mobilised within hydrothermal systems at concentrations many orders of magnitude greater than those predicted by solubility models, there is little consensus over the processes by which colloidal gold suspensions aggregate (i.e., flocculate) to produce ultra-high-grade gold occurrences in hydrothermal veins. Detailed paragenetic observations at the high-grade Brucejack Au-Ag deposit, northwestern British Columbia, Canada, show that ore-stage epithermal carbonate-quartz-electrum veins commonly host bonanza gold where these veins cross-cut earlier mesothermal pyrite veins. This observation, coupled with the discovery of abundant arsenic-rich growth zones in pyrite during LA-ICP-MS and EMP-WDS trace element analyses, offers insight into how such flocculation might occur. We propose that positively charged surfaces on arsenian pyrite, which behave as a p-type semiconductor (due to As substitution in the sulphide structure), promote cationic bridging and electrochemically destabilise colloidal gold suspensions circulating in epithermal carbonate-quartz veins. This destabilisation triggers flocculation (deposition) of the colloidal gold particles and explains why, at Brucejack, many spectacular gold occurrences appear to have been triggered by the intersection of ore-stage epithermal veins with earlier pyrite veins. This genetic model offers a novel explanation for how bonanza gold may form paragenetically late in hydrothermal ore deposits, i.e., after most sulphide mineral deposition has occurred, without the need for phase separation (boiling) or the presence of steep physicochemical gradients. Moreover, it can also be used to predict the potential locations of bonanza gold occurrences from the block-modelled distribution of arsenic based on exploration and resource drill data.

LA-ICP-MS Analysis on Sulfide Minerals from the Higashi-Aogashima Knoll Caldera Hydrothermal Field to Understand the Host Phase of Gold

Tatsuo Nozaki[1, 2, 3], Junji Torimoto[1], Yutaro Takaya[2, 4, 1], Kazuki Kohama[5], Kotaro Yonezu[5], Satoshi Okada[1], Takahiro Kuribayashi[6], Toshiro Nagase[6], Samuel Morfin[7], Keiko Hattori[7]
1. Japan Agency for Marine-Earth Science and Technology (JAMSTEC), Yokosuka, Japan, 2. Univ. of Tokyo, Tokyo, Japan, 3. Kobe Univ., Kobe, Japan, 4. Waseda Univ., Tokyo, Japan, 5. Kyushu Univ., Fukuoka, Japan, 6. Tohoku Univ., Sendai, Japan, 7. Univ. of Ottawa, Ottawa, ON, Canada

The Higashi-Aogashima Knoll Caldera hydrothermal field, 12 km east offshore from Aogashima Island, Japan, was discovered in 2015. This seafloor hydrothermal field occurs at a water depth of ~750 mbsl, and an abnormal gold enrichment was preliminarily reported. The first multi-disciplinary research cruise (KS-21-20) was conducted in 2021 to collect rock, seawater, hydrothermal fluid, and (micro)biological samples.

Bulk Au concentrations of the sulfide/sulfate-rich rock are 22.54 ± 41.34 ppm (n = 34, ± 1SD), ranging from 0.01 to 178.7 ppm. Average Au concentrations at the Central Cone (n = 26), Southeast (n = 5), and East sites (n = 3) are 25.6, 17.4, and 14.1 ppm. The maximum Au concentration of 178.7 ppm was observed at the mound of the Central Cone Site. Au concentrations show positive correlations with those of Zn (r = 0.64), Cd (r = 0.54), Te (r = 0.51), and Pb (r = 0.48). In situ LA-ICP-MS analyses on sulfide minerals (n = 446) yielded average Au concentrations, in ppm, of 12.0 (pyrite; n = 53), 4.62 (sphalerite; n = 176), 10.0 (chalcopyrite; n = 105), 1.33 (galena; n = 58), 3.66 (tetrahedrite; n = 38), 0.10 (stibnite; n = 12), 0.47 (unidentified As-S mineral; n = 3), and 4.20 (enargite; n = 1). Under the microscope, electrum-rich parts are (1) at the edge of sphalerite and (2) on silica (opal) and unidentified clay minerals next to sphalerite. The electrum grains occur as an aggregate of nano- and micro-particles, suggesting that electrum is precipitated after sulfide minerals, trapped in their cracks and defects, and then recrystallized, producing the "apparent" positive correlations between the bulk Au and Zn (and Pb) concentrations. Our observation would be direct evidence of the recently observed nano-particle precipitation of electrum through boiling processes of the seafloor hydrothermal fluid.

Gokona: An Archean Porphyry-Epithermal Au Deposit

Charles Routleff[1], James Lambert-Smith[1], Thomas Blenkinsop[1], Simon Tapster[2], Adrian Boyce[3], Andrew Allibone[4], Joel Holliday[4], Nathan Komarnisky[4]
1. Cardiff University, Cardiff, United Kingdom, 2. NERC Isotope Geosciences Laboratory, Nottingham, United Kingdom, 3. National Environmental Isotope Facility, Glasgow, United Kingdom, 4. Barrick Gold Corporation, Toronto, ON, Canada

Exploration expenditures are increasing as discoveries are in decline; evolving conceptual-genetic models are essential to identify and delineate viable ore reserves. Archean greenstone-hosted Au deposits are generally classified as "orogenic" Au, having formed in structurally complex convergent settings from reduced, low-salinity H_2O-CO_2 fluids across a broad crustal depth profile. The class accounts for ~30% of past production and known gold resources, but the origins of ore-forming fluid(s) remain controversial.

The Tanzanian Lake Victoria Goldfield (LVG) hosts several major Neoarchean Au deposits that are of "relatively late syn-orogenic timing," c. 2660 to 2640 Ma (e.g., Geita, Bulyanhulu, Golden Pride, North Mara). The ~10-Moz Gokona deposit is enigmatic, exhibiting characteristics that are incongruous with the orogenic Au model and LVG mineralization paradigm. At Gokona, mineralization predates regional deformation and occurred in a highly dynamic, near-surface setting. The deposit is associated with intense wholesale K-feldspar alteration, anomalous base metal and telluride enrichments, and cryptic structural control that is only apparent at the camp scale. In this presentation, we synthesise our petrographically constrained paragenesis with ID-TIMS geochronology, fluid inclusion micro-thermometry, and stable isotope data (δD, $\delta^{18}O$, $\delta^{34}S$) to argue that mineralization occurred under similar conditions to Phanerozoic porphyry-epithermal deposits and was driven by a contemporaneous genetically associated calc-alkaline dacitic porphyry complex. Our proposed magmatic-metallogenetic event pre-dates regional orogenic Au mineralization by around 20 m.y., which is broadly coeval with early Au-mineralizing episodes reported in the Yilgarn and Superior cratons. The fortuitous preservation of the entire Gokona stratigraphic sequence permits recognition of this setting.

Our evidence suggests that the Gokona deposit is an oxidised magmatic-hydrothermal system, which may pose significant implications for our understanding of controversial Archean Au deposits and the magmatic contributions to greenstone belt Au inventories. This is exemplified by the increasing acknowledgement of analogous ore-related porphyries spatiotemporally associated with Au deposits.

Sustainable Futures for Small-Scale Gold Mining Communities: Mapping the Socio-Economic, Environmental, and Geological Baseline for Peixoto De Azevedo, MT, Brazil

Maria J. Mesquita[1], James Lambert-Smith[2], Marko Monteiro[1], Rosana Corazza[1], Thomas A. Smith[2], Angelina S. Bellamy[3], Agatha Herman[2], Roberto Greco[1], Marcus Gomes[2], Marcin Piersiak[4], Jerusa Schneider[5], Rubia Quintao[1], Pedro d. Silva[1]
1. Universidade Estadual de Campinas - Unicamp, Campinas, SP, Brazil, 2. Cardiff University, Cardiff, United Kingdom, 3. University of the West of England-UWE, Bristol, United Kingdom, 4. Alliance for Responsible Mining, Medellin, Colombia, 5. Universidade Estadual De Santa Cruz – UESC, Ilheus, BA, Brazil

Researchers from the Universities of Campinas (Brazil) and Cardiff (UK) held an engagement workshop in Peixoto de Azevedo, MT, funded by the Global Challenges Research Fund (GCRF). The aim was to build a working relationship with stakeholders involved in the region's artisanal and small-scale gold mining (ASGM) to co-produce research targets for more sustainable mining practices. Stakeholders came from a wide background, including "garimpeiros" from cooperatives, Ministry of Mines and Energy representatives, junior mining companies, indigenous communities, state and local government members, social organizations, and local institutions academics. The thematic groups discussed geology, environment, economy, social science, and education. The following key points are central to sustainable livelihoods on ASGM in Peixoto: (a) high-quality education and training; (b) local economy diversification, with mining remaining as a vital component; (c) low mining environmental impact; and (d) regulatory environment and social justice improvement. These outcomes provided starting points to develop the second project:

1) Is it possible to farm and produce contaminant-free products in former gold mining areas?

2) What are the environmental impacts of ASGM on soil and water?

3) What model of gold mining certification is the most appropriate for the Peixoto cooperatives?

4) Can new technology replace Hg-amalgamation?

5) How can workplace safety practices be implemented to increase well-being?

The group won the second GCRF Project intending to work with stakeholders in Peixoto to implement new models for ASGM sustainability appropriate to Brazil and relevant to global ASGM sustainability challenges. We are developing the following objectives: (a) map stakeholder interactions within and around mining; (b) explore stakeholders' attitudes toward sustainability and requirements for gold certification; and (c) implement phytoremediation in contaminated areas and test feasibility of Hg substitution. We believe that only community-company-university co-creation programs will effectively meet the community's real needs for sustainable mining in Brazil and elsewhere.

Carbonate U-Pb Geochronology: A Novel Tool for Directly Dating Fluid Flow Events in Mineral Deposits

Catherine Mottram
University of Portsmouth, Portsmouth, United Kingdom

Carbonate veins are ubiquitous in many ore deposits and often are interpreted as a late stage in the ore deposit history. Dating both veins and brittle fault material has been notoriously difficult because of a lack of "datable" material. Using innovative techniques, it is now possible to date carbonate in veins, breccias, and faults with the U-Pb isotope system.

This presentation gives an overview of advances in U-Pb carbonate geochronology and explores the effectiveness of the U-Pb calcite dating technique for providing timing constraints for faulting, fluid-flow, and mineralization events. In similarity with other unconventional geochronometers that have low U content and incorporate common Pb into their structures (such as titanite), calcite is not always straightforward to date and results can therefore be ambiguous to interpret. This method has, however, been successfully applied to a variety of ore deposits, including porphyry, orogenic gold, Carlin-type, and Pb-Zn mineralization.

In a fault-controlled Cu-Au-Mo porphyry system in the central Yukon, Canadian Cordillera, U-Pb carbonate dating has been used to analyze a variety of fault and vein material. Over 50 samples have been dated, revealing a long history of faulting and fluid flow in the deposit spreading over tens of millions of years between ~75 and <20 Ma. The results show the carbonate veins crystallized during the main ore-forming event at ~75 Ma. Subsequently, there was a prolonged period of fault-controlled fluid pulsing that likely concentrated metallic minerals in the deposit. The findings show that carbonate veins are not always late features within ore deposits and are an underutilized resource for understanding the full temporal and fluid evolution of a system. Carbonate U-Pb geochronology is therefore potentially incredibly useful for telling the previously untold and long history of fluid flow in a variety of deposits.

New Approaches to Constraining Cu-Co Mineralisation in the Zambian Copperbelt Using Hydrothermal Rutile and Apatite

Jamie Kelly[1], Simon Tapster[2], Stephen Roberts[1]
1. University Of Southampton, Southampton, United Kingdom, 2. British Geological Survey, Nottingham, United Kingdom

Despite decades of research on the Zambian Copperbelt (ZCB), the timing of mineralisation remains a contentious debate. The long-standing paradigm of multi-episodic mineralisation encompassing syndiagenetic mineralisation through to post-orogenic mineralisation (~800–500 Ma) has been challenged by a restricted 540-490 Ma synorogenic hypothesis. This is primarily based on extensive $^{187}Re/^{187}Os$ molybdenite ages across numerous ore deposits that demonstrate this late temporal affinity. However, the direct relationship between molybdenite and the timing of first copper mineralisation has been questioned. This is due to molybdenite rarely sharing contemporaneous petrographic relationships to Cu and Co sulphides. Here, we address this issue by utilising the widespread occurrence of hydrothermal accessory phases, rutile and apatite. Petrographically characterised vein and ore horizons are used to establish the paragenetic framework and relationship to ore mineral growth and then combined with in situ LA-MC-ICPMS U-Pb analyses of these accessory phases.

Our results indicate two major mineralising episodes: an older ~570-520 Ma period and a younger, previously under-represented ~490-470 Ma period. The older period is attributed to potassic brines mobilising prior to and during peak metamorphism of the Lufilian Orogeny and Hook Batholith emplacement. This event is interpreted as the first mineralising event at Mindola North. We see no clear evidence of dates exceeding 600 Ma. The younger event is interpreted as a second growth event responsible for bornite mineralisation and secondary fluid flow along pre-existing quartz-carbonate veins during post-orogenic uplift of the Lufilian Orogeny. The only previous documentation of this event is by bornite (473 ± 4 Ma) at the Kamoto Cu-Au deposit. Our results demonstrate how our understanding of the temporal evolution of Cu-Co in the ZCB can be enhanced by our approach of uniting the petrographic record with high-spatial resolution U-Pb geochronology which could be applied to other deposits within the ZCB.

Far Field Geochemistry and C-O Isotope Signatures of the Bingham Porphyry Cu-Au-Mo Deposit in Carbonate Wallrocks

Michael J. Kirschbaum[1], Zhaoshan Chang[1], Paul Agnew[3], Adam Pacey[3], Phillip Nickerson[2], Andrew Sasso[2], Schroeder Kim[2], Kathleen Gundy[2]
1. Colorado School of Mines, Golden, CO, USA, 2. Rio Tinto, Salt Lake City, UT, USA, 3. Rio Tinto, Bundoora, VIC, Australia

Far field signals in carbonates around the Bingham Canyon porphyry Cu-Mo-Au deposit may help with vectoring in other projects. Bingham comprises multi-phase Eocene-aged igneous rocks emplaced in folded and thrusted Paleozoic strata, producing proximal porphyry-style mineralization and adjacent skarn deposits in carbonates. Outboard from the porphyry-skarn zones, there are Pb-Zn-Ag vein and carbonate replacement orebodies, and carbonate veins occur in distal carbonate rocks. This study sampled whole rock and veins of the Jordan and Commercial limestones along a 17-km transect between Bingham Canyon and the Stockton porphyry system.

The 182 whole-rock C-O isotope data show that most samples within ~2.5 km of Bingham and Stockton have hydrothermal signatures (<19‰ $\delta^{18}O$; <-1.8‰ $\delta^{13}C$). Samples farther out have marine carbonate signatures, with some exceptions. There is no smooth trend, indicating untreated whole-rock C-O isotope composition is a moderately strong indicator but not a vector. Temperature-corrected data are expected to show smooth trends.

In carbonate whole-rock geochemistry data, Cu, Mo, Fe, and other elements decrease to background levels in <1 km, whereas Ag, As, P, and REEs decrease between ~4 and 6 km. Sr increases, then levels off, over <1 km. Mn, Zn, Cd, and Ba increase to maxima at ~1 km then decrease to background at ~2–4 km. Vein calcite LA-ICP-MS data show decreasing trends in REEs, Fe, Mg, S, and Pb, whereas Ba and P increase outboard to a distance of ~2–8 km, while Mn and Zn increase to ~2-km distance before dropping to background levels ~4 km outboard. Compared with WR data, vein data shows stronger signals. In whole-rock calcite, the trends of Fe and Pb persist (in shorter ranges), whereas other trends disappear, likely because sedimentary calcite contains more S, Mg, P, etc., masking hydrothermal signals. Additional samples along the transect should help refine, strengthen, or verify these trends.

Mineral Petrography, Trace Element, and Pb Isotope Geochemistry of the Black Butte Cu-Co District, Belt Basin, Montana, USA: Constraints on Ore Genesis

Jeffrey Steadman[1], Jerry Zieg[2], Sebastien Meffre[1], Ross Large[1]
1. CODES, University of Tasmania, Hobart, TAS, Australia, 2. Sandfire America, White Sulphur Springs, MT, USA

Petrographic and geochemical analyses of more than 100 samples of pyritic shale from in and around the Cu-Co deposits of the Black Butte district, Belt Basin, Montana, have revealed a complex, multi-stage pyrite paragenesis. Two main pyrite generations—one early (syn-diagenesis) and the other late (syn-hydrothermal)—are preserved across the district. Diagenetic pyrite is typified by its fine-grained, densely packed to sporadic spherical shapes. Hydrothermal pyrite is distinguished from diagenetic pyrite by its euhedral cubic forms, which commonly occur as overgrowths on diagenetic pyrite. Other types of hydrothermal pyrite also occur (e.g., colloform), hosted in either veins or brecciated shale. Diagenetic pyrite typically contains ppm-level concentrations of Co, Ni, Tl, Mn, Ag, Cu, Pb, and Bi, but below-detection levels of Se and Te. Gold has a bimodal distribution in diagenetic pyrite, in that samples above the regional Volcano Valley fault zone have below-detection levels of Au, whereas samples below the fault zone contain ppm-level Au concentrations. Hydrothermal pyrite bears a marked chemical resemblance to diagenetic pyrite, except for Co, As, and Ni, which reach values of up to 5, 3, and 1 wt. %, respectively. Furthermore, the Pb isotope compositions of diagenetic and hydrothermal pyrite enable effective discrimination between the two generations, particularly in samples below the Volcano Valley fault zone, which exhibit more radiogenic Pb compositions than those above it. The deposits of the Black Butte district represent unique examples of Mesoproterozoic sediment-hosted Cu-Co mineralization that do not fit well into any existing models for sediment-hosted Cu deposits. Furthermore, shared textural, mineralogical, and sedimentological affinities with Meso- to Paleoproterozoic sequences in northern Australia may indicate heightened potential for McArthur River analogues in the Black Butte area, which warrants further investigation.

Kyanite Alteration as an Indicator for Saline Fluids and Metal Transport; Case Studies from the Central African Copperbelt

Anna Bidgood[1, 2], Murray Hitzman[2], Helen Twigg[2], David Chew[3], Richard Unitt[4], Paul Slezak[2]
1. Fulbright-UK scholar, Carnegie Institution for Science, Washington DC, DC, USA, 2. iCRAG, University College Dublin, Dublin, Ireland, 3. Trinity College Dublin, Dublin, Ireland, 4. University College Cork, Cork, Ireland

The Central African Copperbelt (CACB) is the world's primary source of Cu and Co, producing about 70% of the metal, which is critical for battery production and other decarbonisation technologies. Highly saline fluids are associated with extensive mineralisation and alteration in the CACB. The intimate association of kyanite-bearing veins and Cu-sulphide mineralisation at low pressure and temperature indicates that unusual thermodynamic conditions facilitated the mobility of traditionally immobile elements such as Al, Ti and Ge, as well as mobilising technology metals such as Cu, Co, and Ni.

Kyanite is traditionally considered a high-pressure and high-temperature Al_2SiO_5 mineral polymorph. Common kyanite-forming reactions such as pyrophyllite or chlorite dehydration take place at ~400°C and >600°C respectively. In this study, kyanite is observed in low-temperature (i.e., sub-greenschist facies) carbonate and carbonaceous siliciclastic lithologies, hosted in post-evaporitic breccia. Kyanite occurs in hydrothermal kyanite-quartz-magnesite-chalcopyrite-monazite veins, within fractures associated with quartz + pyrite, and as porphyroblasts in host rock lithologies. The occurrence of kyanite in low-pressure–temperature rocks indicates unusual conditions of kyanite growth.

Extensive kyanite alteration is also seen throughout the higher-temperature portions of the Copperbelt in northern Zambia, where kyanite-nickel sulphide and kyanite-biotite-copper sulphide assemblages are observed in highly altered regions of intense fluid flow. Cathodoluminescence imaging, laser ablation mass spectrometry, Raman spectroscopy, and U-Pb monazite dating are used in this study to unravel the complex vein paragenesis and characterise the multiple textural generations of kyanite.

Understanding the paragenesis of kyanite in relation to Cu and Ni and the unusual thermodynamic conditions required for low-temperature kyanite growth and Al-mobilisation has implications for exploration of technology metals, as well as metal mobility in highly saline systems. The interpretation that kyanite can be part of an alteration assemblage is also potentially useful as an indicator of highly saline fluids and intense fluid flow.

Potential Sources of Rare Earth Elements (REEs) in Loch Borralan, Northwest Highlands, Scotland: Geochemical and Mineralogical Controls on REE Distribution in Carbonatite

Abdulkadir Mohamed Abdulkadir
UCL, Earth Sciences 5 Gower Place | London WC1E 6BS, London, United Kingdom

This study investigates the Loch Borralan carbonatite as a source for rare earth elements (REEs), which are vital for the green transition. The Caledonian Loch Borralan complex is located in the Assynt district, in the NW Scottish Highlands. Analysis of targeted rocks indicates a diverse group of REE-bearing minerals, including bastnaesite, synchysite, ancylite, monazite, apatite, zirconolite, titanite, and REE-rich ilmenite. There are also significant amounts of sulphide minerals such as pyrite, pyrrhotite, and chalcopyrite. This study aims to constrain REE enrichment processes and isolate different magmatic processes that led to the enrichment of REEs in the carbonatite magma from the effect of hydrothermal processes. The $\delta^{13}C$ (-5.5‰) and $\delta^{18}O$ (+9.1‰) stable isotopes indicate a mantle origin, whereas $^{40}Ar/^{36}Ar$ and $N_2/^{36}Ar$ data are compatible with limited crustal assimilation. REE-bearing minerals in the Loch Borralan carbonatite have experienced a sequence of REE mineralisation where early-formed REE-bearing minerals such as REE-titanite and apatite donated their REE contents to more evolved REE-carbonates and phosphates. Although the size of the deposit is unknown, it could have the potential to contribute to securing a sustainable REE mining industry ad ensuring supply security of these critical elements; establishing a model for REE enrichment will improve understanding of the origin of this deposit, which could have an impact on future targeting of REE-bearing deposits globally.

Mineralogy and Whole-Rock Geochemistry as Exploration Tools for Lithium Pegmatites in Ghana: Examples from the Biriwa-Ewoyaa Pegmatite Fields

Elizabeth Agyekum, Kofi Adomako-Ansah
University of Mines and Technology, Tarkwa, Ghana

Due to its importance for the green energy revolution supported by UN SDG 7: Affordable and Clean Energy, lithium (Li) is attracting interest on a global scale. The worldwide search for Li ore has expanded tremendously as a result of the anticipated huge demand for Li-based technological products. In-depth investigation of the origins, enrichment, and exploration of this rare metal in Ghana has been prompted by the discovery of Li ore in pegmatites. This paper provides preliminary field observations, petrographic studies, and interpretations of the major and trace element concentrations to understand the petrochemical character of the pegmatites and its implications for the evolution and enrichment of Li.

The pegmatites in the Ewoyaa and Biriwa areas, respectively, are associated with Birimian mica schists in the Cape Coast basin, Central Region of Ghana. Although both pegmatites are composed predominantly of quartz, K-feldspar, albite, and muscovite, the pegmatites at Ewoyaa additionally contain spodumene and relatively rare-to-no biotite while the pegmatites at Biriwa contain appreciable amounts of biotite. The geochemical data indicates that both pegmatites are granitic and peraluminous. Standard-normalised diagrams reveal similar distribution patterns for REEs, LILEs, and HFSEs. In contrast to the pegmatites at Biriwa, evolution may have influenced the enrichment of Li, Cs, Ta, Sn, and Rb and depletion of Ba, Cr, Nd, Sr, and REEs in the Ewoyaa pegmatites. Although the source (whether sedimentary or igneous) of Li in the granitic magma is currently not clear, the enrichment of Li and other rare metals in the Ewoyaa pegmatites was most likely caused by fractional crystallisation of Li-rich pegmatitic melt. The differences in elemental concentrations enabled binary plots of elemental ratios that clearly distinguished between mineralised and barren pegmatites, which would be very useful in target mineral exploration programs for mineralised pegmatites in the region and elsewhere in the world.

Structural-Chemical Vectors to LCT-Pegmatites in the Varuträsk Area, Sweden

Joel Andersson[1], Edward P. Lynch[2], Martiya Sadeghi[2], Samuel Lundsten[3], Tobias E. Bauer[1]
1. Luleå University of Technology, Luleå, Sweden, 2. Geological Survey of Sweden, Uppsala, Sweden, 3. Luleå University of Technology, Skellefteå, Sweden

The Varuträsk area in northern Sweden hosts the only historic lithium-caesium mine in Sweden. A small-scale mining operation active between 1936-1946 extracted 90.5 t lithium (petalite, spodumene, lepidolite) and 63.3 t caesium (pollucite). The mineralogy of the Varuträsk pegmatite has been described in detail by several pioneering studies, making the Varuträsk pegmatite a classic example in the pegmatite literature. Despite an exponential interest in battery metals, the Varuträsk area has not been investigated from a broader mineral systems perspective and source-transport-trap mechanisms remain unresolved.

As part of the Horizon Europe project Exploration Information System (HORIZON-CL4-2021-RESILIENCE-01 - n°1010557357 –), we combined structural mapping with whole-rock geochemistry including samples from the Varuträsk pegmatite, regionally occurring pegmatites, and potential source granitoids to identify structural-chemical vectors towards prospective pegmatites in the study area.

Results show that peraluminous leucogranites in the area form a geochemical continuum with highly fractionated pegmatites and may represent source intrusions for LCT pegmatite-forming melts. However, published age data indicate the Varuträsk pegmatite is at least >8 m.y. younger than the candidate source granites, which underlines the need for additional geochronology data to help identify the source(s?) and define timespans of the mineralized system.

Potential source granites are bounded by pre-existing shear zones, and aeromagnetic data indicate fabrics deflect around certain plutons. Quartz/granite/pegmatite veining increases towards major structures, indicating these structures served as conduits for fluid-rich granitic melts. Structural entrapment styles vary in the area. Underground mapping (physical and digital) and geological 3D modelling shows the Varuträsk pegmatite constitutes at least two shallowly dipping fault-bounded sheets that crosscut pre-existing foliation in folded, competent mafic volcanic rocks. Regionally occurring pegmatite veins hosted by metasedimentary rocks are bounded by steep fracture planes with a local fertilization trend (La/Ta vs. Mg/Li) approximately normal to pegmatite trends.

Molybdenite from the Palaeoproterozoic REE-Line in Bergslagen, Sweden: Textural Relations, Trace Element Chemistry, and Re-Os Age Data

Stefan S. Andersson[1], Erik Jonsson[1, 2], Thomas Zack[3], Martiya Sadeghi[1]
1. Geological Survey of Sweden, Department of Mineral Resources, Uppsala, Sweden, 2. Uppsala University, Department of Earth Sciences, Uppsala, Sweden, 3. University of Gothenburg, Department of Earth Sciences, Gothenburg, Sweden

The REE-line in the Bergslagen ore province in Sweden hosts the so-called Bastnäs-type REE deposits, which comprise a suite of skarn-hosted magnetite-dominated iron oxide deposits associated with locally REE-rich and polymetallic mineralisation (e.g., Cu, Co, Bi, Mo, Au). They are generally interpreted as replacements of interlayered carbonate rocks within a volcano-sedimentary sequence by high-temperature Fe-REE-rich fluids at c. 1.9 Ga. The associated polymetallic mineralisation has been known and documented previously but not studied in detail. In this study, we characterise molybdenite from different REE deposits in the REE-line by textural studies combined with trace element chemistry and Re-Os ages by LA-ICP-MS/MS.

Molybdenite is commonly associated with chalcopyrite ± pyrrhotite ± pyrite ± bismuthinite as well as other ore minerals, including Co-rich ones, and occurs in several textural and paragenetic settings: 1) associated with allanite group minerals in irregular fracture fillings, veins or bands in different skarn assemblages, locally crosscutting earlier recrystallised cerite-(Ce) and bastnäsite-(Ce) ore in Nya Bastnäs, 2) within hydrothermally Mg- or Mg-K-altered volcanic rocks, now manifested by metamorphosed mineral assemblages like quartz-anthophyllite or biotite-garnet, 3) within carbonate rock carrying serpentine-dominated pseudomorphs ("ophicalcite") associated with magnetite, or 4) within tremolite-actinolite-magnetite skarn assemblages.

The trace elements Re, W, Se, and locally Te could be identified as being contained in molybdenite, whereas higher concentrations of other elements likely represent inclusions, including Bi-Pb-Cu-Te-Au-bearing and locally REE-bearing phases. Rhenium concentrations range from below 1 to c. 20 ppm. Preliminary LA-ICP-MS/MS data of samples with higher Re concentrations yield Re-Os ages of 1904±17 Ma (Bondgruvan), 1857±17 Ma (Gamla Bastnäs) and 1838±30 Ma (Malmkärragruvan), consistent with published N-TIMS ages. The combined results thus far suggest multiple generations and types of molybdenite mineralisation and associated polymetallic mineralisation during the evolution of the Svecokarelian orogeny.

The Iberian Pyrite Belt as a Strategic Source of Cobalt

Gonzalo Ares, Carmen Conde, Fernando Tornos
Instituto de Geociencias (IGEO, CSIC-UCM), Madrid, Spain

Cobalt is one of the critical elements for the energy transition. It is an essential component of the lithium-ion batteries used in most laptops, mobile phones, and electric vehicles. Cobalt is used in many different sectors, ranging from chemical, metal and industry, and electronics to healthcare.

In recent years, the demand for cobalt has escalated while the global supply chain for this element is still unstable. That is why it is imperative to look for new sources of this element that will help satisfy the demand of the industry.

The Iberian Pyrite Belt (IPB) is the largest concentration of volcanogenic massive sulphide deposits worldwide. Our compilation indicates that several world-class orebodies host large volumes with whole-rock cobalt values greater than 300 ppm. However, this information is very scattered, scarce, and poorly contrasted. This study improves our knowledge on where the Co grades are located at deposit scale, where cobalt is preferentially concentrated (shale-hosted vs. volcanic-hosted deposits and different zones of the massive sulfide vs. stockwork), and in which phases and how it occurs—if it is as discrete phases, nanoinclusions, or solid solution in other phases, very likely pyrite.

The large deposits of Tharsis (shale-hosted), Concepción (volcanic-hosted), and Valverde (shale- and volcanic-hosted) show an average content of 400 to 500 ppm with metre-thick intersections of up to 0.2 wt % Co. pXRF, EPMA, and LA ICPMS analyses of pyrite show that the highest contents are located in the stockwork zones near the contact with the massive sulphides. Co occurs either as grains of cobaltite, glaucodot, and alloclasite or as stoichiometric substitutions in pyrite, which locally has up to 360 ppm of Co, showing a good correlation with As and Au contents.

Study funded by the Horizon Europe EIS project (contract 101057357).

Reassessment of Rare-Metal Potential of the Karagaily-Aktas Deposit (Kazakhstan)

Akmaral Baisalova[1], Yerlan Akbarov[3], Alla Dolgopolova[2], Reimar Seltmann[2]
1. Satbayev University, Almaty, Kazakhstan, 2. Natural History Museum, Centre for Russian and Central Eurasian Mineral Studies (CERCAMS), London, United Kingdom, 3. Committee of Geology of the Ministry of Industry and Infrastructure Development (MIID), Astana, Kazakhstan

The Karagaily-Aktas rare-metal deposit is located in the Terskey Alatau mountains at 2,300- to 3100-m altitude, 20 km southwest of Narynkol, Almaty region. Several stages of exploration were carried out during 1956-1983.

The deposit ("Central"), together with occurrences at Noviy, Lesnoy, Vostochny, and Dalniy, forms an ore field of 12-km extension. Tungsten-tin mineralization is associated with a dike of greisenized granite intruding the Lower Paleozoic shale-carbonate package. The main orebodies amount to about 80-90% of all tin reserves, represented by vein and veinlet zones composed of quartz, topaz, feldspar, and mica. Orebodies with complex shapes are concentrated in the foot and hanging walls of the greisenized granite dike. The orebodies stretch along the strike for 300 to 740 m, thickness varies from 1 to 23.8 m, and vertical extension down the dip is steep to the south for up to 250 m at sub-latitudinal strike. Ores have mica-quartz composition. The main commodity is tin contained in cassiterite, accompanied by tungsten, lithium, feldspar, beryllium, rubidium, caesium, and rare earths.

The total ore reserves amount to 16.4 Mt (medium-size deposit); tin reserves are 78.7 Kt—average grade of 0.48%; tungsten reserves are 19.7 Kt—average grade of 0.12%. Apart from tin and tungsten, the deposit has significant credits of lithium dioxide (reserves >28 Kt) and beryllium oxide (reserves ~5,000 t). Processing tests showed the possibility of obtaining rare metals concentrates. The recoverability of tin was 62-80%, tungsten trioxide 52-68%, and lithium oxide 54-93%.

The ore field around the Karagaily-Aktas deposit is underexplored (total reserves estimated at 50-60 Kt tin; large-size tin deposit). Increase of the reserves of tin ores and accompanying metals can be achieved by exploration of deeper horizons and prospecting in the periphery of the deposit, including thermal waters. Current exploration aims to incorporate new areas and improve the deposit model.

Finding and Quantifying Li-Bearing Mineralisations Using the GeoCore X10 Drill Core Scanner

Mikael Bergqvist[1], Alexander Hansson[1], Mikael Kördel[1], Karin Högdahl[2], Erik Jonsson[3], Magnus Leijd[4], Jaroslaw Majka[2], Pauline Jeanneret[2]
1. Orexplore AB, Kista, Sweden, 2. Dept. Earth Sciences Uppsala University, Uppsala, Sweden, 3. Swedish Geological Survey, Uppsala, Sweden, 4. Bergby Lithium AB, Bollnäs, Sweden

Raw materials from the recent lithium-cesium-tantalum (LCT) pegmatite field discovery in Bergby, in the central part of Sweden, could potentially quickly be put to use in European or even domestic electric vehicle and storage battery production.

When exploring, a challenge has been to actually detect and measure the abundance of light elements, such as Li. Laser induced breakdown spectroscopy (LIBS) is a method with such capabilities, and the task would be to make it practically feasible to apply it in a way suitable for exploration.

In the recently started ULiBS project, LCT pegmatites are studied with LIBS, X-ray computed tomography (CT), and xX-ray fluorescence (XRF), trying to differentiate and quantify Bergby lithium ore minerals such as petalite and spodumene. Moreover, the project aims to study the possibility of quantifying concentrations of quartz, feldspar, and mica, which could potentially be sold as by-products from a future mine at Bergby.

Orexplore's drill core scanner, the GeoCore X10, has been producing x-ray CT and XRF data for tens of thousands of meters of drill core, although LCT-pegmatites have not been scanned previously. In addition to scanning such drill cores, the ULiBS project will utilize a GeoCore X10 version with a modular LIBS system added alongside its other sensors to extend the range of detectable elements and boost both mineral identification and quantification capabilities while maintaining the scanning speed and usability of the original system.

Following the project, the enhanced system and methodologies could be rolled out to the fleet of GeoCore X10 drill core scanners to enable faster discovery of lithium mineralisations and conversion to active mines.

The Yin Deposit, a New REE Discovery in the Gifford Creek Carbonatite Complex, Gascoyne Province, Western Australia

Luke Blais
Dreadnought Resources, Perth, WA, Australia

Dreadnought Resources' Yin deposit, part of the Gifford Creek Carbonatite Complex (GCCC) in the Gascoyne Province of Western Australia, is one of the most significant rare earth discoveries in Australia this millennium. Since identification of outcropping REE-bearing dikes (ironstones weathered from a ferrocarbonatite protolith) in May 2021, Yin has been drilled along an initial 3-km strike, yielding a 14.4 Mt @ 1.13% TREO initial resource, and was the subject of a university study where, through petrography, mineral, and geochemical analyses, REE mineralisation controls have been established.

Since the 1980s, the area has recognised REE mineralisation, but a significant portion of occurrences remained undiscovered until 2021/2022 due to accepted paradigms that all REE mineralisation was located north of the crustal-scale Lyons River Fault. Over the last 18 months, exploration focused entirely south of the Lyons River Fault; Dreadnought Resources have discovered multiple additional REE-bearing dike trends, and a series of carbonatite plutons central to, and related to, the metallogenesis of the region, redefining the mineralised potential of the GCCC and increasing the search space for mineralisation from a ~700-km^2 to a >1,200-km^2 area.

Strong industry-academia relationships have been critical to the advancement of the geological understanding of the GCCC. Studies have found the Yin deposit is defined by two main zones, geologically distinct in mineralogy and chemistry, with the zones interpreted to represent separate pulses during the emplacement history of the carbonatite dikes. The mineralogical complexity between these zones is compounded as subsequent supergene weathering affects the zones independently.

Successful exploration programs which identify potentially economic REE concentrations will be key to sustaining the metals required for a green energy transition. This talk will outline the exploration approach that led to these recent discoveries and discuss mineralisation controls at the deposit, where new geological understandings have significant implications for future exploration.

Investigating Aeration and Filtration as Pre-Treatment Methods for Flowback and Produced Water Prior to Ion Exchange for Lithium Extraction

Bennett Braun, Karthik R. Shivakumar, Daniel Alessi
University of Alberta, Edmonton, AB, Canada

Lithium is crucial for battery production, utilized in electric vehicles and portable electronics. In many cases, Li-bearing flowback and produced water (FPW) requires pre-treatment before entering the ion-exchange process for Li extraction using a highly Li-selective Mn (IV) oxide sorbent. The main obstacle is the reductive dissolution of the Mn (IV) oxide by dissolved organics (DOC) and loss of Mn (IV) as more soluble Mn (II/III). An approach of aeration and filtration can remove DOC to prevent mass loss of the sorbent and improve the Li extraction performance. In Canada, significant concentrations (>75 ppm) of lithium are present in petrobrines near Fox Creek, AB. The FPW samples used for this study have a TDS >157,000 ppm and lithium values from 64-72 ppm. The four aeration experiments used 800 ml of brine, which was aerated with compressed air for 24 hrs at 0.3 L/min and filtered after 24 hours to remove precipitates. Two experiments used raw brine samples and the other two were pH-adjusted to pH 6 and 7. Li extractions were completed on the raw brine, aerated raw brine, and aerated pH-adjusted brine using the Mn (IV) oxide sorbent, and an ICP-MS to measure Li recovery and Mn loss. The pH-adjusted brine displayed a decline in Fe^{2+} concentrations, evidencing the formation of Fe (III) solids, with a concomitant 10-24% decrease in the solution TOC concentrations. Elevated Li uptake of up to 10.8 mg/g was observed in the pH-adjusted brine extractions, compared to 4.5 mg/g in the raw brine extraction. The above trends are proposed to be caused by organics sorbing to iron oxyhydroxides produced through of the oxidation of Fe^{2+} during aeration. These colloids are large enough to be removed by microfiltration. If the pre-treatment methods above are scalable, ion-exchange technology could be feasible for lithium production from petrobrines.

Battery Metal Exploration: Integrating Satellite, Airborne, Drone and Field Data with Machine Learning Outputs in Cornwall

Mark Broadley
CGG, Crawley, United Kingdom

The geothermal brines associated with granitic rocks in Cornwall have been proven to have globally significant lithium grades. These hot-brines are often localized along faults and fractures and alter the surrounding country rocks.

The clay and iron minerals associated with hydrothermal alteration have characteristic spectral signatures that can be identified by satellites. Multispectral and hyperspectral satellites were used to map the alteration and generate target areas for fieldwork. High-spatial and -spectral resolution airborne hyperspectral data and 3D drone imagery were also integrated to enable more precise target detection. Fieldwork follow-up with a field-spectrometer confirmed the presence of alteration.

Combined with the spectral mapping, satellite elevation data sets and airborne LiDAR were used to generate semi-automated fault mapping workflow. Machine learning algorithms and automated workflows resulted in the rapid processing, interpretation, and delineation of possible faulted areas. The outputs from these were integrated with the mineral alteration mapping to geologically contextualize the target areas.

Using these targets, geologists from Cornish Lithium undertook a ground validation campaign as a key element of the iterative approach that the machine learning model took. This allowed for a highly accurate prospectively map to be generated and the improved delineation of field target areas.

Scientific Drilling at a REE-Mineralised Carbonatite

Sam Broom-Fendley[1], Kathryn Goodenough[2], Michael Marks[3], John Spear[4]
1. University of Exeter, Penryn, United Kingdom, 2. British Geological Survey, Edinburgh, United Kingdom, 3. Eberhard Karls University, Tübingen, Germany, 4. Colorado School of Mines, Golden, CO, USA

Carbonatites host the world's largest and highest-grade rare earth element (REE) deposits and are critical for the global transition from fossil fuels to a low-carbon society. Many compositional and structural aspects of these complexes remain poorly understood, such as the relationship between carbonatites and associated alkaline rocks, how carbonatites evolve through magmatic-hydrothermal processes to form economic REE enrichment, how mineralising fluids vary with depth and time, and the importance of local structures for fluid-rock interaction.

Carbonatites are commonly preferentially weathered, and so there are remarkably few global localities where they can be studied in three dimensions. Scientific drilling represents an alternative way to study processes occurring at depth within a mineralised carbonatite complex. In May 2023 we will hold an International Continental Drilling Programme (ICDP)-funded workshop to plan deep drilling that will study the 3D structure of a mineralised carbonatite complex and the influence of these unusual rocks on groundwater chemistry and deep life. We are targeting the Songwe Hill carbonatite complex in Malawi as it represents an excellent example of a sub-volcanic REE-mineralized alkaline-carbonatite complex, with an extensive catalogue of surface data, existing shallow drilling (to 300 m), and a range of geophysical data to support projections of the carbonatite body to depth.

In this presentation we will summarise discussions at the workshop and outline the geological basis of our plan to drill at Songwe Hill. The current interpretation of Songwe Hill comprises a REE-mineralised carbonatite and phonolite neck emplaced between 1 and 5 km, enveloped by carbonatite-rich breccias and altered phonolite and passing outwards through a halo of carbonated nepheline syenite into fresh nepheline syenite. Drilling will enable us to understand the relationship between the carbonatite and the alkaline igneous rocks and the structures that have influenced fluid flow over time.

Mineralogy and Composition of Vanadium-Bearing Shales at the Van Property, Mackenzie Mountains, Northwest Territories

Stefanie M. Brueckner[1], Turner Green[1], Merilie A. Reynolds[2], Daniel Gregory[3]
1. Department of Earth Sciences, University of Manitoba, Winnipeg, MB, Canada, 2. Northwest Territories Geological Survey, Yellowknife, NW, Canada, 3. Department of Earth Sciences, University of Toronto, Toronto, ON, Canada

Vanadium is a vital component in new battery technologies related to energy storage and is hence listed as a critical metal by several countries. Shale-hosted V deposits are a type of V resource that is underexplored. To account for an increasing V demand, the mineralogy of V-bearing phases in and chemistry of V-hosted shales need to be better understood.

The Van property, in the Mackenzie Mountains, Northwest Territories, is a shale-hosted V prospect. Samples from the V-enriched Duo Lake Formation (OSD) and the younger Prevost Formation were collected for mineralogical and compositional analyses including lithogeochemistry, X-ray diffraction, and scanned electron microscopy.

The OSD has average $C_{graphite}$, C_{org}, and S concentrations of 4.0±2.1, 7.8±3.2, and 0.8±0.2 wt %, respectively. Vanadium, Zn, and Cu are enriched especially within the Lower Cherty Mudstone Member of the OSD, with maximum concentrations of several thousands of ppm. In contrast, samples from the Prevost Formation have lower $C_{graphite}$, C_{org}, and S concentrations, are depleted in V, Zn, and Cu, and can have elevated BaO concentration (1.2 wt %).

Vanadium-rich samples of the OSD are dominated by quartz with minor graphite-muscovite and trace sulfides, illite, and oxides. Vanadium does not form own its mineral phases but is hosted in Ti-bearing oxides that are spatially associated with clays but not with graphite.

The relatively flat REE pattern with a negative Ce anomaly and low Th/Sc ratio (<0.6) indicates a mafic sedimentary source and deposition along a passive margin, similar to other critical metal-enriched shales in the Selwyn Basin. The association of V with Ti-oxides indicates a moderately oxidized state (e.g., V^{4+}). However, it is unknown if the enrichment of V in the shales was a product of direct V scavenging from seawater, syngenetic or epigenetic hydrothermal activity, and if organic matter played a role.

The Mineralogy and Mineral Associations of Platinum-Group Elements and Base Metal Sulfides in the Down-Dip Critical Zone at Sandsloot, Northern Bushveld Complex

Kate R. Canham[1], David Holwell[1], Iain McDonald[2], Hannah Hughes[3], Andy Lloyd[4], Lara Du Preez[4], Kofi Acheampong[4]
1. Centre for Sustainable Resource Extraction, University of Leicester, Leicester, United Kingdom, 2. School of Earth and Environmental Sciences, Cardiff University, Cardiff, United Kingdom, 3. Camborne School of Mines, University of Exeter, Exeter, United Kingdom, 4. Anglo American South Africa, Johannesburg, South Africa

The extraction of new resources of PGEs, Cu, Ni, and Co is critical to satisfy increased demand driven by the energy transition. Developing genetic models for ore deposits aids in the generation of exploration targets, whilst understanding the geometallurgy of individual deposits is paramount to efficient processing of ores. The Northern Limb of the Bushveld Complex is renowned as one of the Earth's largest resources of PGEs and also contains important resources of Cu, Ni, and Co. The Critical Zone of the limb hosts the world-class Platreef PGE-Ni-Cu-(Co) deposit, which thickens and increases in grade down-dip, reaching >10 g/t (4E) over tens of meters. In the Sandsloot area of the Northern Limb, recent exploration of the down-dip Critical Zone has identified multiple high-grade PGE-rich zones, as well as thick base metal sulfide (BMS)-rich zones up to several hundred meters below. To date, these down-dip zones at Sandsloot have received no detailed mineralogical study.

The upper PGE-rich zone has around 1 to 2 vol % sulfides, with a typical assemblage of pyrrhotite-pentlandite-chalcopyrite, with lesser cubanite. The most common platinum-group minerals (PGMs) are Pt-Fe alloys, Pt-Pd-Pb alloys, and Pt-Pd bismuthotellurides. In contrast, the lower BMS-rich zone contains lower-tenor semi-massive to massive sulfides, composed dominantly of pyrrhotite, with lesser pentlandite and minor chalcopyrite. Relative to the PGE-rich zone, this zone contains much lower PGE grades, and notably an IPGE-dominant PGM assemblage of Ir arsenides, Ir-(Pt,Rh,Rh) arsenosulfides, and Ru sulfides (62%), with lesser Pt arsenides (18%), Pt-Pd bismuthotellurides (9%), and Pd-Pb alloys (4%).

In this ongoing study, we integrate PGM and BMS variation throughout these zones, with lithological and alteration styles established from hyperspectral and mineralogical analysis. This will establish how magmatic, contamination, and hydrothermal processes have affected the geometallurgy and metal budget of each zone.

Stream Sediments as an Exploration Tool for Spodumene Pegmatites

Lawrence C. Carter[1, 2], Julian Menuge[3], Ben J. Williamson[2], Teimoor Nazari Dehkordi[3], Gavyn K. Rollinson[2]
1. LC Geoscience, Redruth, United Kingdom, 2. Camborne School of Mines, Cornwall, United Kingdom, 3. University College Dublin, Dublin, Ireland

Stream sediment geochemistry has long been used in exploration for gold and other chemically resistant minerals, typically at district or larger scale. Spodumene pegmatites contain a variety of distinctive heavy minerals, and common minerals with distinctive chemistry and/or texture, many of them resistant to weathering. As part of the GREENPEG project, which is developing toolkits for buried pegmatite exploration, we have applied this technique on the prospect scale (25-50 km^2) in two areas of known spodumene pegmatite subcrop beneath thin (<1 m) soil: the Moylisha prospect, southeast Ireland, and the Wolfsberg prospect on the Carinthia-Styria border, Austria. In Moylisha, ~400 Ma spodumene pegmatites were emplaced along the contact between ~410 Ma granites intruding Ordovician mica schists; in Wolfsberg, ~ 270 Ma spodumene pegmatites were emplaced into Variscan mica schists and amphibolites, and later metamorphosed up to eclogite facies at around 90 Ma.

Samples were collected mainly from first- and second-order streams, washed, and sieved onsite into three size fractions. Heavy mineral fractions were separated by shaking table from the 125/150- to 500-µm sieved samples. An aliquot of each heavy fraction was resin mounted, polished, and analysed by QEMSCAN automated mineralogy.

In both Moylisha (Ireland) and Wolfsberg (Austria), we demonstrate relatively high abundances of spodumene, beryl, and kaolinite (amongst other minerals) in streams draining known subcropping spodumene pegmatite intrusions. Abundances of these minerals are markedly lower in samples draining other lithologies. Further, we show that certain normalised mineral associations (calculated by QEMSCAN as a percentage of what touches what in the sample), such as spodumene with beryl and spodumene with quartz, are highest in streams draining pegmatite intrusions. These findings highlight the potential of QEMSCAN of stream sediments as a vectoring tool at the district to prospect scale at early stages of spodumene pegmatite exploration campaigns.

Mapping Critical Minerals in Mineral Systems with the Critical Minerals in Ores (CMiO) Database

George N. Case[1], Evgeniy Bastrakov[2], Garth Graham[3], Albert Hofstra[3], David Huston[2], Christopher Lawley[4], Vladimir Lisitsin[5], Bronwen Wang[1]
1. U.S. Geological Survey, Alaska Science Center, Anchorage, AK, USA, 2. Geoscience Australia, Canberra, ACT, Australia, 3. U.S. Geological Survey, Geology, Geophysics, and Geochemistry Science Center, Denver, CO, USA, 4. Geological Survey of Canada, Ottawa, ON, Canada, 5. Geological Survey of Queensland, Brisbane, QLD, Australia

Base- and precious-metals are the primary focus of most mineral systems, yet many critical minerals (CMs) occur as co- or by-products. Understanding the distribution of CMs in mineral systems is crucial to determine if they are concentrated in ores, waste, or unmined parts of the system. The Critical Minerals in Ores (CMiO) database, compiled by Geoscience Australia, the U.S. Geological Survey, and the Geological Survey of Canada as part of the Critical Minerals Mapping Initiative, contains high-quality, geospatial geochemical data from ore deposits around the world. Each deposit has been classified using a consensus nomenclature, facilitating global comparisons.

Preliminary investigation of porphyry and iron oxide-Cu-Au/iron oxide-apatite (IOCG/IOA) systems highlights differences in CM abundance within mineral systems, and spatial variations within individual deposits. The database confirms that porphyry Cu ± Au ± Mo systems are enriched in rhenium (Re), platinum (Pt), palladium (Pd), selenium (Se), and tellurium (Te). In the Pebble Cu-Au-Mo porphyry deposit (Alaska, USA), Re and Se abundances are broadly coincident with the East and West Cu-Au-Mo orebodies, while Te enrichment is highest immediately outside of the orebodies. A similar pattern is observed in the CuMo porphyry deposit (Idaho, USA), in which mineralization is zoned from Mo-Cu in the core to peripheral Te. Iron oxide-Cu-Au systems are enriched in a broader range of CMs that include cobalt (Co) and light rare earth elements (LREEs). Overall, cobalt is enriched in magnetite-dominant IOCG systems such as Ernest Henry and the E1 Group (Queensland, AUS), whereas LREEs are enriched in IOA and hematite-dominant IOCG systems like Olympic Dam (South Australia, AUS). However, IOA mineralization beneath or adjacent to Cu-Au orebodies in both types of IOCG systems can also be enriched in LREEs. The results demonstrate the utility of the CMiO for characterizing CM distribution in ore deposits.

Cobalt Mineralization in the Central African Copperbelt: Preliminary Insights from Petrography and LA-ICP-MS Analyses

Malena Cazorla Martínez[1], Lingli Zhou[1, 2], Murray Hitzman[1]
1. iCRAG - SFI Research Centre in Applied Geosciences, University College Dublin Belfield, Dublin, Ireland, 2. Department of Earth Sciences, Vrije Universiteit Amsterdam, De Boelelaan, Amsterdam, Netherlands

Cobalt is a key metal for the manufacture of rechargeable batteries, and there is currently intense interest in exploration and development of new genetic models for cobalt-bearing deposits. This study investigates trace element distribution in sulfide minerals in the sedimentary rock-hosted Cu-Co Kwatebala deposit in the Tenke-Fungurume Mining (TFM) district of the Central African Copperbelt. Eleven mineralized samples from very high to low-grade cobalt zones were selected for petrography and trace element geochemical analyses. Two main mineralization styles were identified from microscopy work: 1) stratiform-to-lenticular and disseminated sulfides in dolomitic shales, and 2) bed-parallel and crosscutting sulfide-bearing quartz-carbonate veins. Carrollite ($CuCoS_2$) is the main cobalt mineral and occurs in both mineralization styles. The associated copper sulfides are bornite (Cu_5FeS_4), chalcopyrite ($CuFeS_2$), and chalcocite (Cu_2S). Carrollite is observed as (1) anhedral, elongated grains parallel to the bedding, with abundant quartz, carbonate, bornite, and muscovite inclusions, (2) anhedral overgrowths with bornite, quartz, and carbonate inclusions, (3) subhedral grains with carbonate inclusions, and (4) subhedral-to-euhedral inclusion-free grains. High-grade cobalt zones are characterized by abundant stratiform-to-lenticular, inclusion-rich carrollite, with local inclusion-poor overgrowths. Bornite and chalcocite occur in both high-grade and low-grade cobalt zones, whilst chalcopyrite only is present in high-grade cobalt zones. LA-ICP-MS analyses (n=107) reveal that carrollite is depleted in most trace elements, except for Ni (150-1,700 ppm), As (150-1,300 ppm), Se (10-90 ppm), Mo (1-100 ppm), Sb (5-1,000 ppm), and Bi (0.1-100 ppm). A significant difference in Ni content is recognized between carrollite in stratiform-to-lenticular (150-550 ppm), bed-parallel vein (900-1,700 ppm), and cross-cutting vein (350-550 ppm) styles. Results suggest multiple generations of carrollite precipitation with evidence for variations in the geochemistry of mineralizing solutions. Further work will expand the data set to more mineralized samples from additional deposits in the TFM district and trace element geochemical analyses of copper sulfides.

REE-Enriched Migmatites and Gneisses of the Son Valley Crystalline Area Sonbhadra, Uttar Pradesh: Can it be a Potential REE Deposit in India?

Sakthi S. Chinnasamy, Abhishek Verma
Indian Institute of Technology Bombay, Powai, Mumbai, MH, India

Rare earth elements (REEs) are critical for many contemporary technologies, including electronics, renewable energy, and defense application. However, the supply chain for REEs is frequently intricate and influenced by geopolitical and commercial issues. In the present study, we investigated the migmatite and gneisses of different parts of the Son Valley Crystalline (SVC) area for its REE potential. The study area is part of the Chhotanagpur Granite Gneissic Complex (CGGC), which is exposed along Son Valley in the Sonbhadra district of Uttar Pradesh, India. The area is highly deformed and metamorphosed. Petrography of migmatite and gneiss from the Kudri, Jarha, Barwatola, and Jaurahi areas was carried out. It shows a higher concentration of REE-bearing phases viz. zircon, apatite, titanite, monazite, and at times xenotime, and garnet. The REE-bearing phases, particularly zircon, monazite, and xenotime were found as inclusions within biotite, whereas apatite and titanite were found exclusively as inclusions within feldspar. To quantify rare earth oxide (REO) concentration, we did an electron probe microanalysis of REE-bearing phases such as apatite, titanite, monazite, and xenotime. The mineral chemistry of these phases reveals a higher concentration of total rare earth oxides (TREO; 0.84-1.91%, av. 1.37%) and monazite with TREO (44.54-68.51%, av. 58.98%). This study highlights the potential of REE mineralization in the SVC area, which warrants a detailed exploration. The chemical age of monazite from Kudri migmatite is 799 to 1056 Ma, av. 928 Ma (n=10), and Jarha gneiss is 837 to 1339 Ma, av. 1088 Ma (n=13). These chemical ages of monazite show its origin from Mesoproterozoic to Neoproterozoic age, which is a quite lot of variation in the formation age. Considering the total REE contents and abundance of REE-bearing minerals, SVC has huge potential to host REE reserves. However, rigorous geological and exploration studies are warranted.

Integration Of Multiscale Hyperspectral Data, Geochemistry, and Mineralogy for Assisting Mineral Exploration: The Brownfield Punta Corna Co-Ni Vein-Type Deposit (Italy)

Rita Chirico[1], Nicola Mondillo[1, 3], Carsten Laukamp[2], Jens Najorka[3], Diego Di Martire[1]
1. University of Naples Federico II, Naples, Italy, 2. CSIRO Mineral Resources, Perth, WA, Australia, 3. Natural History Museum (NHM), London, United Kingdom

This study aims to investigate how the integration of hyperspectral field-based to hyper- and multispectral satellite data with geochemistry and mineralogy can support in recognizing the signature of alteration potentially associated with Co-Ni–bearing mineralization. The test site is the Punta Corna brownfield exploration prospect, owned by AltaMin Ltd., located in the Servin Valley in the Western Alps of Italy, an area well-known for hosting several post-orogenic hydrothermal ore deposits exploited in the past. The Punta Corna deposit is characterized by “five element” hydrothermal polymetallic vein-type orebodies composed of zoned Fe^{2+}-rich carbonates and Co-Ni mineralization, hosted by metabasite in greenschist facies. Laboratory-based hyperspectral spectroscopy, mineralogical (XRPD), and micro- and bulk geochemical (FESEM-EDS and ICP-AES/MS) analyses were performed to define target alteration minerals and validate the results obtained from optical satellite image processing. The absorption features at 900, 2,200, and 2,250 nm were considered diagnostic for identifying and/or characterizing the compositional variation of the main alteration minerals detected (i.e., Fe-oxy-hydroxides, white mica, chlorite). The results show that the strongest hydrothermal alteration consists mainly of heavy sericitization in the immediate proximity of the ore occurrences, associated with Fe-oxy-hydroxides after weathering of hypogene Fe-bearing carbonates. Both were recognized with laboratory-based hyperspectral data and satellite remote sensing. By combining hyperspectral and geochemical data from hand specimens, instead, a more widespread chloritization was identified. Even though mineralogical and geochemical systematic patterns with regards to the Mg-rich chlorite and its relationship with the Co-Ni mineralization were observed, the background regional greenschist facies metamorphism and deformation must be considered for the formation of chlorite. The integration of distinct data sets allowed identifying distinct hydrothermal alteration types at different observation scales, including sericitization and secondary supergene alteration, as well as a possible chloritization, which can be useful to support exploration in the area.

Melt–Fluid and Fluid–Fluid Immiscibility in Na_2SO_4–SiO_2–H_2O System and Its Implications for the Formation of Rare Earth Deposits

Hao Cui[1], Richen Zhong[1], Yuling Xie[1], Xiaolin Wang[2], Huan Chen[1]
1. School of Civil and Resource Engineering, University of Science and Technology Beijing, Beijing, China,
2. State Key Laboratory for Mineral Deposits Research & Institute of Energy Sciences, School of Earth Sciences and Engineering, Nanjing University, Nanjing, China

Liquid–liquid immiscibility has crucial influences in geological processes, such as magma degassing and formation of ore deposits. Sulfate, as an important component, associates with many kinds of deposits. Two types of immiscibility, including (i) fluid–melt immiscibility between an aqueous solution and a sulfate melt, and (ii) fluid–fluid immiscibility between two aqueous fluids with different sulfate concentrations, have been identified for sulfate–water systems. In this study, we investigate the immiscibility behaviors of a sulfate- and quartz-saturated Na_2SO_4–SiO_2–H_2O system at elevated temperatures, to explore the phase relationships involving both types of immiscibility. The fluid–melt immiscibility appeared first when the Na_2SO_4–SiO_2–H_2O sample was heated to ~270 °C, and then fluid–fluid immiscibility emerged while the sample was further heated to ~450 °C. At this stage, coexistence of one water-saturated sulfate melt and two aqueous fluids with distinct sulfate concentrations was observed. The three immiscible phases remain stable over a wide pressure–temperature range, and the appearance temperature of the fluid–fluid immiscibility increases with increasing pressures. Considering sulfate components occur extensively in carbonatite-related deposits, the fluid–fluid immiscibility can result in significant sulfate fractionation and provides implications in understanding the formation of carbonatite-related rare earth deposits.

Magma Mixing in the Huangshandong Ni-Cu Sulfide-Bearing Mafic-Ultramafic Complex, NW China

Yu-Feng Deng[1], Sheng-Xuan Cao[1], Feng Yuan[1], Xie-Yan Song[2]
1. Hefei University of Technology, Hefei, China, 2. Institute of Geochemistry, Chinese Academy of Sciences, Guiyang, China

Magma mixing occurs in some magmatic Ni-Cu-(PGE) sulfide deposits and is interpreted to play an important role in attachment of magma sulfide saturation, but the magma-mixing process remains unclear. In this study, we use petrology, mineral compositions, and LA-ICP-MS trace element mapping to better understand the magmatic mixing process in the Huangshandong sulfide-bearing mafic-ultramafic complex, eastern Tianshan, NW China.

A sulfide-bearing layer developed at the boundary between the bottom hornblende gabbro and upper pyroxenite in the Huangshandong Ni-Cu sulfide deposit. There are hornblende gabbro inclusions enclosed in the pyroxenite, and the contents of the xenocrystal plagioclases in the hybrid layer increase closer to the hornblende gabbro. The average An contents of plagioclases from hornblende gabbro (67.2%) are similar to those from the hybrid layer (67.6%), suggesting that the plagioclases of the hybrid layer come from the hornblende gabbro. The plagioclases in hornblende gabbro show a zonal structure in backscattering image, and their An contents range from 43.9 to 73.9. LA-ICP-MS trace element mapping data shows that An content of the plagioclases in the hornblende gabbro have a trend of first decrease and then increase from core to rim, suggesting a more mafic magma may have been added into the hornblende gabbro when the plagioclases crystallized. When replenishing pyroxene-rich crystal mush intruded at the top of the plagioclase-rich crystal mush residing in a chamber, the plagioclase crystals would rise into the pyroxene-rich crystal mush due to buoyancy, while the sulfide in the pyroxene-rich crystal mush would settle down and percolate into the bottom hornblende gabbro .

Textural and Mineralogical Characterization of One Lithium Deposit from the Barroso-Alvão Aplite-Pegmatite Field: Preliminary Study

Filipa Dias[1, 2], Alexandre Lima[1], Encarnación Roda-Robles[3], Tânia Martins[4]
1. Universidade do Porto - Faculdade de Ciências, Porto, Portugal, 2. Institute of Earth Sciences (ICT), Porto, Portugal, 3. Universidad del Pais Vasco/EHU, Bilbao, Spain, 4. Manitoba Geological Survey, Winnipeg, MB, Canada

The AL56 deposit is located in the Barroso-Alvão aplite-pegmatite field in the north of Portugal and has outcrops along an area of approximately 500 × 300 meters. The western side of the area has several small bodies (4–5 meters wide) of aplite-pegmatites where lithium is majorly in the form of spodumene and/or petalite. The bodies tend to be N130° dipping 30° to 50° to the NE. They also have a preferred type of zonation from top to bottom: 1) up to 0.5cm of border zone; 2) up to 10 cm of a pegmatitic zone constituted by quartz and feldspar (crystals up to 2 cm) and occasionally white mica; 3) aplite-pegmatite alternations with thickness typically varying from 4 to 10 cm each; 4) Li-mineralization in the shape of fine spodumene and/or petalite in the center of the body; 5) last meter without Li-mineralization that can either be a layered aplite or pegmatite with megacrystals of feldspar (up to 10 cm).

In the area of the AL56 deposit, Li-mineralization occurrence and distribution seem to be related to deformation, which was still active during the crystallization of these bodies. Zones with spodumene > petalite seem to be aligned with zones of greater stress. It is possible that while the aplite-pegmatites were cooling and entering the zone of stability of spodumene, zones with greater stress led to the dissolution of petalite and precipitation of fine spodumene and quartz around other minerals in zones of lesser stress. The zones of lower stress can correspond to the zones where petalite can still be observed or to the zones where the breakdown to spodumene and quartz occurred, preserving in a great extent the outer shape of the petalite crystals.

This work was financially supported by FCT - Foundation for Science and Technology (UIDB/04683/2020, UIDP/04683/2020 and 2020.05534.BD).

Controls on the High-Grade Gold-Cobalt Mineralisation at Rajapalot, Finland

Lily Dickson, Taija Torvela, Robert C. Chapman
University of Leeds, Leeds, United Kingdom

Rajapalot is a gold-cobalt deposit owned by Mawson Gold in the Peräpohja Schist Belt (PSB), Northern Finland. It is an example of an orogenic gold deposit (OGD) with an atypical metal association, of which the PSB hosts several types: (e.g., Au-Cu, Au-U) [1].

OGDs with an atypical metal association commonly occur in Paleoproterozoic-aged rocks, for example in the Transvaal Basin, South Africa, and Tennant Creek, Northern Australia [2]. The deformation of evaporite-hosting intracratonic basins releases unusually saline fluids, which can transport base metals [2]. It is possible that Au and Co are transported together in an anomalously saline orogenic fluid [2,3], however this has not yet been demonstrated at Rajapalot. Co-transport of Au and Co would probably generate covariance within the deposit, but instead, Co-only, Au-only, and Au-Co zones have been identified, each with distinct whole-rock trace element signatures [4].

The gold-only mineralisation stage has been dated between 1.75 and 1.78 Ga [5], however no dating of the primary- or late-phase Co in the Rajapalot prospect has been conducted, and the deposit model explaining Co distributions is not currently understood.

Chemical mapping has been undertaken using core micro-XRF analysis at Bruker Nano Analytics (Berlin). This information will inform subsequent subsampling to generate thin sections for routine mineralogical and paragenetic characterisation of ore samples, which will also seek to identify minerals that could underpin geochronological studies. Trace element distribution in sulphides will be established using LA ICP-MS to investigate the evolution of the entire mineralised system to establish the genetic relationships of Au and Co throughout the deposit.

References:

[1] Eilu (2015). In: *Mineral Deposits of Finland.*

[2] Goldfarb et al. (2001). *Ore Geology Reviews,* 18

[3] Williams-Jones and Vasyukova. (2022). *Economic Geology,* 117(3)

[4] Raič et al. (2022). *Solid Earth*, 13(2)

[5] Molnár et al. (2016). *Mineralium Deposita*, 51(5)

The Shavazsay Deposit (Uzbekistan) – a Large Caldera-Hosted Volcano-Sedimentary Lithium Deposit

Alla Dolgopolova[1], Reimar Seltmann[1], Rustam Mirkamalov[3], Farid Divaev[3], Rustam Khalmatov[2]
1. Natural History Museum London, Centre for Russian and Central EurAsian Mineral Studies (CERCAMS), London, United Kingdom, 2. University of Geological Sciences / Centre for Advanced Technologies, Tashkent, Uzbekistan, 3. Institute of Mineral Resources / Centre for Advanced Technologies, Tashkent, Uzbekistan

Compared to the lithium deposits of volcano-sedimentary origin of Miocene age such as Thacker Pass (McDermitt caldera, Nevada) and Jadar (Serbia), little is known about similar formations in the Central Asian Orogenic Belt.

The Shavazsay deposit is located in the Western Tienshan (Uzbekistan), as part of the Chatkal ridge, Middle Tienshan continental arc, with subduction completed in the late Carboniferous. Extensive Permian post-subduction collisional magmatism includes volcano-plutonic complexes of leucogranite, gabbro syenite-syenite-granosyenite, leucogranite and sub-/volcanic equivalents of early rhyolite-trachyrhyolite, trachybasalt-trachyandesite-trachyte and late rhyolite-trachyrhyolite, respectively. These complexes extend over 100 km, exhibiting several calderas and volcano-tectonic depressions reflecting extensional tectonics.

The Chilten caldera hosts several lithium-bearing exploration objects: the Shavazsay rare metal deposit (Li, B, Rb, Cs, REE +/- Au, Ag, As, Mo, Sr, U) and the North Shavaz and Kamyshly occurrences. Upon Devonian and Carboniferous volcanics and intrusives forming the caldera basement, the intra-caldera volcano-sedimentary suite comprises the followig from bottom to top: Oyasay complex with basal conglomerate and gravellites; rhyolite lavas, tuffs and tuffobreccias; volcanomictic aleurolites and argillites (Li mineralization is confined to a productive zone of up to 350 m with up to 4% carbonaceous to bituminous matter as trap). It is overlain by basalt seal. Both complexes are crosscut by dikes of topaz rhyolites ("ongonite," Li-F microgranite) and trachyrholites.

Stratiform Li mineralization is hosted mainly by mica (polylithionite, tainiolite, phengite) and clays (montmorillonite). Grade varies 0.3-0.8 wt % Li_2O (average 0.57 wt % Li_2O), exceeding 2-3 wt % Li_2O in mica metasomatites. Confirmed combined indicated and measured resources ($B+C_1+C_2$) are reported as 21.1 Mt of ore and 123.4 Kt of Li_2O (State Geoinform Centre of Uzbekistan).

Whole-rock K-Ar geochronology of polylithionite metasomatites gave an age of 286+10 Ma. With its Permian age of mineralised host lithologies, Shavazsay can be considered the oldest known volcano-sedimentary lithium deposit.

The Successful Application of Stream Sediment Geochemistry as an Exploration Tool for As-Bi-Co-Cu-Fe-Ni Mineralisation in Volcano-Sedimentary Settings

Adam Eskdale[1], Sean Johnson[2], Amy Gough[1]
1. Department of Earth Sciences, Royal Holloway University of London, Egham, United Kingdom, 2. School of Earth Sciences, University College Dublin, Dublin, Ireland

Cobalt (Co) is essential in the development of batteries for carbon-neutral technologies, such as plug-in hybrid electric vehicles. Despite this criticality, primary sources of Co are limited, typically sourced as a by-product within other critical and base metal deposits across a range of metallogenetic conditions. Understanding Co ore formation styles and developing rapid-assessment exploration workflows, making use of low-cost, high-reward tools, is crucial to the identification of future new deposits. Within this remit, we utilize the UK Geochemical Baseline Survey of the Environment (G-BASE) stream sediment geochemical data set to demonstrate its successful applicability as a lithogeochemical mapping and exploration tool for As-Bi-Co-Cu-Fe-Ni mineralisation in the UK Lake District.

Using a combined unsupervised-supervised statistical approach, we reduced the dimensionality of the G-BASE data, resulting in geochemical maps that correlate strongly with regional, 50k-scale geological boundaries for each of the major lithological groups: Skiddaw Slates, Borrowdale Volcanic Group (BVG), and Ordovician-Devonian Plutonics. These maps provide average lithogeochemical values for major and trace elements. Using these, we identified combined anomalies for As-Co-Cu-Ni, correlated these with available exploration-related data, identifying 10 prospective areas for Co-bearing mineralisation, four of which align with previously recorded Co mineralisation.

Upon visiting six of the prospective sites, base metal sulphides were identified as disseminations in quartz-chlorite veins or the local volcano-sedimentary host rocks. Bulk geochemical concentrations of host rock and vein material were then measured. Comparison of these values with the modelled G-BASE values indicates relatively similar concentrations of major and trace elements, with slight overestimations of Bi, Co, Cu, Pb, S, and Zn in the model but still within acceptable boundaries. Furthermore, detailed mineral characterisation of three sites (Scar Crags, Dale Head North, Ulpha) link the mineralisation to "Five-Type" vein systems, demonstrating the applicability of this workflow to exploration in similar, more economic regions globally.

LREEs, Zr, and Th Enrichment of Autometasomatic Alkaline Dikes in Mamuju, West Sulawesi, Indonesia: Constraints from Petrology and Whole-Rock Geochemistry

Fadlin Fadlin[1], Shaban Godang[2], Arifudin Idrus[3], Hartaja M. Wicaksono[4], Wildan N. Hamzah[1], Rinaldi Ihkram[1]
1. Akita University, Japan, Japan, 2. Geochemistry tectonic modeling, REEs Researcher, Jakarta, Indonesia, 3. Universitas Gadjah Mada, Yogyakarta, Indonesia, 4. Geological Agency of Indonesia, Bandung, Indonesia, 5. Jenderal Soedirman University, Purwokerto, Indonesia

The alkaline intrusive complexes and their weathering products are one of the main sources of REEs. Therefore, a better understanding of the petrogenesis of the alkaline rocks related to the enrichment of REEs and other critical elements is an essential future scientific goal. In this study, we present optical microscopy, XRD, SEM-EDS, and whole-rock geochemistry (XRF; ICP-MS/OES) studies of alkaline dikes to elucidate the petrogenesis and characteristics of the enrichment in LREE-Zr-Th. The alkaline dike can be categorized into Fitzroyite lamproite rocks, which mainly consist of leucite, phlogopite, and pyroxene (ferrosilite; augite-aegirine), with accessory minerals such as zircon and apatite. The alkaline dike is a part of the Adang volcano complex, which is related to metasomatic undersaturated silica rock formed within plate continental extension. Whole-rock geochemistry of alkaline dikes varies from 50 to 52 wt % SiO_2. The ultra-potassic series contains very high K_2O (7–8 wt%) and K_2O/Na_2O (3–6), with moderate Mg# values (33–56). Therefore, they can be classified as alkaline series and belong to the metaluminous rocks, with A/CNK ratios between 0.67 and 1.04. Furthermore, the large-ion lithophile elements (LILEs) are enriched, followed by Ti, Sr, and P depletion, with insignificant Eu anomalies. The alkaline dikes are enriched in LREEs (901–1,558 ppm), Zr (968–3083 ppm), Th (152–408 ppm), U (30–37 ppm), Nd (164–282 ppm), and Pb (148–460 ppm). The REE-bearing minerals include zircon, barioperovskite, romanechite, rhabdophane, and Si-Fe-Al-Zr-Ti-Ca (Ce) mineral. The enrichment in LREE-Th-Zr of alkaline dikes is related to primary enrichment processes (fractional crystallization) and secondary upgrading by deuteric processes (autometasomatisme), indicated by the pseudomorphism of primary minerals. The parent magmas are likely derived from a metasomatized sub-continental lithospheric mantle (SCLM) origin.

Magmatic Glass as a Source of Lithium: A Case Study of the Cerro Galán Ignimbrite of Northwest Argentina

Conor Farrell[1], Alison Rust[1], Edward Bunker[2], Stephen Edwards[2], Naomi Thompson[2], Joseph Emmings[2]
1. University of Bristol, Bristol, United Kingdom, 2. CGG, Crawley, United Kingdom

Lithium (Li) is a metal of strategic and societal importance due to its use in high-energy-density batteries. Global demand for Li is projected to increase nearly ten-fold by 2050 in response to the electrification of vehicles and transformation of energy production and storage. To meet this demand, better understanding of the sources, transport, and sinks of Li is fundamental to ensuring effective mineral exploration and discovery of new deposits of this critical metal.

Roughly 58% of known Li reserves are hosted within evaporative saline basins, such as the salars of the Atacama Desert in South America. Several Andean salar basins contain dissolved Li concentrations exceeding 1,000 mg/L, representing highly economic prospects for Li extraction, whereas others have relatively low Li concentrations of <10 mg/L and are considered uneconomic. A key control on Li enrichment in salar brines is the presence of Li sources within the catchment, which are principally believed to be Li-bearing silicic volcanic glasses in ignimbrites that release their Li during weathering.

Here, we investigate ignimbrite textures, compositions, and physical properties in terms of controls on Li mobility and enrichment in salar systems. The Cerro Galán ignimbrite is recognised as a major Li source for the Salar de Hombre Muerto in northwest Argentina and is considered an archetype source rock in salar systems. A suite of samples from Cerro Galán were studied to better constrain the magmatic and volcanic processes leading to enrichment of Li. We identify and explore the key physio-chemical characteristics that facilitate liberation of Li into solution and frame our findings in the context of the design and implementation of regional fertility assessments in the search for undiscovered Li brine deposits.

Hydrothermal Alteration and the Formation of an Iron-Oxide-Apatite (IOA) Type Rare Earth Element (REE) Deposit at Kwyjibo, Québec, Canada

Gary Fung, A.E. Williams-Jones, David Martineau
McGill University, Montreal, QC, Canada

Kwyjibo is an unusual iron-oxide-apatite (IOA) type deposit, with a concentration of light and heavy REEs sufficient to constitute a potential economic resource of 6.92 million tons grading 2.72% TREO with an HREE proportion of 33%. Magmatic processes led to the crystallization of magnetite and fluorapatite, and hydrothermal processes led to the mobilization of REEs from the fluorapatite. The latter processes resulted in an enrichment of the REEs in andradite, allanite, and particularly in britholite. The last of these minerals (britholite) formed as a replacement of the fluorapatite through the coupled substitution of REE^{3+} for Ca^{2+} and SiO_4^{4-} for PO_4^{3-}, with the source of the REEs from the fluorapatite and the Si from hydrothermal fluid. The andradite is interpreted to have formed from the Fe^{3+} in the magnetite, Ca^{2+} and heavy REE^{3+} in the fluorapatite, and Si^{4+} from the fluid. Allanite replaced andradite with light REE^{3+} substituting for Ca^{2+}, and charge balance was achieved by the addition of hydroxide from the fluid. Hydrothermal activity terminated with the precipitation of fluorite as a result of the mobilization of Ca^{2+} and F^- into the aqueous fluid.

A reconnaissance study was conducted on fluid inclusions trapped in fluorapatite, andradite, and fluorite to gain insight into the nature and evolution of the hydrothermal system that mobilized the REEs. Two-phase (liquid + vapor), liquid-rich inclusions were the only fluid inclusions present. Primary inclusions, however, were observed in andradite, providing a record of the REE-mineralizing fluid. Secondary fluid inclusions with similar liquid/vapor volume ratios are preserved in fluorapatite and likely record the passage of this fluid. The fluid inclusions trapped in fluorite have higher liquid volume ratios consistent with the evolution of the fluid to lower temperatures. Future work will characterize the microthermometric behavior of the fluid inclusions to reconstruct the physicochemical controls of REE mineralization.

Geochemical and Petrographic Effects of Halo Formation in Host Rocks of Spodumene Pegmatites in Southeast Ireland

Elena J. Geiger[1, 2], Teimoor Nazari-Dehkordi[1], Julian Menuge[1, 2]
1. University College Dublin (UCD), Dublin, Ireland, 2. SFI Research Centre in Applied Geosciences (iCRAG), Dublin, Ireland

The unexposed but extensively drilled spodumene pegmatites of South Leinster, southeastern Ireland, intruded the East Carlow Deformation Zone extending along the eastern margin of the ~400 Ma Tullow Lowlands pluton of the Leinster Granite. Metasomatic geochemical halos developed in the mainly granitic country rocks around spodumene pegmatites. However, the mineralogy and texture of halo minerals remains largely unknown, limiting their use in exploration. Whole-rock geochemical analyses of 117 country rock samples and petrographic observations from 141 thin sections are available to address this question and draw first conclusions. Certain characteristics of the system complicate the evaluation and interpretation of the whole-rock data. The most problematic of these are i) "invisible" pegmatites not cut by the drilling, creating pronounced halo mineralization in adjacent samples; ii) halo mineralization bound to locally confined fluid pathways, causing mixed signals in the respective samples, and iii) a likely unrelated alteration event, causing chloritization of biotite and seritization of feldspar, which potentially redistributed fluid-mobile elements like Li and Cs. In a cleaned data set (74 of the 117 samples), excluding chloritized, weathered, and immediate contact samples, Li contents are between 238 and 1,070 ppm. Plotted against the distance to the nearest mineralized pegmatite, Li, Sn, Mn, Cs, Rb, and Ta are distributed over their full compositional range in samples close to pegmatites, while they are restricted to lower values farther away. Plotting concentrations against the ratio of distance to the thickness of the nearest pegmatite shifts former outliers towards the above-mentioned distribution. Preliminary conclusions drawn from the results are that i) a careful petrography is necessary to evaluate and interpret whole-rock geochemical data, and ii) to track the geochemical halo signature, an in situ approach might be more fruitful as it will avoid possibly mixed and uninterpretable signals from whole-rock geochemistry.

Mineralogy and Critical Metal Deportment of Fe-Ni-Co Laterite Ores from Sebuku Island, SE Kalimantan, Indonesia

Michele Giorno[1], Alkiviadis Kontonikas-Charos[2], Doreen Ebert[1], Joachim Krause[1], Robert Möckel[1], Dirk Sandmann[3], Mathias Burisch[4], Ernowo Ernowo[5], Arifudin Idrus[6], Yoseph Swamidharma[7], Yudhi Krisnanto[7], Max Frenzel[1]
1. Helmholtz-Zentrum Dresden-Rossendorf, Helmholtz Institute Freiberg for Resource Technology, Freiberg, Germany, 2. School of Earth and Atmospheric Sciences, Queensland University of Technology, Brisbane, QLD, Australia, 3. ERZLABOR Advanced Solutions GmbH, Freiberg, Germany, 4. Department of Geology and Geological Engineering, Colorado School of Mines,, Golden, CO, USA, 5. National Research and Innovation Agency, Banten, Indonesia, 6. Gadjah Mada University, Yogyakarta, Indonesia, 7. Sebuku Iron Lateritic Ore, Sebuku, Indonesia

Laterite deposits in Indonesia are a major source of Ni and Co. Here, we present new geological data on the Sebuku laterite deposit (Island of Sebuku, SE Kalimantan, Indonesia), with a JORC-compliant resource of ~390 Mt at 42.5 wt.% Fe, 0.9 wt.% Ni, and 0.15 wt.% Co. The laterites are mostly limonitic, oxide-dominated Fe-Ni-Co(±Sc)-rich horizons, which formed by weathering of Jurassic-Cretaceous ophiolitic units. Although the deposit is under production since 2006 (primarily for Fe), there is little mineralogical and geochemical data available, which would allow optimizing beneficiation and recovery of Ni, Co, and Sc.

Typical laterite profiles at Sebuku consist of 1) weathered bedrock composed of serpentinized dunites and harzburgites overlain by 2) a 0.2- to 7-m-thick saprolite zone, 3) a 2- to 8.5-m-thick yellow limonite zone, and 4) a 1- to 3.5-m-thick red-limonite zone.

Preliminary X-ray fluorescence (XRF), X-ray diffraction (XRD), and mineral liberation analysis (MLA) data show a decrease in Mg and Si and an increase in Fe moving upwards through the laterite profile, corresponding to a transition from silicate- to oxide-rich mineralogy. Oxides and (oxy)-hydroxides comprise goethite, maghemite, hematite, magnetite, chromium spinel, gibbsite/bayerite, and various Mn-minerals, whereas silicates consist of serpentine, chlorite, talc, quartz, pyroxene, olivine, clay minerals, and "garnierite"-like minerals. Ni is hosted by various minerals, which include goethite, Mn-oxides, serpentine, and garnierite, whereas Co is mainly hosted by Mn-oxides and garnierite. It is still unclear in which minerals Sc is primarily hosted.

Mineral chemical analyses (EPMA) are planned to further understand critical metal variability and distribution within the host minerals and throughout the deposits. The first results will be presented here. Our ultimate goal is to characterize and quantify the distribution of Ni, Co, and Sc in order to develop more efficient beneficiation processes.

New Lithium Search Spaces – A Mineral Systems Approach

Howard C. Golden, William Heins, Paul Jackson
Getech Group plc, London, United Kingdom

Lithium, a critical mineral for the future of energy storage, is forecast to experience a 40-fold demand increase by 2040 by the International Energy Agency in their Sustainable Development Scenario. The increased pressure to discover new lithium deposits will increase commensurately. In this presentation we lay out how global data sets and sophisticated analytics can be applied in concert to discover new and covered search spaces to provide currently unknown exploration targets for future lithium supply.

Lithium can be extracted from brines or salars, pegmatites, and sedimentary deposits. Exploration for lithium traditionally employs direct detection of lithium and is well established. Common techniques for exploration of pegmatite deposits include remote sensing, airborne geophysics, geological mapping, soil geochemistry, and drilling. For brines and salars, data such as water geochemistry and geothermal plant locations are commonly employed. Sedimentary lithium deposits, though under cover and less energy intensive than pegmatite deposits and higher grade than brines and intrusive deposits, are less well known and explored for less frequently.

New lithium exploration search spaces require a fresh approach to predictive exploration technology. Specialists at Getech have identified several genetic factors which, when present at the same time in the same location, provide compelling sedimentary lithium exploration targets. These genetic factors comprise criteria such as felsic volcanic and volcaniclastics as a source of lithium, regularly recharged groundwaters to leach lithium in a long-lived endorheic basin, evaporation and precipitation required to produce a brine/evaporitic accumulation, tectonic activity for upwelling of lithium-rich fluids, minimal runoff to encourage surface leaching, rivers eroding igneous source rocks into the basin, and basinward wind from an igneous source to allow lithium-rich ashfall tuff to enrich basin fluids.

We present results from analysing the criteria above or their proxies to produce play maps that generate new regional targets for lithium exploration.

Battery Metals in Diverse Geological Environments of the Lesser Caucasus, Armenia: An Overview for Future Exploration

Samvel Hovakimyan[1], Robert Moritz[2], Marianna Harutunyan[1], Karen Hambaryan[1], Kima Charatanyan[1]
1. Institute of Geological Sciences of the National Academy of Sciences of the Republic of Armenia, Yerevan, Armenia, 2. Department of Earth Sciences, University of Geneva, Geneva, Switzerland

The Lesser Caucasus belongs to the central part of the Tethyan metallogenic belt, between the Anatolian and Iranian segments. Mineral deposits and prospects of this segment belong to two metallogenic belts, recording diverse geological and geodynamic settings located between the Arabian and Eurasian plates, and ranging from Mesozoic island-arc to Cenozoic collisional and post-collisional settings. Critical and battery metal targets in Armenia include porphyry Cu-Mo, epithermal Au and base metal, Cu-rich and Ba-rich polymetallic type, volcanogenic Mn, and Kiruna-type magnetite-apatite deposits, mine waste and tailings, as well as various magmatic and metamorphic complexes. The existing mines generate huge volumes of waste rock and tailings. They contain significant potential resources of critical metals, including REEs, Cr, Ni, Co, Li, PGEs, Re, Se, Te, Bi, Ge, Ti, W, V, Zr, Th, etc., which are by-products and are typically not recovered during mining.

Cenozoic alkaline and nepheline syenite complexes are considered one of the major potential sources for REEs and represent potential targets for battery metals, including Li, Ni, and Co in Armenia. The Li potential of granitic pegmatite and tourmaline-bearing felsic intrusions of the highly mineralized Cenozoic Meghri-Ordubad pluton in southernmost Armenia remains underexplored. Clay minerals in hydrothermally altered rhyolite, as well as brines of salt mines near Yerevan, might be potential targets for Li resources. Mn prospects in Late Cretaceous volcano-sedimentary rocks along the Somkheto-Karabagh belt of the Eurasian margin in northwestern Armenia remain to be evaluated. Potential targets for REEs and battery metals are carbonatized rocks located along the Jurassic-Cretaceous ophiolite suture zones. The Neoproterozoic metamorphic basement rocks are considered potential targets for REEs and battery metals. Further investigations are required for the assessment of critical metals enrichment of various deposit types, magmatic complexes, and metamorphic basement rocks.

A Fresh Look at the World-Class "Five-Element" Vein Deposit, Cobalt, ON, Canada: Constraints from Fluid Inclusion and Stable Isotope Systematics and Mineral Paragenesis

Joshua Jackman[1], Jacob Hanley[1], Daniel Kontak[2], Richard Stern[3], David Lewis[4]
1. Department of Geology, Saint Mary's University, Halifax, NS, Canada, 2. Harquail School of Earth Sciences, Laurentian University, Sudbury, ON, Canada, 3. Department of Earth & Atmospheric Sciences, University of Alberta, Edmonton, AB, Canada, 4. Kuya Silver Corporation, Vancouver, BC, Canada

The world-class "five-element" (Ag-Co-Ni-As±Bi) vein deposits of the Cobalt-Gowganda district (Ontario, Canada) have not been rigorously studied for decades. We present results of a fully integrated study involving mineralogical and mineral chemical, fluid inclusion (FI), and in situ O-isotope data to reassess fluid and metal sources relevant to ore-forming processes in this high-tenor, five-element vein setting. Our ongoing work uses fluid inclusion assemblage (FIA) protocol and a range of analytical methods (microthermometry, cathodoluminescence imaging, Raman spectroscopy), in situ (SIMS) $\delta^{18}O$ quartz analysis, and Fe-sulphide trace element mapping (LA-ICP-MS).

Early-stage (pre-Co-Ni-Ag-As-Bi ore) and late-stage (syn- to post-ore) vein quartz preserves two dominant fluid types in both homogeneously and heterogeneously trapped FIAs. These comprise (at room T) single-phase carbonic fluid (CH_4±N_2), single-, two-, or three-phase aqueous fluids (liquid±vapour±halite), and coentrapped carbonic and aqueous fluids. Salinity of aqueous FI varies between 0.2 and 50 wt.% ($NaCl_{equiv.}$) but with minimal variation within FIAs. The FIAs are of secondary and pseudosecondary origin and preserve numerous episodes of brine and carbonic fluid interaction and mixing with a more dilute aqueous phase.

Cathodoluminescence imaging of early- and late-stage quartz shows pronounced primary growth zonation and basal-terminated fracture plumes, indicating pervasive dissolution-recrystallization related to incursion of the aqueous-carbonic fluid mixtures. In situ $\delta^{18}O$ analyses of primary growth zones (early-stage quartz 11.9-14.8%; late-stage quartz 14.5-15.3%) and secondary features hosting fluid inclusions (early-stage quartz 12.2-16.1%; late-stage quartz 14.7-15.8%) suggest <100°C of cooling throughout early vein mineralization. This, in tandem with fluid mixing, was a major control on mineralization.

Elemental mapping of accessory Fe-sulphides reveals complex enrichment/depletion patterns across primary zones and discordant overprints, indicating highly variable fluid enrichments in metals, decoupling of Ag from Ni-Co-As, and oscillations between arsenide/sulpharsenide-dominant (Ni-Co-As), Ag, and base metal (Cu-Zn-Pb) sulphide mineralization stages. This is inconsistent with traditionally documented geochemical paragenesis for five-element vein-type mineralization.

Green Transition-Critical Metal and Mineral Resources in the European Nordics

Erik Jonsson[1, 2], Tuomo Törmänen[3], Jakob K. Keiding[4], Terje Bjerkgård[5], Pasi Eilu[6]
1. Geological Survey of Sweden (SGU), Uppsala, Sweden, 2. Department of Earth Sciences, Uppsala University, Uppsala, Sweden, 3. Geological Survey of Finland (GTK), Rovaniemi, Finland, 4. Geological Survey of Denmark and Greenland (GEUS), Copenhagen, Denmark, 5. Geological Survey of Norway (NGU), Trondheim, Norway, 6. Geological Survey of Finland (GTK), Espoo, Finland

The rapidly increasing demands for specific metals and minerals to fuel expanding green transition technologies are pushing a Europe that is overly dependent on imported raw materials towards necessary changes in their sourcing (cf. the European Critical Raw Materials Act of 2023). Notably, many of these metals and minerals are already classified as critical to the European industry. In this context, the Nordic countries, excepting Denmark and Iceland, represent one of the most fertile regions for exploration and mining of different metals and minerals in Europe. The geology of the Nordics is very diverse and comparable to abundantly mineralised shields globally, as it ranges from Archaean nuclei via Proterozoic mobile belts and accreted terrains, into the Phanerozoic, also featuring different young magmatic complexes and platform cover units. This extended geological evolution and the diverse processes involved provided favourable conditions for the formation of a wide variety of mineral deposits. We have recently synthesised available data from the relevant Nordic countries (Greenland, Norway, Sweden, and Finland) to assess the potential for critical raw materials (CRMs) specifically relevant for green technologies, i.e., cobalt, graphite, hafnium, lithium, niobium, platinum-group metals (PGMs), rare earth elements (REEs), silicon, tantalum, titanium, and vanadium, describing the most important deposits and their geological settings and resource characteristics. Of these, a number are very relevant in a European and, in several cases, also a global context. Among important resources for green technologies, REE-(±Nb-Hf) deposit projects in Greenland as well as Sweden and Norway stand out; Finland has significant cobalt and PGMs, while Norway has major graphite and titanium resources. Additionally, Sweden, Finland, and Greenland have very large vanadium resources, while several lithium-(tantalum) projects are underway in Finland and Sweden. Overall, we conclude that the exploration potential for most CRMs is significant in these Nordic countries.

Project BATTRACE - Using Geo-Based Fingerprinting for Battery Raw Material Traceability

Harri Kaikkonen, Quentin Dehaine, Mari Kivinen
Geological Survey of Finland, Espoo, Finland

Traceability is one of the ways to increase transparency along the battery raw materials supply chain. Ideally, traceability of raw materials would not be required if there was global trust that all mining companies act in a socially, environmentally, and economically sustainable way. When in place, sustainability and responsibility standards should ensure transparency and trust along the mining supply chains.

However, as the global demand for battery minerals is expected to grow exponentially, the increased demand may introduce financial incentives for some companies or actors with poor performance to adopt unsustainable business practices. The technological ability to trace raw materials back to their geographical origin would work as a barrier against any actors trying to claim false origins for their raw materials.

BATTRACE is a research project where the Geological Survey of Finland has been developing a novel method for battery mineral traceability, called geo-based fingerprinting. In geo-based fingerprinting, a material's origin can be traced back to its original source based on certain material characteristics, such as its mineralogical and elemental compositions.

There are currently many ongoing initiatives to introduce traceability to battery mineral supply chains. Geo-based fingerprinting could be used as a method for traceability for several raw materials used in batteries. As the geological and elemental characteristics of a sample cannot be altered, the geo-based fingerprinting method developed in BATTRACE has unique potential in working as a standalone traceability technology or as a complementary verification method with any other traceability system.

Market Drivers and Barriers for Battery Raw Material Traceability

Harri Kaikkonen, Mari Kivinen
Geological Survey of Finland, Espoo, Finland

Transport electrification and digitalization have created an increasing demand for raw materials needed for battery production. The supply chains involved are global, complex, non-transparent, and sometimes associated with unsustainable production practices. Customer awareness and regulations are pushing battery manufacturers to make their supply chains more transparent and traceable, yet the application is still lacking from this industry.

Traceability has already become an integral element for many global industries with complex supply chains. The relevant maturity of these supply chains from a logistical standpoint has enabled them to develop differential value through traceability, which cannot be said for battery raw materials. Although the need for traceability is coming from consumers and regulatory agencies, the same is not yet evident from equipment manufacturers (OEMs). Increased costs are a clear hindrance to the adoption of traceability, but there are more pressing considerations for the OEMs. If the supply gap for the battery raw materials grows too large, OEMs may have to buy their materials where they can get them and might not want to disclose their sources.

Widespread adoption of traceability is dependent on financial incentives for the manufacturers and raw material producers. Already, the producers of sustainable and low-emission raw materials have identified the commercial benefit of tracing the greenhouse gas emissions of their products along the supply chain. This may be further highlighted with future CO_2 pricing regulations, such as the EU Carbon Border Adjustment Mechanism.

There may be additional fracturing for the commodities market in the future with different pricing for sustainable raw materials. Still, as the price for sustainability is still revolving around the raw materials' CO_2 emissions, these developments may not help with the ethical considerations surrounding cobalt mining, for example. The ethical considerations of consumers are still valued by the manufacturers through their bottom line.

Tellurium: Geometallurgy of a Critical Element at the Perseverance VMS Deposit, Quebec, Canada

Filip Kasprowicz[1], Katharina Pfaff[1], Thomas Monecke[1], David Diekrup[2], Samuel Pierre[4], Mark Hannington[3]
1. Center to Advance the Science of Exploration to Reclamation in Mining, Department of Geology and Geological Engineering, Colorado School of Mines, Golden, CO, USA, 2. Geological Survey, Department of Industry, Energy, Technology, Government of Newfoundland and Labrador, St. John's, NL, Canada, 3. Department of Earth and Environmental Sciences, University of Ottawa, Ottawa, ON, Canada, 4. Samuel Pierre Geoservices, Bern, Switzerland

Tellurium is a metalloid needed in numerous high-tech applications, most notably in cost-effective Cd-Te photovoltaic panels. The role of Te in green technologies and limited domestic sources led to Te being listed on the 2022 USGS critical mineral list. Tellurium is recovered as a by-product from anode slimes generated from processing porphyry copper and volcanogenic massive sulfide (VMS) ores. Tellurium can occur as traces in sulfides and sulfosalts, as native Te, or as discrete tellurate and telluride minerals.

VMS deposits can host significant Te and currently produce ~17% of global tellurium. However, there is a knowledge gap regarding the occurrence and mineralogical sequestration of tellurium in VMS deposits. To narrow the gap, samples from the Perseverance VMS deposit, located in the Matagami district of the Abitibi greenstone belt, Quebec, Canada, were analyzed to better understand the occurrence and sequestration of Te. Perseverance was mined between 2008 and 2013, with pre-mining reserves of 5.1 Mt of ore grading at 15.8% Zn, 1.24% Cu, 29.4 g/t Ag, and 0.4 g/t Au. Deposits of the Matagami district are notable for elevated Te concentrations; however, there is no systematic description of the occurrence and distribution of Te. Tellurides were identified and analyzed using field-emission scanning electron microscopy, scanning electron microscope-based automated mineralogy, and electron probe microanalysis.

In total, 17 discrete telluride minerals were identified. Spatial relationships studied using reflected light microscopy and electron microscopy methods show that tellurium-bearing phases form multi-mineral grains of up to four distinct phases, inclusions in highly recrystallized major ore minerals, and along grain boundaries. Microtextural evidence suggests telluride sequestration may be a direct result of regional greenschist facies metamorphism and post-depositional structural modification of the deposit. This research provides new insights that may impact ore processing methods and improve our understanding of tellurium sequestration in the VMS environment.

"Multi-Step Distillation" Anatexis as a Mechanism for Generating LCT Pegmatites

Lot Koopmans[1], Tania Martins[2], Robert Linnen[3], Richard M. Palin[1], Catriona Breasley[4], David Silva[4], Nicholas J. Gardiner[5], Lee A. Groat[4], Laurence J. Robb[1]
1. Department of Earth Sciences, University of Oxford, Oxford, United Kingdom, 2. Manitoba Geological Survey, Winnipeg, MB, Canada, 3. Department of Earth Sciences, Western University, London, ON, Canada, 4. Department of Earth, Ocean, and Atmospheric Sciences, University of British Columbia, Vancouver, BC, Canada, 5. School of Earth and Environmental Sciences, St Andrews, United Kingdom

Lithium-caesium-tantalum (LCT) pegmatites are an important subset of highly differentiated intrusions that retain geochemical signatures, implying that a component of recycling of Earth's crust is required for their formation. Deposits of LCT pegmatites account for roughly half of global lithium output and the majority of global caesium and tantalum production. Despite their fundamental and growing importance to critical metal supplies, their petrogenesis remains debated.

Two main petrogenetic models have been proposed: (1) extreme fractionation of granitic intrusions, leading to a melt concentrated in incompatible elements, or (2) low-degree partial melting of host rocks during prograde metamorphism, enriching the anatectic component in incompatible elements. Although both models have been invoked empirically in different terranes, several pegmatite fields fail to fit either model.

Here, we present a novel method to quantitatively assess the fate of trace elements during anatexis to suggest a third petrogenetic model of economic pegmatite deposit formation, which we have termed "Multi-Step Distillation." Through innovative petrological modelling, we show that water content and degree of partial melting significantly affects the lithium content of the melt component; however, a single step of melting is unable to generate economic grades. Alternatively, partial melting of leucosomes or leucogranites that were formed as part of an earlier melting event is highly effective at enriching the melting package in lithium, concentrating lithium into the melt by a factor of ~20.

Our models show that this mechanism is able to generate pegmatites of economic interest and may explain the petrogenesis of several pegmatite fields (such as the Maine pegmatite field and pegmatites of the St-Sylvestre complex) around the world.

Exploration Model for REE Deposits Associated with Nepheline Syenite Occurrences in Tropical Areas: Case Study of the Eboundja Nepheline Syenite (Cameroon)

Kenneth K. Kwanang, Louise Marie Ngo Bidjeck, Yannick Ngono Onana
University of Yaounde I, Yaounde, Cameroon

Our ambitious goals for a green future could be challenged by the global supply of critical metals. In order to meet the growing demand for critical metals, more exploration projects are needed. In this context, nepheline syenite is an interesting exploration target. These are undersaturated alkaline rocks generally enriched in HFSEs, rare earths, and other relatively rare elements (such as Li, Zn), and have the potential to form significant economic deposits of these elements. Previous work shows that the mineralising processes behind rare earth deposits associated with nepheline syenite are of three types: magmatic differentiation, post-magmatic hydrothermal events, and meteoric weathering. The Eboundja nepheline syenite is an intrusion dated at about 590 Ma. It belongs to a group of syenitic plutons found along the southern Atlantic coast of Cameroon, at the margin of the Congo craton. It was evaluated for possible rare earth mineralization in 1985. Petrographic observations show that the Eboundja syenite consists of albite, nepheline, microcline, cancrinite, sodalite, amphibole, biotite, sulphur, and monazite. Monazite appears to be the main rare earth ore. But the REE concentration in the rock is low, as is the monazite content. The agpaitic index indicates a metaluminous to peralkaline character. The favourable climatic factors observed in this region (average annual rainfall of 1,700 mm) allow sufficient weathering of this rock. The present work reveals total rare earth contents on the order of 1,000 ppm in the weathering materials (higher than the generally accepted limit of exploitability for rare earth deposits in regolith). Furthermore, all rare earths are enriched in the weathering materials relative to their content in the source rock. These results allow us to consider weathering as the main REE enrichment process in this case. In addition, we used all the exploration data to build a 3-D model.

Source to Sink Modelling of Salar-Hosted Lithium Brine Systems

Christopher Lambert, Jean-Jacques Ferguson, Esme Whitehouse, Edward Bunker, Alexandra Scarlet, Jorge Gines
CGG, Crawley, United Kingdom

The integration of remote sensing-derived data sets and robust geological maps is a powerful predictive tool for the prospectivity modelling of salar-hosted lithium brine systems. Lithium in these systems is most likely sourced from ignimbrites through the process of devitrification and then transported in soluble form by surface and ground waters. Rivers, catchments, and basins, as well as a wide array of other slope and aspect derivatives, may be extracted from continental-scale, high-resolution Digital Elevation Models (DEMs) to provide accurate proxies for transportation and sink features.

Understanding the geographical extent, morphological geometry, and connectivity of these features linked to the suitability of lithium-bearing source rocks is a critical step towards understanding the mineralisation potential and dilution or enrichment factors of surface brine deposits. Similarly, working with these features in a palaeo-environmental context may indicate the presence of subsurface reservoirs in the form of palaeo-salar deposits.

Here we apply these geospatial relationships to a regional study of brine potential for lithium and associated minerals in a High Andean setting. Systematic calculations across a variety of input parameters and processed geospatial data sets have defined a ranked order of salar, and palaeo-salar, prospectivity of lithium and associated brine-related minerals by watershed. Holistic interpretation of the results may be used to inform critical exploration and investment decisions.

Critical Metal Occurrences in Historic Mining Waste in Sweden

Alexander Lewerentz[1], Johan Camitz[1], Patrick Casey[1], Dick Claeson[1], Roger Hamberg[1], Erik Jonsson[1, 2], Daniel Larsson[1], Stefan Persson[1], Gunnar Rauséus[1], Helge Reginiussen[1], Johan Söderhielm[1], Björn Wiberg[1]
1. Geological Survey of Sweden, Uppsala, Sweden, 2. Uppsala University, Uppsala, Sweden

In 2023, the Geological Survey of Sweden completed a project focused on sampling and characterisation of occurrences of historic mining waste in Sweden. Sampling methodology focused on both finding compositional anomalies and enabling the calculation of approximated average compositions. Analyses included whole-rock chemistry and mineralogy.

Waste rock (mine dump material) was sampled from 34 historic mines, in total comprising some 31 Mt of waste rock material. Average concentrations for battery and critical metals are highly variable and include 2–2,930 ppm Ni, 3–371 ppm Co, 77–8,200 ppm REE+Y, and 6–6,960 ppm Cu. Assuming that these average concentrations are representative for the entire waste rock volumes, the sampled dumps contain some 1,200 t Ni, 850 t Co, 10,000 t REE+Y, and 24,100 t Cu.

Tailings were sampled from 23 historic tailings dams, comprising some 121 Mt of overall material. Average concentrations include 1–114 ppm Li, 2–124 ppm Ni, 1–321 ppm Co, 54–2,680 ppm REE+Y, and 5–3,000 ppm Cu. Assuming that the average concentrations are representative for the entire tailings, these contain some 2,200 t Li, 2,600 t Ni, 4,900 t Co, 49,000 t REE+Y, and 60,000 t Cu.

Pyrometallurgical slags from both Fe and Cu smelters were sampled from six historic occurrences. Average concentrations include 1–3,275 ppm Co, 98–3,087 ppm REE+Y, and 6–21,800 ppm Cu. The total tonnages of remaining slag material are unknown.

These results should be seen as order-of-magnitude estimations and by no means as qualified mineral resource estimates. Neither do they weigh in whether the metals are extractable or not from technical and/or economic standpoints. Nevertheless, the results clearly highlight historic mining wastes as a potential source for battery and critical metals, especially so if extraction is combined with primary mining of similar resources.

Petrogenesis and Ni-Cu Sulfide Mineralization of Early Permian Mafic-Ultramafic Complexes in the Beishan Area, Xinjiang

Hao Lin, Dongyang Zhang
China University of Geosciences (Wuhan), Wuhan, China

Many small ultramafic-mafic intrusions, accompanied with series of economically magmatic Ni-Cu sulfide deposits, e.g., Qixin, Podong, Poyi Ni-Cu deposits, were discovered in the Beishan area, southern margin of the Central Asian Orogenic Belt (CABO). Sparsely disseminated sulfide mineralization occurs in the Poyi and Qixin lherzolite, whereas densely disseminated sulfide occurs in the Podong orthopyroxene-bearing ultramafic-mafic intrusions. Zircon U-Pb geochronological dating suggests these mafic-ultramafic intrusions were formed during the early Permian (294-273 Ma). These mafic-ultramafic samples show low La/Yb ratios, LILE enrichment, HFSE depletion (e.g., Nb and Ta), and high $\delta^{18}O$ (5.36-9.26‰), implying that they were derived from the shallow sub-lithospheric mantle metasomatized by subducted slab-derived fluid. The generation of these early Permian mafic-ultramafic intrusions associated with Ni-Cu sulfide mineralization in the Beishan area may be attributed to post-orogenic extension and a ~280 Ma mantle plume event. The Podong mineralized units show relatively lower $\varepsilon_{Nd}(t)$ (-1.00 - +0.51) compared to those of the Poyi and Qixin mineralized samples ($\varepsilon_{Nd}(t)$= +4.29 - +6.97), suggesting the addition of more contaminated material in the former. Furthermore, in the fO_2 calculation of parent magma between the olivine and Cr-spinel compositions, the Podong mineralized units show relative lower ΔQFM (-0.5 - +1.5) and olivine Ni contents (200 - 3427 ppm), resulting in the effective sulfide immiscible process. In contrast, the relatively higher ΔQFM (+0.3 - +2.5 and +2 - +3.5) and olivine Ni contents (2,771-6,747 ppm and 1,097-4,244 ppm) for Poyi and Qixin mineralized samples suggest the parent magma was characterized by relatively oxidized conditions, and sulfide segregation may be limited or occur in the deep magma chamber. The above observation confirms the orthopyroxene-bearing portions (e.g., Podong olivine gabbronorite and gabbronorite) may be a potential exploration target for Ni-Cu deposits in the Beishan area.

A Mineral Systems Approach to the Classification of Lithium Deposits

Robert Linnen[1], Wilson Bonner[2], Kurt House[2]
1. University of Western Ontario, London, ON, Canada, 2. KoBold Metals, Berkeley, CA, USA

A mineral systems approach to classifying lithium deposits needs to consider the sources of Li, tectonic setting, magmatic processes, transport and emplacement of magma, hydrothermal processes, weathering, and preservation. Mantle processes exert little direct control on the formation of Li deposits. The source of lithium is the crust, where it is concentrated via partial melting, fractional crystallization, then once magmas are exposed at surface, through weathering and enrichment of Li in clay minerals. The tectonic environment is in collisional settings and initial Li enrichment is in late syn-tectonic to early post-tectonic melts. Li is a moderately incompatible element, and melting of pelitic sedimentary rocks and intermediate to felsic plutonic rocks produces Li-bearing peraluminous melts. During the crystallization of these melts, Li is initially concentrated in micas, and with enough crystallization, a Li ore mineral crystallizes. Structures at terrane boundaries are important as conduits for the emplacement of magma at higher crustal levels. After the emplacement of peraluminous magma, extension and subsidence can form closed basins, in which brine deposits can develop.

Li deposits can be broadly classified into those associated with intrusive rocks and those associated with extrusive rocks, with some overlap between the two classes. The intrusive class consists of pegmatite deposits, which can be either zoned or poorly zoned, and granites, which can contain disseminated mineralization or Li-bearing veins. The source of the Li in at least some of the brine deposits is from remobilization from granites, and the host can either be in sedimentary basins or in granites. This potentially includes Jadar-type deposits. The extrusive class consists of Li in geyserites and volcanic glasses. This Li can be remobilized into evaporite-associated or geothermal brines. Similarly, the re-distribution of Li from Li-rich volcanic rocks can generate clay deposits. Lastly, bauxite-hosted Li deposits form in weathered environments.

SEG 2023 Conference: Resourcing the Green Transition

Using Scanning Electron Microscope-Based Automated Mineralogy to Better Understand Magnetic and Conductive Anomalies in Ni-Cu-Co-PGE Exploration Targets in W-Greenland

Stefanie Lode[1, 2], Björn H. Heincke[2], Nynke Keulen[2], Bo Møller Stensgaard[3]
1. Norwegian University of Science and Technology (NTNU), Trondheim, Norway, 2. Geological Survey of Denmark and Greenland (GEUS), Copenhagen, Denmark, 3. Bluejay Mining Plc, London, United Kingdom

The West Greenland Basalt Province hosts economically significant exploration targets for Ni-Cu-Co-PGE mineralization. Bluejay Mining PLC is actively exploring in the province and funded the current study. The presence of magnetic and conductive minerals (native iron, pyrrhotite, graphite, and magnetite) in the volcanic rocks complicates the understanding of the origins of magnetic and conductive anomalies observed in data from geophysical surveys.

Detailed optical and electron microscopical and petrophysical studies were carried out on drill core material from Qullissat on the north of Disko Island intersecting an intrusion and over- and underlying sedimentary sequences. The body consists of two basaltic blocks with distinct mineralogical and petrophysical signatures that could explain observed geophysical anomalies. On the same samples, automated mineralogy (ZEISS Mineralogic) and petrophysical properties (magnetic susceptibility, remanence, resistivity, and induced polarization) were acquired.

Minerals of interest for conductive and magnetic properties include native iron, magnetite, pyrrhotite, and graphite, as well as Ni-Co-Cu-mineral phases (NiFeCo-phosphides, pentlandite, chalcopyrite). The Ni-Co-Cu-minerals are predominantly observed in the lower block.

The results of this study demonstrate that a) stronger magnetic anomalies observed in the lower block are a result of magnetite and not, as initially thought, caused by the presence of native iron; b) native iron and slightly higher occurrences of pyrrhotite in the upper block explain its more conductive behavior compared to the lower block; c) graphite disseminated throughout both blocks, previously thought a major cause for conductivity, does not appear to have any significant influence on the electric properties; and d) Ni-Co-Cu-bearing minerals occur in higher quantities in the lower block and correlate also with higher amounts of Ti-bearing minerals, which therefore, may represent a set of minerals with potential as indicator minerals.

Automated mineralogy, therefore, can complement petrophysical measurements and geophysical surveys to better define and characterize potential exploration targets for Ni-Cu-Co-PGE mineralization.

Geology of the Paleoproterozoic Järkvissle Li-Sn-Ta Pegmatite System, Central Sweden

Edward P. Lynch[1], Joel B. Andersson[2], Martiya Sadeghi[1], Irma Bečelytė[3], Peter Hedin[1], Tobias E. Bauer[2]
1. Geological Survey of Sweden, Uppsala, Sweden, 2. Luleå University of Technology, Luleå, Sweden, 3. Lund University, Lund, Sweden

The Järkvissle Li-Sn-Ta prospect in central Sweden is a key example of an LCT-type granitic pegmatite system in the country. Historical exploration at Järkvissle identified pegmatite dikes containing spodumene, petalite, cassiterite, and columbite group minerals, and produced a pre-regulatory resource of c. 600 kt grading 0.45% Li, 0.07% Sn, and 80 ppm Ta. In this presentation, we give an overview of the geology of the Järkvissle prospect based on new field mapping, lithogeochemistry and geophysical data, and 3D geomodelling. Results form part of the larger EIS project* which aims to develop new digital exploration tools for several critical metal-bearing mineral systems.

The oldest rocks at Järkvissle are metamorphosed siliciclastic sedimentary and intercalated volcanic rocks that form part of a Paleoproterozoic (c. 1.9 Ga) continental margin termed the Bothnian Basin. The sedimentary-volcanic package records at least two episodes of ductile deformation as evidenced by bedding-cleavage relationships, large- to mesoscale folds, crenulation cleavages, shear bands, and boudin structures. Two main intrusive rock types also occur; a foliated biotite-muscovite granodiorite-granite, and a foliated garnet-bearing granite with unzoned "simple" pegmatite parts.

At Järkvissle, a set of subvertical, mainly NNE- to NNW-trending, spodumene-bearing pegmatite dikes occur over a c. 2- × 7-km area. The pegmatites range from c. 0.2 to 2.0 m in thickness, are c. 10 to 800 m in length, have planar to somewhat wavy forms, and are internally zoned. They occur as either individual dikes or grouped "sheeted" bodies that preferentially intrude the eastern limb and hinge zone of a large-scale, steeply SSE-plunging asymmetrical fold. Pegmatite orientations follow those of pre-existing planar fabrics in metasupracrustal host rocks, indicating earlier-formed structures were pathways and/or traps for pegmatite-forming volatile-rich melts. Several pegmatites also display folded forms, suggesting localized ductile deformation accompanied pegmatite emplacement.

*Exploration Information System (Horizon Europe grant 1010557357)

Emplacement Mechanisms of Rare Element Pegmatites, Separation Lake Greenstone Belt, Northwest Superior Province

James A. Macdonald, Alfredo Camacho, Garnet J. Ching
University of Manitoba, Winnipeg, MB, Canada

The rare element pegmatites of the Superior Province host a variety of critical metals (Li, Cs, Rb, Be, Ta, Nb, Ga, Tl, and Ge) essential for developing sustainable technologies. The prevailing theory regarding the Li-bearing pegmatites of the Separation Lake greenstone belt is that the dikes intruded in one event at ~2640 Ma and were pervasively overprinted by a ductile shear fabric. However, recent investigations suggest that these rocks intruded over a ~50 Ma period (ca. 2649–2601 Ma) and that the fabric is related to emplacement processes. In the current study, a variety of analytical methods such as electron backscattered diffraction (EBSD), in situ U-Pb geochronology, LA-ICP-MS, and micro-XRF scans are used to correlate the mechanisms and timing of pegmatite emplacement within the tectonic framework of the western Superior Province. Our new data show that S-type granitic magmatism in the greenstone belt had commenced by ca. 2690 Ma and that quartz and feldspar in granite bodies that share an orientation with the rare element pegmatites were deformed and recrystallized during regional deformational events. Quartz and plagioclase crystallographic axes within the domains of the Li-bearing dikes that have attributes most suggestive of deformation show very little rotation from their primary orientations (<5° misorientation) and lack a crystallographic preferred orientation, supporting the interpretation that these features were generated during dike emplacement. Micro-XRF scans of polished slabs of the rare element pegmatites show Ca enrichment at the country rock interface that decreases towards the centre of the dike and may reflect diffusion of elements from the country rock into the pegmatite. In our model, emplacement of the pegmatites operated under a mechanism similar to fluidization, in which metasomatism coincident with melt emplacement allowed for localized deformation to be imparted on the country rocks upon intrusion.

Origin of Alkaline-Carbonatite Rocks and Related Mineral Deposits at the NW Margin of the West African Craton: Inferences from Field and Petro-Geochemistry

Cheikh Elwali Malainine[1], Muhammad Ouabid[1], Otmane Raji[1], Jean-Louis Bodinier[1, 2]
1. Geology and Sustainable Mining Institute, Mohammed VI Polytechnic University, 43150, Benguerir, Morocco, 2. Geosciences Montpellier, Université de Montpellier & CNRS, 34095, Montpellier, France

The pericratonic terrains of the Reguibat shield in the NW part of the West African Craton host several occurrences of carbonatites, whose mineral potential deserves consideration due to the growing global demand for critical mineral resources. In the present work, we report a comprehensive field inspection and a petrographic and geochemical study of three annular structures in the western Oulad Dlim massif (southern Morocco), namely Twihinate, Lamlaga, and Lahjayra. These structures are partially to completely covered by silica breccia and ferruginous crusts. The carbonatites are Cretaceous sövites, evolving from primary magmatic paragenesis to hydrothermally reworked facies hosting REE- and Nb-bearing phases, mainly REE-fluorcarbonate and pyrochlore. These occurrences were previously thought to be isolated carbonatite structures, lacking coeval silicate alkaline rocks. However, our study revealed the existence of silicate alkaline rocks with ultrapotassic affinity, mainly composed of K-feldspars and characterized by high K/Na ratios. The occurrence of such rocks in close association with the calciocarbonatites suggests the derivation of the igneous complexes from a parental carbonated silicate melt, probably formed by low-degree partial melting of a phlogopite-bearing mantle source.

The iron oxide cap rocks host the highest concentrations of REE-Nb and igneous phosphates and locally preserve relics of primary carbonatite paragenesis. They exhibit normalized REE and multi-trace element patterns consistent with those of associated carbonatites, indicating that their geochemical signatures are primarily inherited from their carbonatitic protoliths. Similarly, the siliceous breccia facies show a trace-element signature comparable to that of the carbonatites. However, they have locally preserved the same sedimentary stratification as observed in the sandstone and quartzite country rocks. Their trace-element signature is attributed to intense hydrothermal activity during the carbonatite's emplacement. In light of these results, the studied structures are considered suitable sites for better understanding of the genesis of carbonatite-related mineral deposits through hydrothermal/supergene enrichment processes.

Lithium in Manitoba: Production and Exploration Potential

Tania Martins
Manitoba Geological Survey, Winnipeg, MB, Canada

Lithium continues to be in the spotlight due to the continue demand for batteries in tandem with the advent of decarbonization of our economy and the fossil fuel transition. Lithium has other important physical and chemical properties that are suited to a wide range of applications including pharmaceuticals, glass and ceramics, and aerospace technologies. In addition, lithium is considered a critical element by the governments of Canada, the United States, and the European Union.

The Canadian province of Manitoba has a long and rich history of lithium exploration and mining associated with granitic pegmatites. Granitic pegmatites contain the largest known resources of lithium in Manitoba. The most prolific region is the Winnipeg River-Cat Lake pegmatite field, which hosts the world-class Tanco lithium cesium-tantalum deposit, along with numerous other pegmatite occurrences. The Tanco mine was until very recently the sole producer of spodumene concentrate in Canada. The Winnipeg River-Cat Lake pegmatite field is Archean in age and is part of the Superior province. Elsewhere in the Archean Superior province of Manitoba, lithium-bearing pegmatites occur at Red Sucker Lake, Gods Lake, Cross Lake, Red Cross Lake, and McLaughlin Lake.

A second prolific lithium-hosting region in Manitoba is the Wekusko Lake pegmatite field, located just east of the town of Snow Lake. The Snow Lake area in Manitoba is well known for the world-class Flin-Flon-Snow Lake mining camp and its production of gold, copper, zinc, and nickel. The Wekusko Lake pegmatite field is of Paleoproterozoic age (ca. 1800 Ma). The main lithium-bearing pegmatite mineral found in the Wekusko Lake field is spodumene. However, more recently, highly fractionated pegmatite dikes with lepidolite were reported, making this area even more attractive for pegmatite exploration projects.

The REDOX Setting of the Stratabound V Mineralization at the Van Property, Northwest Territories

Chelsey C. Merrick
University of Toronto, Toronto, ON, Canada

This project will investigate two aspects of stratbound critical metal (V) mineralization on the Van property, Northwest Territories. First, it will use sequential leach extractions to understand how V is held within the metalliferous carbonaceous shales. This will be useful for two reasons. First, knowing what phases (mineral or organic matter) the V is held within will help understand the likely depositional mechanism, fluid that transported the metals, and trap mechanism, which will be instrumental in developing a refined geologic model for this type of mineralization. Second, one of the main impediments to mining these types of deposits is the poor recovery using traditional metallurgic techniques. The sequential leaches we will conduct will give initial data to aid engineers in determining more efficient ways to process the ore. In addition to the sequential leach analyses, we will conduct pyrite framboid size analysis on the pyrite framboids from the metalliferous shales and the immediate hanging wall and footwall. This will help determine whether the host rocks for the mineralization were deposited in an oxic/anoxic or euxinic water column. This is important for economic geology because it will aid in developing a more accurate geologic model for these potential new sources for critical metals—important resources as we transition to a green economy. This project received one of the 2022 SEG Canada Foundation Undergraduate Scholarships.

Timing and Linkage of the Mesoproterozoic Ree-Y-Zr-Nb Mineralized Strange Lake and Flowers River Deposits to the Neoarchean Nekuashu-Aucoin Magmatic Event, Northern Canada

Nadia Mohammadi[1], Anne-Aurélie Sappin[2], David Corrigan[3], Neil Rogers[1], Nicole Rayner[1]
1. Geological Survey of Canada, Ottawa, ON, Canada, 2. Geological Survey of Canada, Quebec City, QC, Canada, 3. Geological Survey of Canada (Retired), Ottawa, QC, Canada

The subcircular Mesoproterozoic Strange Lake Peralkaline Complex (SLPC: 1240±2 Ma) is a rare-metal-enriched (Nb-Zr-REE-Yb-Be) multi-phase intrusion that occurs near the Nekuashu Suite and transects the Pelland Suite within the Mistinibi-Raude Block (MRB) of the southwestern Churchill Province (Quebec and Labrador, Canada). The Flowers River Igneous Suite (FRIS: 1281±3 Ma) comprises several discrete peralkaline ring intrusions and coeval volcanic rocks, also enriched in REEs and rare elements, and intrudes the Aucoin Complex in the Nain Province in Labrador. Although the SLPC and FRIS have attracted considerable economic interest and have been the target of numerous detailed studies since the mid-1980s, lingering questions on their overall tectonomagmatic setting and source of metal endowment remain.

Our new crystallization age of 2551±8 Ma (SHRIMP U-Pb zircon) for the Nekuashu Suite broadly coincides with the emplacement age of the Pelland Suite (re-evaluated age of 2519±35 Ma), the Aucoin Complex (2573±8 Ma, 2580±8 Ma, and 2567±4 Ma), and the newly identified Mikuasheunipi Complex (2569±4 Ma). These findings suggest a regional igneous event that is larger than previously recognised, herein termed the Nekuashu-Aucoin event (ca. 2.58–2.52 Ga). All Nekuashu-Aucoin intrusions are characterized by alkaline to calc-alkaline, quartz-poor to quartz-absent, but mostly high-temperature rocks that range in composition from ultramafic to mafic, intermediate, and to a lesser extent, felsic.

Given the close spatial relationship between the SLPC and the MRB, and the FRIS and the Aucoin Complex, we propose that these Late Neoarchean intrusions generated from a hydrous metasomatized mantle source that may have acted as a precursor REE enrichment event. Subsequent remelting/reworking processes associated with additional plume-related magmatism and/or far-field tectonic activities/orogeny likely caused the later Mesoproterozoic, rare-metal-enriched peralkaline magmatism of the SLPC and FRIS.

Battery Minerals in the Democratic Republic of Congo: Overcoming Challenges to Discover a New World-Class Lithium Deposit in the Kasese-Lulingu Pegmatitic Belts (KLP)

Rub'son H. N'nahano, Huan Li, Mohammed A. Ibrahim
Central South University, School of Geosciences and Info-Physics, Department of Geological Engineering and Geological Resources, Changsha, Hunan, China

Battery metals have experienced a surge since last year because of the need for the globe to transition to a green and low-carbon envrionment by creating electric vehicles, despite COVID-19's effects on the mining industry. The greatest outstanding reserves of critical metals, such as cobalt and lithium, which are crucial for the energy transition, are found in the Democratic Republic of the Congo, making it a strategically important country for the energy transition. The nation produced 115,371.91 tons of cobalt in 2022 as opposed to 93,010.53 tons the year before in 2021. There is a proven resource of around 44.6 million metric tons of lithium in the Manono deposit. The richest mining firms presently travel to the DRC to acquire mining permits of cited minerals for the aforementioned reasons. The Katanga area in the country's southeast is home to the battery mineral deposits that have been researched and recorded along with huge exploitation activities. Due to the increased demand for licenses, issues and disagreements have been taking part of the business in the country. Exploring and finding new deposits for transition metals for green energy thus becomes urgently necessary in DRC. Here, we describe a prospective larger pegmatitic zone that is located in South-Kivu province. The KLP belt was the world's greatest tantalum and third-largest niobium producer in the 1950s. According to early exploration results, the spodumene-bearing pegmatite in the KLP belt is significantly richer than the spodumene pegmatite in Manono. The granite-pegmatite rocks extend to over 500 km in the belt with the accessory minerals of pegmatite being 60 to 70% spodumene and amblygonite. The chance of finding a new world-class lithium deposit in this zone is still plausible, but the absence of infrastructures along with technological and environmental difficulties keep these enormous lithium reserves hidden and unexplored.

Application of Soil Geochemistry for Pegmatite Exploration: Spodumene Pegmatites in Ireland and Austria, and High-Purity Quartz-Bearing NYF Pegmatites in Norway

Teimoor Nazari-Dehkordi[1], Julian Menuge[1, 2], Lawrence Carter[3], Ben Williamson[3], Axel Müller[4, 5]
1. University College Dublin, Dublin, Ireland, 2. iCRAG, the SFI Research Centre in Applied Geosciences, Dublin, Ireland, 3. Camborne School of Mines, University of Exeter, Cornwall, United Kingdom, 4. Natural History Museum, University of Oslo, Oslo, Norway, 5. Natural History Museum, London, United Kingdom

We investigate the applicability of soil geochemistry in exploration for buried rare metal pegmatites at prospect scale (<25 km^2) as one of a range of exploration tool sets being developed for such deposits within the EU H2020 GREENPEG project. We compare the geochemistry of Ah- and C-horizon soil samples underlain by spodumene pegmatites from the 400 Ma Moylisha prospect, Ireland, and the 270 Ma Wolfsberg prospect, Austria, with those of niobium-yttrium-fluorine (NYF) pegmatites from the granite-hosted 1800 Ma Tysfjord prospect, Norway. At each locality, samples were collected along several parallel transects perpendicular to the long axis of a known buried pegmatite, or pegmatites. All soils were <1 m thick.

We show that the spatial distributions of common indicator elements for spodumene and NYF pegmatites (Li, Cs, Ta, Nb, Be) can be used to accurately locate the known buried pegmatites. Moreover, from principal component analysis (PCA), there are strong correlations among these indicator elements, especially in C-horizon samples. This horizon is also preferentially enriched in rare earth elements (REEs). C-horizon samples above the NYF pegmatites, which are typically rich in REEs, appear to be more enriched in HREEs, whereas Ah-horizon samples tend to host most of the LREEs. These patterns suggest that the comparatively immobile indicator elements and HREEs remain in the C-horizon closest to the buried pegmatites, whereas LREEs and other relatively mobile elements (e.g., Mg, Fe, Ni, Cr, Co, Pb) concentrate predominantly in the Ah horizon. In the Tysfjord NYF pegmatites, there are high U and Th concentrations in both the C and Ah samples that are likely to be detectable in the latter using airborne gamma ray spectrometry. From these results, soil geochemical data and PCA can be effectively used to explore for buried rare metal pegmatites.

Implications of the Magma-Mush Concept for Lithium Enrichment Processes in LCT Pegmatites

Louis R. G. Penfound-Marks[1], Ben J. Williamson[1], Julian F. Menuge[2], Lawrence C. Carter[1]
1. Camborne School of Mines, University of Exeter, Cornwall, United Kingdom, 2. iCRAG, the SFI Research Centre in Applied Geosciences, and School of Earth Sciences, University College Dublin, Belfield, Dublin 4, Ireland

LCT pegmatites are the world's premier hard-rock source of Li, which is used in rechargeable batteries and an increasing range of other high-tech and green technologies. Despite this, the mechanisms that concentrate Li in pegmatites are poorly understood, hampering the development of exploration models. Processes advocated by other authors include fractional crystallization, constitutional zone refining, and melt-melt-fluid immiscibility under supercritical conditions. Although these could cause short-lived saturation with respect to minerals such as spodumene and petalite, their crystallization would stop as Li melt concentrations fall. These processes cannot therefore explain sustained crystallization of Li phases through almost the entire volume of certain pegmatites, especially of spodumene in most spodumene pegmatites.

As part of the EU H2020 GREENPEG project, we are studying the implications of the magma-mush concept for the enrichment of Li in pegmatite-forming systems, initially based on unzoned to weakly zoned spodumene pegmatites in the Leinster belt of south-east Ireland. From theoretical considerations, we propose that Li in the magma source variably diffuses from solid phases into melts and fluids which then migrate through crystal-mush pathways and evolve to more Li-rich compositions due to percolative reactive flow. Fluids are sourced from a relatively larger volume than melts and move more rapidly through the crystal mush due to their comparatively low viscosity and surface tension. As Li melt concentrations fall due to the crystallization of Li-bearing phases, so the flow of Li-rich fluids through the mush refertilises the melts due to diffusion of Li along chemical potential gradients. Macrotextural evidence in support of the presence of fluid-crystal mush and percolative fluid flow includes aplite and quartz (originally fluid) mingling/flow channels, chaotic and jumbled pegmatite structures, and multiple generations of magmatic and hydrothermal spodumene. The results of ongoing micro-textural and chemical studies will be discussed.

The McDermitt Deposit (Oregon, US): Preliminary Results from a Li-Clay Volcano-Sedimentary System

Francesco Putzolu[1], Robin Armstrong[1], Rebecca Ball[2], Paul Meyer[2], Richard Herrington[1]
1. Natural History Museum, London, United Kingdom, 2. Jindalee Resources Limited, Reno, NV, USA

The McDermitt Caldera (western US) is host to one of the world's largest Li-clay resources, including the Thacker Pass (Nevada) and McDermitt (Oregon) deposits. The McDermitt project (Jindalee) holds indicated and inferred resources of 3 billion tonnes at 1,340 ppm Li, making it the largest lithium deposit in the US. In this contribution, we report on preliminary findings dealing with the geology, lithogeochemistry, and ore mineralogy of the system. The Li-clay resource sits in extensional sub-basins formed during caldera collapse caused by eruption of the McDermitt tuff. Post-collapse resurgence of the caldera created the geomorphology for the development of hydrologically closed lakes. Volcanic glass-rich air falls are common across the orebody, while welded tuffs occur at the bottom of a former lake system. High Li concentrations are found in the clay-rich beds, in association with geochemical anomalies indicating that smectite formed under low Eh conditions. Post-formation processes that altered the initial nature of the system include (1) uppermost oxidation-bleaching producing a deposit-scale V-Zn-Fe anomaly; and (2) late volcanism producing silicification and Mo-Sb-S anomalies in the basal mineralised section. The breakdown of volcanic glass is the main trigger of the formation of ±Li-rich smectites. The main Li-smectite formation events were disrupted by the authigenesis of zeolites and K-feldspar. This paragenesis is consistent with the pH increase occurring at the latest stages of closed hydrologic system diagenesis as observed in other alkaline lakes worldwide. The preliminary study of the clays indicates that whilst these are described as "hectoritic ores," the smectite assemblage includes other mineral species (e.g., saponite, ferrosaponite, etc.) not normally described as potential Li repositories. Based on these findings, future research efforts will be focused on understanding the relationship between hectorite and other clays to constrain the aging of smectites and thus the pathway of Li fixation in volcano-sedimentary systems.

Relationship Between Granitic Intrusions and Hydrothermal Veins in the Argemela Tin and Lithium Mine

Ricardo Ribeiro[1,2], Rui Moura[1,3], Alexandre Lima[1,2], Charles Gumiaux[4,5]
1. Faculty of Sciences - University of Porto, Porto, Portugal, 2. Institute of Earth Sciences, University of Porto Pole, Porto, Portugal, 3. INESC TEC, Robotics and Autonomous Systems, Instituto Superior de Engenharia do Porto, Porto, Portugal, 4. Institut des Sciences de la Terre d'Orléans (ISTO), Université d'Orléans, Orléans, France, 5. Bureau de Recherches Géologiques et Minières (BRGM), Orléans, France

The Argemela Tin and Lithium mine is located in Central Portugal, within the Central Iberian Zone (CIZ). The quartz-rich hydrothermal veins are mainly subvertical and are hosted in metasedimentary country rocks. The main lithium-bearing mineral is montebrasite, which is associated with cassiterite (Sn-bearing mineral). The genesis of this lithium mineralization is still debated. Authors say that this mineralization is related to non-outcropped granite, as happens in other lithium-rich quartz vein mineralization in the CIZ. In the late 70's, three drill holes were made by the Serviços de Formento Mineiro in order to understand the mineralizations. At the end of drill hole 1, granitic rocks were found, and at the end of drill hole 3, spotted schist was found that can be related to a non-outcropped granitic intrusion.

To study this genetic origin, we performed a gravimetric campaign in the Argemela area. A complete Bouguer anomaly (CBA) was obtained, to which were applied four filters highlighting different anomalies: tilt, analytic signal, and vertical and horizontal derivates. An inversion was also performed in order to delimit the anomalies at depth.

On the CBA map, it is possible to observe that the mineralization occurrences are related to a zone of transition of higher to lower values, from SE to NW. Similar to CBA, the tilt and the vertical derivative are also associated with a transition between higher (SE) to lower values (NW). The analytic signal and the horizontal derivative are correlated with higher values. The inversion shows low-density values at depth on the W side of the Argemela mine.

The gravimetric results in conjunction with the drillholes, made in the past, indicate that at the West of the Argemela Mine, there is a non-outcropping granitic body that could be the source of the hydrothermal fluids that originated this mineralization.

Structural Controls on the Origin of Regolith-Hosted Rare Earth Elements Deposits at the Coastal Cordillera, Central-Southern Andes (35°–37°S)

Tomás Roquer, Paulina Vergara, Nicolás Bustos, Carlos Marquardt
Departamento de Ingeniería de Minería, Pontificia Universidad Católica de Chile, Santiago, Chile

In the coastal range in the southern-central Chilean Andes (35°–37°S), regolith-hosted rare earth element (RH-REE) deposits have been discovered in recent years. These deposits develop on pediplanes where in situ weathering of REE-bearing host rocks and minerals has occurred. However, the structural controls on the genesis of these deposits remain unknown and may be dual. First, regional brittle fault zones localize fracture networks in damage zones adjacent to fault cores, increasing host-rock potential for weathering. Second, neotectonic faults may shape the landscape by generating morphologies favorable for weathering. Our preliminary results indicate the existence of two groups of faults and shear zones: the Pichilemu-Arauco faults (PiAF) and the Andean transverse faults (ATF). The PiAF are N- to NE-striking transpressional synmagmatic shear zones, which govern the emplacement of Carboniferous batholiths. The ATF are NW- to WNW-striking faults with sinistral and sinistral-normal kinematics and control the occurrence of NW-SE–oriented Triassic continental sedimentary basins and hypabyssal stocks. Both the PiAF and ATF show a late episode of transtensional brittle reactivation resembling neotectonics of specific ATF segments at the northern termination of the study area. Comparison of the structural data to published geomorphological models reveals several areas where regional faults limit regional pediplanes (e.g., 14 × 8 to 5 × 2 km^2), suggesting a neotectonic control in the formation of the landscape. Overall, results suggest that faults and shear zones increase the prospective potential for RH-REE deposits, especially when they limit pediplanes and affect host rocks with adequate REE-bearing mineralogy.

Occurrence and Geological Features of LCT Pegmatite in the Kalba-Narym Belt, East Kazakhstan

Gleb Smirnov[1], Azam S. Dehnavi[1], Reimar Seltmann[2]
1. Nazarbayev University, Astana, Kazakhstan, 2. The Natural History Museum, Earth Sciences Department, London, United Kingdom

The Great Altai of Eastern Kazakhstan, part of the Central Asian Orogenic Belt, consists of four major metallogenic zones (Dyachkov et al., 2018) in which the Kalba-Narym Belt is famous for the occurrence of rare metals. The Great Altai accretion process is a result of the collision between the Kazakhstan and Siberian continents in the Late Carboniferous, which induced the large-scale formation of the Kalba-Narym batholith. The batholithic body is made up of several complexes, including Kunush plagiogranites and tonalites of 325 to 285 Ma, Kalguta granodiorites and leucogranites of 286 to 272 Ma, Kalba granites, aplites, and pegmatites of 300 to 275 Ma, and Monastery leucogranites and pegmatites of 285 to 271 Ma (Kotler et al., 2015). The latest fractionation stage of the Kalba-Narym batholith resulted in the formation of pegmatites, some of which are mineralized and can be classified as LCT pegmatites. These pegmatites are discovered in several regions of East Kazakhstan, including the Asubulak deposit cluster. The mineralized pegmatites are classified based on mineralogical and geochemical characteristics and are divided into three categories: a) microcline-albite with pollucite and petalite (Ta, Cs, Be, Sn), b) microcline-albite with spodumene (Ta, Nb, Cs, Li, Be, Sn), and c) spodumene-albite (Li, Ta, Nb, Sn) (Dyachkov et al., 2021).
At the deposit scale, the structurally controlled mineralization occurs at the intersection of the EW-oriented major Asubulak fault with NS minor structures. At the greater regional scale, the orientation of pegmatitic bodies is aligned with the NW-SE direction of the Irtysh Shear Zone. This research documents the geological overview of the Kalba-Narym Belt in relation to mineralogical and structural variations within known and prospective regions. The preliminary research on various tourmaline types and their geochemical composition is undertaken for obtaining data on fluids responsible for the formation of mineralized pegmatites.

Rare Earth Element Recovery from Apatite

Julia Sordyl[1, 2], Jarosław Majka[1, 2], Karin Högdahl[2], Maciej Manecki[1, 2]
1. AGH University of Science and Technology, Krakow, Poland, 2. Uppsala University, Uppsala, Sweden

Rare earth elements (REEs) are considered critical elements integral to advanced energy production, transmission, and storage. Growing demand and supply risks have significantly escalated the problem of their availability in recent years. Exploring methods to recover REEs from both primary and secondary sources becomes crucial. Swedish deposits of apatite iron oxide ores are the major source of iron in the European Union and host REE-rich apatite ($Ca_5[PO_4]_3[OH,F,Cl]$) that has been disposed of in large quantities for over a century. Regarded as waste for a long time, most of the apatite was dumped together with the gangue in large tailing dams rather than being recovered and reused.

We propose a new method for REE recovery that has been tested at a laboratory scale on different types of material containing REE-rich apatite as a source of REEs. The essence of our approach is the chemical conversion of REE-rich apatite into a crystalline precipitate of REE-rich lead phosphates through a sequence of dissolution-precipitation processes resulting in rapid and efficient removal (97-100%) of REEs from pregnant solutions. Leaching, precipitation, and transformation into a marketable product are relatively simple and environmentally sustainable. The method does not require energy-intensive initial mechanical pre-concentration of heavy minerals or reactions at elevated temperatures. A crystalline precipitate can be easily separated from the suspension. In the final steps, Pb is separated from REEs and recycled in the process, while the final product is REE oxalate. Due to its compatibility with existing technologies and high REE recovery even from low-content sources, the method holds promise for circular and resource-efficient use of existing mines and waste streams.

Geological Analysis of Graphite Deposits in Capim Grosso (BA), Brazil

Thales F. Tandel
UNESP, Rio Claro, SP, Brazil

Natural graphite is an ore that has several industrial uses, from refractories to the manufacture of electric batteries and graphene. The purpose of this work is the geological and economic analysis of graphite rocks found in the surroundings of Capim Grosso (BA), Brazil. The municipality of Capim Grosso is located in the center-east portion of the State of Bahia; inserted in the geological context of the São Francisco craton. The São Francisco craton is a geological unit of great mineral importance in the Brazilian territory, being the target of prospecting for various mineral inputs. The shear zone known as the Salvador-Curaçá Orogen has lithologies belonging to the Tanque Novo-Ipirá Complex, known for the occurrence of mineralized graphite. In the area, both gneisses and mafic and ultramafic rocks rich in graphite of the flake type were recognized. The study was carried out through field work, opening trenches and carrying out rotary diamond drill holes and chemical analysis of samples. The mineralization occurs within an alteration zone subject to a sinistral shear zone. Chemical analyses reveal graphite carbon contents of an average of 14%, with anomalies reaching 41%. The results show that the analyzed lithology presents high levels of graphite, with ample capacity for mineral exploration. The study area also presents the presence of sulphides inserted in the alteration zone, indicating the possible presence of secondary minerals. The analyses also reveal Ni, Cu, Cr, and V anomalies. In addition, preliminary studies suggest that the rocks present at the site could be used as soil remineralizers. The prospective methods of trenches and drilling proved to be efficient, revealing the economic potential of the studied area.

Assessing Vanadium Speciation in Hyper-Enriched Black Shales to Identify the Geochemical Controls Involved: Insights from X-Ray Absorption Near-Edge Spectroscopy

Ozgur Can Tekin[1], Daniel D. Gregory[1], Anthony Chappaz[2], Danielle McGill[1], Daryll B. Concepcion[1], Merilie A. Reynolds[3], Chelsey Merrick[1], Daisy Zhang[1], Stefanie M. Brueckner[4]

1. University of Toronto, Toronto, ON, Canada, 2. STARLAB - Earth and Atmospheric Sciences - Central Michigan University, Mount Pleasant, MI, USA, 3. Northwest Territories Geological Survey, Yellowknife, NW, Canada, 4. University of Manitoba, Winnipeg, MB, Canada

The demand for Vanadium (V) is increasing rapidly every year due to its critical role in supporting the green energy revolution. One of the major drivers of this demand is vanadium redox flow batteries. Additionally, V is used as a steel alloy to improve the lifespan of steel products while reducing energy consumption. The ultimate goal is to provide a deeper understanding of V enrichment, which will aid in the exploration of additional V resources and contribute to the green energy revolution.

To meet the demand for V, it is necessary to explore new V deposits. One potential source of V is hyper-enriched black shales (HEBS) found, and widely spread, in the Selwyn Basin, Yukon. The source of this enrichment is still debated, with two scenarios being direct precipitation from seawater (top-down) or seafloor hydrothermal activity (bottom-up). While the oxidation state of V can range from –III to +V, only +III, +IV, and +V occur in natural settings.

Synchrotron-based spectroscopy techniques can precisely determine the speciation of V, such as its oxidation state, coordination, molecular environments, and hosting phases. A selection of samples was analyzed using μ-XANES at the Advanced Photon Source (BL-13-IDE). Our preliminary results indicate the presence of two V species: (a) V(+IV) bound to sulfur atoms and (b) V(+III) bound to oxygen atoms. The V(+IV)–S species is predominant and suggests the top-down scenario might control V burial. This study is still ongoing work, and additional geochemical analyses like isotope chemistry, whole-rock, and scanning electron microscope analyses are being conducted.

The Deeper Platreef at Sandsloot: New Insights Into Controls on Ni-Cu-PGE Mineralisation in the Bushveld Complex

Erin Thompson[1], David Holwell[1], Iain McDonald[2], Marc Reichow[1], Thomas Blenkinsop[2], Lara Du Preez[3], Kofi Acheampong[3], Andy Lloyd[3]

1. Centre for Sustainable Resource Extraction, University of Leicester, Leicester, United Kingdom, 2. School of Earth and Environmental Science, Cardiff University, Cardiff, United Kingdom, 3. Anglo American South Africa, Johannesburg, South Africa

Platinum group elements (PGEs) are vital components in internal combustion engines and hydrogen fuel cells and play an increasingly important role in battery technology. The northern limb of the Bushveld Complex, South Africa, is widely regarded as one of the world's largest resources of PGEs. Brown-field exploration at sites within the northern limb, such as downdip of Sandsloot, offers exciting new targets for PGEs, whilst further illustrating the complexities present within the Bushveld Complex.

Within the northern limb, the Platreef forms the major PGE-Ni-Cu-Co deposit and is part of a complex package of Critical Zone (CZ) rocks differing from the wider Bushveld in that high-grade PGE mineralization is spread over much greater thicknesses compared to the Merensky Reef. Additionally, it lies north of the Thabazimbi-Murchison lineament and rests directly on the metasedimentary Transvaal Supergroup and accompanying footwall structures. Therefore, footwall interaction is likely to have had a significant impact on the emplacement and magmatic pathways of the northern limb magmas, as well as their contamination histories and resultant stratigraphy.

In this study, the petrology and bulk geochemistry, including Nd-Sr-Hf-Pb isotopic compositions, of samples from Sandsloot are examined, with the aim of establishing the primary magmatic stratigraphy of the newly discovered deep Platreef, downdip of the Sandsloot open pit. The Platreef is thought to have formed from discrete magmatic units, and several, including a barren zone, PGE-reef, and base metal zone, are identified at Sandsloot. Varying CaO/Al_2O_3 and Eu/Eu* ratios, as well as Pb and Sr anomalies and the presence of enigmatic hybrid textures, are interpreted to represent local dolomite contamination and hydrothermal alteration of the magmatic units. Thickness variations are proposed to be related to the changing topography of the underlying footwall architecture, and magmatic erosion during intrusion of Main Zone magmas post-Platreef emplacement.

SEG 2023 Conference: Resourcing the Green Transition

Multi-Scale Insights Into the High-Grade Mt Weld REE Deposit from 2D and 3D Active Seismic Survey Data

Greg Turner[1], Tom Bell[1], Mohamed Megebry[1], Ganesh Bhat[2], Brad Knell[2], Nigel Miller[3]
1. HiSeis Pty Ltd, Subiaco, WA, Australia, 2. Lynas Rare Earths Ltd, Laverton, WA, Australia, 3. HiSeis Pty Ltd, Subiaco, United Kingdom

The Mt. Weld mine, which is located 35 km southeast of Laverton in central Western Australia, is one of the world's highest-grade rare earth elements (REE) mines. The mine currently extracts ore from the weathered material sitting at the top of a carbonatite intrusion.

In 2022 seismic investigations were carried out to:

- Explore for potential additional REE resources and groundwater resources in proximity to the existing mine; and
- Better inform the structural setting of the current Mt. Weld REE deposit and the underlying carbonatite intrusion.

Both 2D and 3D seismic reflection and refraction surveys were performed together with reprocessing of a deep-crustal seismic line acquired by Geoscience Australia.

The refraction data showed a strong correlation between deeper near-surface zones of low seismic velocity and both high-grade REE concentrations and high porosity. These deeper low-velocity zones correlate well with where existing drilling indicates deeper weathering and highlight additional deep zones as yet untested by drilling.

The reflection data highlighted the existence of a series of steep faults in the fresh carbonatite, which appears to play a key role in the location of deeper weathering by potentially focussing groundwater flow and migration.

The reprocessed deep-crustal line highlighted a zone of low seismic reflectivity extending to at least 9 km below the surface. We interpret this to represent the multi-phase carbonatite intrusive pathway, breaking through Archean sedimentary country rock. The edge of the low-reflectivity zone has a complex shape, suggesting the carbonatite was preferentially squeezed out into different parts of the stratigraphy and along structures. There are also indications of lenses of rafted Archean stratigraphy within the carbonatite body.

Overall, the combination of the seismic data sets over multiple different scales has provided vital input into further understanding potential groundwater and REE sources.

Controlling Factors in Regolith-Hosted Rare Earth Element Deposit Formation at the Coastal Cordillera, Central-Southern Andes of Chile

Paulina Vergara, Tomás Roquer, Nicolás Bustos, Carlos Marquardt
Departamento de Ingeniería de Minería, Pontificia Universidad Católica de Chile, Santiago, Chile

In the central-southern Andes, in the Coastal Cordillera, where tropical to mid-temperate climatic conditions prevail, regolith-hosted rare earth element (RH-REE) deposits have been discovered in the last decade. Understanding conditions of pediplanes in which these deposits develop is still matter of research. Remote sensing, geophysical surveys, field data, petrographic studies, and geochemical and U/Pb dating analysis were carried out to characterize areas with the potential to contain RH-REE deposits. Remote-sensor geomorphological analysis was used to determine zones with low slope-and-roughness index in order to find areas with regoliths associated with pediplanes and zones with high weathering and low erosion. Climatic conditions over the last thousands of years have allowed the development of regoliths in these areas. Complementary field mapping and petro-geochemical analysis allows us to understand the favorable lithologies for RH-REE deposit formation. The Carboniferous granitic plutonic rocks (330–310 Ma, U/Pb zircon ages), with a volcanic-arc and syn-collisional geochemical signature, seem to be the most favorable lithological unit. Triassic (230–215 Ma, U/Pb zircon ages) subvolcanic rocks also seem to be favorable with an intraplate signature. We find a relation between the %SiO_2 and REE content: more basic compositions in Carboniferous intrusive rocks exhibit a higher total REE content, while more acidic compositions in Triassic subvolcanic rocks show a higher total REE content. Additional preliminary geophysical and structural analysis suggests that regional faults may control pediplane formation. From this, we conclude that the main controlling factors in RH-REE deposit formation are as follows: 1) favorable lithologies, like granitic and subvolcanic rocks, most likely to be REE-bearing host rocks; 2) climatic tropical to mid-temperate conditions that led to in situ weathering of the rocks; and 3) the existence of geomorphologic and structural features, over which regoliths preferentially develop.

Cobalt Distribution and Enrichment in Skarn Iron Deposits: A Case Study of the Zhuchong Skarn Fe-Cu-Co Deposit, Eastern China

Fangyue Wang, Xian Liang
School of Resouces and Environmental Engineering, Hefei University of Technology, Hefei, China

Skarn Fe deposits are important sources of Co reserves, accounting for approximately 15% of such deposits. Here we present a comprehensive investigation into the spatial distribution and occurrence of cobalt in a typical skarn deposit, the Zhuchong Fe-Cu-Co deposit, located in Eastern China. This deposit contains 50 Mt of Fe @50.5%, 4.3 Mt of Cu @1.15%, and ~9,700 t of Co metal with a grade of 0.019%. The trend of Co contents in Zhuchong ore and gangue minerals is as follows: pyrite> pyrrhotite> sphalerite> magnetite > diopside > chalcopyrite> garnet. Mass balance calculations revealed that pyrite is the most significant Co carrier in Zhuchong, accounting for ~5,000 tonnage of Co (average Co ~7,000 ppm). A negative correlation between Co and Fe in pyrite indicated that the dominant substitution for Co in pyrite is $Fe^{2+} \leftrightarrow Co^{2+}$. Although magnetite has low Co content, it is the most abundant ore mineral in Zhuchong, accounting for approximately half of the total Co resources. Depth profile analysis found that the Co-enriched pyrite is mainly developed in the contact zone between the diorite pluton and the skarn iron orebody, with pyrite far from the pluton having lower Co contents ranging from several ppm to thousands of ppm. Co in pyrite is temperature dependent. The Co/Ni ratios of pyrite in Zhuchong range from 0.001 to 10,000, indicating that Co/Ni ratio is not a reliable indicator for distinguishing hydrothermal and sedimentary pyrite. Overall, this study demonstrates that estimating the amount of cobalt resources in each mineral phase of skarn iron deposits combined with economic availability is crucial for accurately evaluating the economic value of the critical metal Co in such deposits. Recovery and utilization of Co as a by-product from skarn iron deposits has a broad future, especially given current demand for cobalt.

Characteristics and Origin of the Li-Rich Hydrothermally Altered Granite: A New Type of Clay Li Deposit?

Zhiqiang Wang, Feng Yuan
Hefei University of Technology, Hefei, China

Lithium (Li) is an energy-critical element due to the development of sustainable energy technologies. The discovery of new lithium resources is required to meet the growing global demand. Clay Li deposits are important Li resources (7% of the world's Li resources), commonly hosted in volcanic rocks and volcanic sediments, such as the McDermitt deposit in Nevada, USA, and the Sonora deposit in Mexico. Recently, a new type of clay Li deposit was discovered in eastern China, hosted in the hydrothermally altered Jingde granite. Li contents in the Li-rich altered granite range from 405 to 2,050 ppm with an average of 1,245 ppm. Li is structurally bound in cookeite in the Jingde Li deposit rather than hectorite and illitic clay, the main host minerals of the McDermitt and Sonora deposits. The altered mineral assemblage is cookeite + illite + kaolinite + chamosite. Calcite veins and fluorite veins are widely distributed in Li-rich altered granites, suggesting that CO_2 and F are the critical transport carriers for Li. Granodiorite and monzonite, the main rock type of the Jingde pluton, are lowly evolved and have relatively low Li contents (~50 ppm), making them unlikely to be the source of Li-rich hydrothermal fluid. It is worth noting that highly evolved aplite and pegmatite are widely distributed in the Jingde pluton. They (146.2–147.8 Ma) are contemporary with granodiorite and monzonite (147.2–149.8 Ma). As a moderately incompatible element, Li is theoretically enriched in the highly evolved granite. On the contrary, the highly evolved aplite and pegmatite are unusually depleted in Li (1–20 ppm) compared with the typical highly evolved granites (100-1000 ppm). This indicates that abundant Li has been released from the highly evolved melt during fluid exsolution. The exsolved Li-rich hydrothermal fluid from the deep, highly evolved melt is responsible for forming the Li-rich altered granite.

Ewoyaa Lithium Project - Rediscovery to Resource

Isaac I. Williams[1], Lennard Kolff van Oosterwijk[1], Abdul-Razak S. Ballah[1], Kathryn Goodenough[2], Richard Shaw[2], Marieke Van Lichtervelde[3]
1. Atlantic Lithium Ltd., Sydney, NSW, Australia, 2. British Geological Survey, Edinburgh, United Kingdom, 3. Institut de Recherche pour le Developpement, Toulouse, France

The Ewoyaa Project is situated in the Central Region of Ghana, some 100 km west of Accra, and is being explored and developed by Atlantic Lithium Ltd. (AIM: ALL, ASX: A11, OTCQX: ALLIF).

The Ewoyaa deposit comprises several coarse-grained, spodumene-rich pegmatite intrusive bodies, which currently contain a JORC mineral resource estimate of 35.3 Mt at 1.25% Li_2O (28 Mt in measured and indicated categories). The pre-feasibility study delivered a 12.5-year open-pit mining operation with an 8:1 strip ratio, producing clean spodumene concentrate via dense medium separation. The project has an enviable location proximal to existing infrastructure, being 1 km from the coastal highway, electricity crossing the tenement, and only 110 km by metalled road to the deep-water port of Takoradi.

The spodumene pegmatites were rediscovered during research targeting the Birimian S-type granites, following up references to spodumene in the Saltpond area by the Ghana Geological Survey during the 1960s. Early exploration proved challenging due to the dearth of outcrop and deep tropical weathering, combined with spodumene's instability and lithium's high mobility in this environment. A combination of outcrop, airborne radiometrics, and soil geochemical anomalies provided targets for pitting, trenching, and auger drilling to map subsurface pegmatite bodies. Drilling of the pegmatites below the deep weathering confirmed spodumene mineralisation.

The typical mineralogy of the pegmatites comprises quartz, feldspar, spodumene, muscovite, and phosphate minerals. Spodumene is the only significant Li-bearing mineral. The pegmatites are notable in being predominantly free of albitisation, and although potassic alteration does exist, it is structurally controlled and localised. Initial geochronology suggests an Eburnean age of intrusion (2040.7 ± 17.1 Ma).

The origins of the Ewoyaa pegmatites remain enigmatic and require further research, with both anatexis during orogenesis and a fractionated magma source being evaluated.

Timing of Cu-Mo Mineralization in the Deep Extension of the Giant Chuquicamata Deposit, Northern Chile

Claudio A. Araya, Antonio Arribas
The University of Texas at El Paso, El Paso, TX, USA

The giant Chuquicamata Cu-Mo deposit has a documented complex origin and evolution, like most world-class porphyry copper-type deposits. This deposit evolved within the Chuquicamata intrusive complex through the superimposition of several magmatic and hydrothermal events over a period of ~5 million years that started at ~35 Ma with the intrusion. First, an early event (~34-30.4 Ma) of potassic alteration was developed, which transitioned laterally to propylitic alteration. This was followed by a second magmatic-hydrothermal event of intense potassic alteration, Qz-Mo veining (~32.9-31.7 Ma), and white mica alteration (~31.8-30.5 Ma). This second pulse appears to be responsible for the bulk of metallic mineralization in the deposit. Finally, the upper part of the hypogene alteration/mineralization was altered by a younger Miocene supergene alteration and mineralization enrichment event at ~19 to 15.2 Ma.

This study will consider all previously available geochronological data for Chuquicamata, including U-Pb in zircon from intrusive host rocks, $^{40}Ar/^{39}Ar$ in biotite, K-feldspar, and white mica for potassic and sericitic alteration stages, and Re-Os in molybdenite from Qz-Mo veins. The objective of this research is to better understand the temporal evolution of hydrothermal alteration within the Chuquicamata complex, with emphasis on the precise timing and duration of the main stage of Cu-Mo mineralization in the Deep Extension Chuquicamata deposit, which has not been dated yet. This will be achieved through radiometric dating analyses ($^{40}Ar/^{39}Ar$ and/or K-Ar in biotite and white mica, and Re-Os in molybdenite) of pure mineral separates selected from a collection representative of the various alteration and mineralization stages within the Deep Extension deposit.

Geology, Mineralization, Geochemistry, Geochronology, and Ore Potential of Cu Deposits, Central African Copperbelt: A Review

Kevin Igor Azeuda Ndonfack[1, 2], Jilin Zhang, Zhiming Yang
1. JCHX Maning Management CO., LTD, Beijjng, China, 2. Chinese Academy of Geological Science, Beijjng, China

The Central African Copperbelt is the most mineralized Neoproterozoic basin on Earth, preserving a spectacular scale of Cu mineralization with 8.7 billion tons of ore reserve grading at 2.6% Cu. Deposits occur within stratigraphic layers of Mines-Kitwe and Mwale subgroups, as stratiform and vein-type deposits with ore zoning consisting of chalcocite + bornite + chalcopyrite outward to chalcopyrite and then to pyrite. Chalcocite is common within the Congolese orebodies, whereas chalcopyrite is the main ore mineral of the Zambian deposits. The $\delta^{34}S$ sulfide values (-28 to +29‰) of Cu sulfides display lighter sulfur isotopes in the Congo part (-14 to +2‰) and become heavier within Zambian deposits (-16 to +23‰). The heavy $\delta^{34}S$ values (-2 to 25‰) are consistent with the $\delta^{34}S$ data of Cu-veins (-4 to +23‰) and host granitoid (+3.5 to +10.9‰). The $\delta^{13}C$ and $\delta^{18}O$ values (-23.1 to +8.7‰, -1.8 to -28.0‰) of carbonate from ores are consistently lighter than sedimentary host carbonate values, implying the oxidation of organic matter during carbonate deposition. $^{206}Pb/^{204}Pb$, $^{207}Pb/^{204}Pb$, and $^{208}Pb/^{204}Pb$, with μ between 9.44 and 10.66, suggest multisource of ore. The dating data fall within a ca. 670-490 Ma window, yielding two major spectra at ca. 670-645 Ma and 540-490 Ma for early stratiform and late vein stages of Cu mineralization.

Virtually, the Lufilian orogeny (ca. 600-465 Ma) may have strongly influenced the Cu deposits in the Central African Copperbelt, particularly in the Zambian deposits. The Copperbelt displays a prograde metamorphic trend varying from the prehnite-pumpellyite facies in Congo to greenschist-amphibolite facies in Zambia. The metamorphic process during orogeny may have led to metal redistribution characterizing the Zambian deposits. Common features include large ore reserves but relatively low grades (averaging 1.9% Cu versus 2.7% Cu in Congo), chalcopyrite as the main Cu sulfide, and heavy $\delta^{34}S$ values for sulfides.

Tectonic Controls on Porphyry Copper Formation in the Northern Great Basin (Nevada-Utah-Idaho), Western USA

Hero Bain[1], Thomas Lamont[1], Dan Bevan[2], Frances Cooper[3, 1], Tim Elliott[1], Adam Gorecki[4]
1. University of Bristol, Bristol, United Kingdom, 2. University of Western Australia, Perth, WA, Australia, 3. University College London, London, United Kingdom, 4. Narwhal Exploration LC, Tucson, AZ, USA

The prevailing Andean model for the formation of porphyry copper deposits (PCDs) along convergent continental margins requires a prolonged period of compression and crustal thickening, which stalls juvenile mantle wedge-derived magmas in the lower crust. While stalled, fractionation gradually causes their volatile content and thus ore-forming potential to increase. Upon relaxation of compressional stress, these deep fertile magmas ascend to form PCDs in the shallow crust. Tectonic models are becoming increasingly important in defining new exploration areas for buried, undiscovered deposits in the western USA. However, it is unclear to what extent the Andean model applies in this region. Understanding the controlling factors for mineralisation requires comparison of copper-rich districts such as the Laramide belt in Arizona (~80–50 Ma) to tectonically similar areas of less extensive mineralisation, such as the Northern Great Basin (NGB), where there are fewer PCDs and they tend not to be of Laramide age (generally >100 Ma, or <40 Ma).

This study tests two predictions of the Andean model for PCDs in the NGB: (1) they are emplaced directly after compression ends; and (2) they are most prevalent in areas with the greatest magnitude and longest duration of compression. To do this, we focus on three metamorphic core complexes, which represent exhumed remnants of the mid-lower crust (the Raft River-Albion-Grouse Creek Range, the Ruby Mountains, and the northern Snake Range). Preliminary in situ Rb-Sr dating of syntectonic micas from contractional structures within the Raft River-Albion-Grouse Creek Range yields ages of ~45 to ~100 Ma, providing new constraints on the timing of compression in this area. Further dating of structures within all three core complexes, combined with thermobarometric constraints to estimate crustal thicknesses, will provide a spatial-temporal database for comparison with PCD mineralisation both within the NGB and in copper-rich Arizona.

Alteration Products of Seafloor Massive Sulfides: A New Source of Copper?

Christian S. Bishop[1], Bramley J. Murton[1], Stephen Roberts[2], Anna Lichtschlag[1]
1. National Oceanography Centre, Southampton, United Kingdom, 2. University of Southampton, Southampton, United Kingdom

Deep-sea mineral deposits could make a major contribution to future global raw material supply. Requirements to reduce carbon emissions may result in deep-sea mining, as these deposits have increased ore grades, little to no overburden, and cause significantly less deforestation compared to terrigenous deposits.

Extinct seafloor massive sulfide (SMS) systems comprise the largest and only viable means of potentially mining these deposits. However, the fate of metals during seafloor weathering remains poorly constrained. It is uncertain if metals contained in sulphide minerals, such as Zn, Cu, and Au, are lost from the primary minerals during seafloor weathering forming barren systems, or are potentially stored in alteration products such as Fe-oxyhydroxides (FeOOH), forming supergene gossans.

Forty-two samples of FeOOHs and massive sulfide were collected at Semyenov hydrothermal field, an extinct, ultramafic-hosted SMS system at the Mid-Atlantic Ridge. The average Cu content of FeOOHs collected (excluding metal depleted chimneys) is 2.5 wt.% (n = 29), which lies at cut-off grade for Fe-Mn nodules of 2.5 wt.% Ni and Cu, illustrating the economic potential of FeOOHs. However, massive sulphides can contain up to 13.4 ppm Au and 10.6 wt.% Cu with an average of 4.8 wt.% Cu (n = 5), thus resulting in significant metal loss during oxidation.

Despite seafloor weathering resulting in metal loss from the sulphides, some of these metals remain within the Fe-oxyhydroxides, retaining a significant metal tenor. Furthermore, FeOOHs have the benefit of containing little to no sulphur, removing the need for energy-intensive smelting during processing and are thus more sustainable. Seafloor exploration at SMS deposits largely ignores FeOOHs; however, this research demonstrates the potential of Fe-oxyhydroxides for possible future exploitation. Further work will include acid leaching studies to determine the amount of recoverable Cu in FeOOHs.

Application of Alteration and Geochemical Vectors for Blind Porphyry Cu-Au-Ag-Mo Exploration: An Example from Camp Creek, Thorn District, British Columbia, Canada

Farhad Bouzari[1], Shaun Barker[1], Robert G. Lee[2], Cassady Harraden[1], Gary Thompson[3], Christina Anstey[3], Brock Riedell[4]
1. MDRU-Mineral Deposit Research Unit, Vancouver, BC, Canada, 2. BHP, Vancouver, BC, Canada, 3. Brixton Metals, Vancouver, BC, Canada, 4. Consulting Geologist, West Vancouver, BC, Canada

Improving exploration success for strategic metals such as copper will be required to meet increasing global demand. Blind porphyry deposits present a unique exploration challenge, and advancements in exploration methods will become increasingly necessary. Integrating field and core logging with lithogeochemical (4-acid-ICPMS), shortwave-infrared spectroscopy (SWIR), and micro-XRF data can identify host rock, alteration, and trace element zonation patterns, providing insights to improve blind porphyry exploration.

The Late Cretaceous Camp Creek Cu-Au-Ag-Mo property is located near the western margin of the Stikine terrane in northwestern British Columbia, 90 km ENE from Juneau, Alaska. Zones of high-sulfidation Au-Cu and sediment-hosted Au occur, such as those at Oban, Glenfiddich, and Outlaw. Recent drilling by Brixton Metals beneath zones of advanced-argillic alteration at Camp Creek identified blind porphyry mineralization below an oxidized pyrite-rich zone. New Re-Os dates of quartz-molybdenite veins in Camp Creek and in Oban breccia clasts have similar ages (86-87 Ma), suggesting a link between porphyry and epithermal-type mineralization.

The mineralization at Camp Creek occurs with biotite and K-feldspar alterations that are overprinted by chlorite and sericite. Trace element composition, particularly Zr, Ti, and Sc, led to the identification of five intrusive units (Z, Y, W, X, and V). The X porphyry, characterized by lower Zr concentrations and phenocrysts of plagioclase, biotite, quartz, and hornblende, is the main host to the mineralization. SWIR data shows vertical alteration from pyrophyllite near-surface, which transitions to illite and then muscovite with depth. Sericite crystallinity increases with depth, and its composition becomes phengitic. Sericite is white near the surface but transitions to pale-green and green with depth. Pyrite:chalcopyrite ratio decreases with depletion in Na-Ca, whereas the MDRU porphyry index increases with depth. These geochemical and mineralogical variations provide effective tools to characterize host rocks and alteration, and vector toward blind porphyry mineralization in BC and elsewhere.

The Potential Porphyry Cu-Mo Deposit on the Patacual Claim, El Limon-La Cuesta Locations, Municipality of Tecpan De Galeana, Guerrero State, Southern Mexico

Lizbeth G. Carrillo-Marrodan[1], Jesus Castro-Mora[1]
1. Autonomus University of Hidalgo State, Pachuca de Soto, HID, Mexico, 2. Geoconsulting Ingenieros S.C., Ciudad de Mexico, CMX, Mexico

The Patacual claim represents a prominent greenfield project and will be very important to be considered in the short-midterm according to the increasing global demand for copper ore in the near future. The Patacual claim hosts a potential porphyry copper-molybdenum (Cu-Mo) and epithermal base metals quartz veining mineral deposit, located inside the Santa Lucia major claim, Guerrero State. The local geological setting consists of country rock dominated by a series of granitic stock outcrops exposed all along the studied area. Main stocks are intruded by mafic dikes showing a very clear crosscutting relationship, which defines the existence of postmagmatic late intrusion activity, apparently related to epithermal quartz veining mineralization.

The field examination was done with the purpose of getting an exploration program related to four types of mineralisation: a) copper-molybdenum disseminated mineralisation hosted in the granitic stocks (low grade); b) irregular supergene copper compounds, mostly malachite and azurite, from leaching processes on the granitic stocks; c) epithermal quartz veining base metals mineralisation enriched with molybdenite related to the emplacement of the mafic dikes; and d) base metals in the mafic dikes occurring in microfractures on the hanging wall with the quartz veins.

Magmatic-Hydrothermal Crystal Mush within the Porphyry-Forming Environment – Spatial and Temporal Constraints

Lawrence C. Carter[1, 2], Ben J. Williamson[2], Simon Tapster[3], Gavyn K. Rollinson[2]
1. LC Geoscience, Cornwall, United Kingdom, 2. Camborne School of Mines, Cornwall, United Kingdom, 3. BGS, Keyworth, United Kingdom

Porphyry-type deposits are associated with large, long-lived magmatic systems, where mineralising fluids are mostly derived from mid- to shallow crustal magma or possibly crystal mush reservoirs. For the latter, what remains to be established is their role in the evolution, transport ,and focussing of fluids into relatively narrow and shallow (ca. 2- to 5-km-deep) zones of mineralisation. Even more problematic, however, is a paucity of first-order textural evidence for the presence of crystal mush in magmatic systems, including those that host porphyry-type deposits.

To address this, we have studied uniquely exposed cupolas and root zones of magmatic-hydrothermal systems in Saginaw Hill, Tuscon, Arizona, and in the archetypal Yerington District, Nevada. From field- to micro-scale (QEMSCAN, SEM-EDX-CL, EPMA, and TitaniQ) textural and geochemical studies, particularly observations of vermiform quartz in miarolitic cavity- and vein-mineralised aplitic dikes, we evidence the development of crystal mush-fluid pathways at the magmatic-hydrothermal transition. These are shown to be in close spatial association with unidirectional solidification textures and massive silica caps, indicative of magmatic undercooling and first-type boiling, and lie just below the main zone of porphyry mineralisation. By integrating our observations with zircon U-Pb CA-ID-TIMS and molybdenite Re-Os geochronology, we place spatial and temporal constraints on the formation of these features at the deposit to district scale and suggest that their presence should be considered in any new exploration models for porphyry- and similar-type deposits.

The Orion Cu-Au-Ag-Co-Zn Discovery in the West Kimberley Region, Western Australia – a New Mineralisation Style in a Greenfields Terrane

Nicholas Chapman
Dreadnought Resources Ltd., Perth, WA, Australia

The Orion deposit, discovered in July 2021 by Dreadnought Resources at the Tarraji-Yampi Project in the West Kimberley region of Western Australia, is a massive sulphide body consisting of a structurally hosted hydrothermal, polymetallic assemblage of Cu-Au-Ag-Co-Zn metals. The Orion discovery is hosted in Proterozoic metamorphic rocks within the Wunaamin-Miliwundi Orogen, and is the first documented occurrence of this style of mineralisation in the region.

Polymetallic deposits with similar metal endowment are found in the Mt Isa-Cloncurry districts in northwest Queensland, which are host to an array of iron sulphide-copper-gold (ISCG) and iron oxide-copper-gold (IOCG) deposit styles (Ernest Henry, Eloise, etc.). Mineralisation is typically massive to vein-like with brecciated textures and composed of chalcopyrite-pyrrhotite-quartz-carbonate, strongly analogous to Dreadnought's Orion deposit.

From 1978 to 2013, a significant area of the western Wunaamin-Miliwundi was off-limits to all exploration activity due to the establishment of a military training area. In 2013, exploration restrictions were lifted, but there has been minimal to no drill-testing. Black soil plains, and the nature of the regolith profile, are not amenable to surface geochemical techniques, as there is poor to no dispersion of elements through that type of transported cover, hindering exploration efforts.

Modern exploration techniques, comprising high-powered airborne and ground-based geophysics, and auger drilling, can penetrate through the transported regolith cover in Kimberley and unlock a new exploration space in highly prospective geology. These techniques have been used to great effect in the Mount Isa-Cloncurry districts, which host significant deposits that are heavily endowed in the same metal associations as discovered at Orion—Cu-Au-Ag-Co-Zn-Pb.

Operating in an environmentally sensitive region has required a strong relationship with the Dambimangari Traditional Owners and Department of Defence, critical thinking in relation to drill-pad preparation and rehabilitation, and progressive ideas to generate targets for future base metal discoveries.

The Role of Exhumation in Generating High-Grade Copper Deposits: Insights from the Central Andes, Northern Chile

Frances J. Cooper[1, 2], Byron A. Adams[1, 2], Simon I. Dahlström[2], Todd A. Ehlers[3, 7], Matthijs C. van Soest[4], Kip V. Hodges[4], Brian R. Jicha[5], Bradley S. Singer[5], Jaime Cortes Yañez[6]
1. University College London, London, United Kingdom, 2. University of Bristol, Bristol, United Kingdom, 3. University of Tübingen, Tübingen, Germany, 4. Arizona State University, Tempe, AZ, USA, 5. University of Wisconsin-Madison, Madison, WI, USA, 6. BHP Copper, Tucson, AZ, USA, 7. University of Glasgow, Glasgow, United Kingdom

Copper is a vital part of the global effort to transition to a greener, low-carbon economy. Today, most copper derives from magmatic-hydrothermal porphyry copper deposits (PCDs), which are typically associated with subduction zones. However, there are two problems: (1) finding PCDs is increasingly difficult because the relatively easy surface deposits have already been discovered; and (2) they are notoriously low grade (≲1% Cu), so large volumes of rock must be removed to retrieve a tiny amount of copper. With global demand set to double by 2035, mining companies must not only seek more elusive, buried deposits, but also those that will have the least environmental impact. To achieve this means finding regions likely to host the highest-grade deposits (≳1% Cu) that require the smallest amount of extraction. Determining the exhumation history of an area could provide a critical tool to assist in this effort by significantly narrowing down potential search space.

The exhumation history of a PCD plays an important role in determining its copper grade. During the initial hypogene stage of mineralisation, rapid exhumation is required to advect heat towards the surface, allowing metal-carrying magmas and fluids to transport their cargo into the shallow crust. To achieve maximum ore grade, exhumation must then slow considerably so the deposit can linger close to the surface, where supergene enrichment by meteoric water takes place. We present an example from the Middle Eocene-Early Oligocene copper belt in northern Chile, which hosts the world-class Chucquicamata and Escondida deposits. By combining U-Pb zircon geochronology, Al-in-hornblende geobarometry, low-temperature thermochronology, and thermal-kinematic modelling, we track spatial-temporal patterns of exhumation. Our results suggest that both the rate and timing of exhumation are critical in determining the potential of an area to host high-grade PCDs and should be considered when scoping out new targets for exploration.

Rapid Cu-Porphyry Indicator Mineral Characterization by µXRF: Automated Detrital Indicator Mineral Analysis of British Columbian Cu-Porphyry Exploration Properties

Ben M. Eaton, Shaun L. Barker, Lindsey J. Heagy
University of British Columbia, Vancouver, BC, Canada

As the global population rises and the world economy transitions towards decarbonization, electrification upgrades to energy, transportation, and industrial infrastructure will result in an increase in copper demand by upwards of 350% worldwide by 2050. Continued exploration for Cu-porphyries is critical to meeting this demand; however, the remaining undiscovered deposits are in regions mostly or entirely obscured by post-mineral surficial cover. Effective, timely, and economically feasible exploration approaches must be developed to meet the rise in copper demand and address challenges associated with exploring for obscured deposits. Indicator minerals (IMs) are minerals that contain textural or chemical information indicating the presence of specific mineralization in bedrock from which the minerals were derived, and they are commonly used to vector towards and/or assess the fertility of a potential deposit. In Cu-porphyry exploration, IMs have been widely used in regions of extensive surficial cover to explore for obscured deposits.

The project's research goal is to improve, quantify, and expedite the identification of porphyry copper IMs by investigating rapid, cost-effective analytical technologies and approaches. In this study, Cu-porphyry IM identification methods were developed utilizing benchtop micro-X-ray-fluorescence (µXRF) corescanners and automated scanning-electron-microscopes with energy-dispersive-detectors (ASEM-EDS). Heavy-mineral-concentrates (HMCs) of stream sediment samples from Northwest Copper's Cu-porphyry exploration properties in central British Columbia were analyzed using µXRF core scanners and ASEM-EDS. The IM mineralogy of the HMCs was characterized using the Bruker AMICS automated mineralogy software to identify IMs from µXRF and ASEM-EDS data. HMC mineral characterization and IM identification by µXRF core scanners and ASEM-EDS is presented, and the opportunities and challenges associated with applying µXRF core scanners to rapid Cu-porphyry IM characterization are discussed. Benchtop µXRF core scanners appear to be a promising IM analytical tool; however, relatively low spatial resolution and resulting mixed mineral spectra may complicate identification of fine minerals or textures.

Petrology and Geochronology of Pre- to Post-Ore Igneous Rocks from the Asarel Porphyry Cu(±Au±Mo) Deposit, Central Sredna Gora Zone, Bulgaria

Stoyan Georgiev[1], Irena Peytcheva[1], Elitsa Stefanova[1], Atanas Hikov[1], Albrecht von Quadt[2], Marcel Guillong[2], Desislav Ivanov[3]
1. Bulgarian Academy of Sciences (BAS), Sofia, Bulgaria, 2. ETH, Zürich, Switzerland, 3. Assarel Medet JSC, Panagyurishte, Bulgaria

The Panagyurishte ore region of the Central Sredna Gora Zone, Bulgaria, is a part of the Upper Cretaceous Apuseni-Banat-Timok-Sredna Gora magmatic-metallogenic arc belt (ABTSB). In the frame of the AGIMERA project, we focus on potential of porphyry copper deposits (PCDs) as sources of Cu and critical raw materials. The application of novel geochemical methods and precise geochronology in the study of the Asarel magmatic-hydrothermal system may result in improvement of the deposit model and provide additional knowledge on metal concentration. The ore formation is closely related to subsequent intrusion of pre-, syn-, and post-ore shallow magma bodies and dikes. The magmas are calk-alkaline to high calc-alkaline with diorite to granodiorite composition. The presence of small mafic enclaves, together with some slightly reverse zoning in the plagioclases and various εHf_{90} of zircons (-3.80 to 1.22), suggests mafic magma recharge and mixing. The early amphibole fractionation in water-saturated conditions and delay of plagioclase crystallization took part in the evolution of the magmas and facilitated transporting metal-enriched fluids. The crystallization of amphiboles shows shallow crustal depths (8.13–6.36 km), temperatures of 882° to 839°C, and H_20 (in the melt) of 5.5 to 5.2 wt %. Different redox conditions were found during the zircon growth from pre- to syn-ore stages and syn- to post-ore magmatic stages, indicating that the first ones are more oxidized and in favor of metal precipitation. The depletion of Ce* may additionally partly result from concurrent growth of allanite. The precise CA-ID-TIMS geochronology distinguished the age of porphyries from the pre- to syn-ore stages at 90.29±0.42 Ma and from the syn- to post-ore stages at 89.56±0.11 Ma, consequently a short magmatic-hydrothermal life of the system (0.73 Ma).

Aknowledgements: The study is a contribution to the Horizon Europe AGEMERA project (Agile Exploration and Geo-modelling for European Critical Raw Materials), grant no. 101058178.

Competency-Controlled Copper Mineralization at the Onganja Mining District, Namibia

Matthew P. Hales[1], Judith Kinnaird[1, 2], Paul Nex[1]
1. School of Geosciences, University of the Witwatersrand, Johannesburg, South Africa, 2. School of Mines, Exeter University, Exeter, United Kingdom

The Onganja Mining District, Namibia, comprises several subvertical breccias and stratiform massive sulphides. These orebodies are reported to have formed during the late phases of the Pan-African Damara Orogen (ca. 517 Ma); a period contemporaneous with significant copper deposition within the African Copper belt.

The ore and alteration assemblages bear some similarity to iron oxide-copper-gold deposits; however, geochemical analyses and monazite age-dating data suggest that biotitisation was related to metamorphic processes, while REE enrichment significantly postdates the sulphide mineralisation. Likewise, magnetite trace-element chemistry and whole-rock chemistry are distinct from that of typical IOCG deposits. In addition, based on field and petrographic analyses, the controls on mineralisation appear to be related to specific lithologies and structures associated with D_3 folding.

Both the subvertical breccias and massive sulphides have spatial relation to a N-S–trending quartz-albite vein system. The subvertical breccias, however, are limited in strike along the system but show regional occurrences along lithological strike. Furthermore, old workings have a near-ubiquitous occurrence of biotite-scapolite schists with lenses of amphibolite as country rock to both the breccias and massive sulphides.

Structurally, three phases of folding are recognised within the mining licence. Shearing occurred along an axial-planar foliation that developed with the third phase of folding, while the fold hinges of F_1 and F_2 are preserved within lenses enveloped by the shears. Jointing is restricted to these lenses and has comparable orientations to that of the vein system. Furthermore, veins show brecciation and displacement when intersected by these shears, with breccias in more pelitic units being barren.

As such, lithological and structural data suggest that the ore-forming fluids migrated through the axial-planar shears, mobilising sulphides from the amphibolites, before escaping upwards through structures that formed in the more competent veins and schist lenses.

Porphyry Cu Fertility of Late Mesozoic-Cenozoic Magmatism Along the East Anatolian Magmatic Belt Based on Zircon Trace Element Compositions

Ali İmer[1], Rui Wang[2], Kaan Sayıt[1], Ersin Koralay[3]
1. Department of Geological Engineering, Middle East Technical University, Ankara, Turkey, 2. State Key Laboratory of Geological Processes and Mineral Resources, Institute of Earth Sciences, China University of Geosciences, Beijing, China, 3. Department of Geological Engineering, Dokuz Eylül University, İzmir, Turkey

The East Anatolian Magmatic Belt (EAMB) is a product of the closure of the Southern Neotethys ocean and subsequent collision between the Tauride-Anatolide and Arabian blocks during the Mesozoic-Cenozoic. Cretaceous to Eocene subduction-related and Late Oligocene-Miocene collisional to post-collisional igneous rocks of the belt display whole-rock geochemical signatures broadly consistent with derivation from subduction-modified sources. Nevertheless, economic porphyry Cu±Mo±Au systems at EAMB are rare and appear to be restricted to two narrow time intervals: ~48 to 44 Ma (Çöpler and Mavidere) and ~25 to 24 Ma (Cevizlidere).

In order to constrain porphyry magma fertility signatures along the EAMB, we investigated zircon trace element compositions of selected Late Cretaceous-Miocene intrusive suites. Zircons from mineralized Eocene and Oligocene intrusions display weak negative Eu anomalies on chondrite-normalized plots, whereas barren intrusions have prominent negative Eu anomalies. Mineralized intrusions are also distinguished by lower Ti contents (<7 ppm) as well as by lower Dy/Yb (<0.25) and higher Eu/Eu* (>0.3), Ce/Nd (>3), and (Ce/Nd)/Y (>0.003) ratios compared to most other intrusive suites.

Calculated temperatures based on Ti-in-zircon thermometry revealed relatively lower crystallization temperatures of <730°C (avg.= 708°C) for mineralized intrusions compared to barren intrusions (704–817°C; avg.= 758°C). Additionally, an increase in magma fO_2 is evident from the Late Cretaceous (ΔFMQ: -0.6 to +0.8) to middle Eocene and Oligocene-Miocene intrusive systems (ΔFMQ: +0.4 to +2.0).

Zircon trace element data suggest an evolutionary trend for EAMB magmatism with increasing magma oxidation states and water contents from the Late Cretaceous to the middle Eocene-Oligocene period, a change likely triggered by crustal thickening prior to the onset of Eocene magmatism. Mineralized igneous suites are dominated by early amphibole fractionation and higher degrees of crustal interaction, which are hereby suggested as key factors for magma fertility.

Copper in the Guiana Shield

Salomon B. Kroonenberg[1]
1. Anton de Kom Universiteit van Suriname, Paramaribo, Suriname, 2. Delft University of Technology, Delft, Netherlands

Two types of copper deposits predominate in the rainforest-clad Guiana shield. Volcanogenic deposits associated with orogenic-type gold mineralisation are concentrated in the Paleoproterozoic Maronian greenstone belt stretching over 1,500 km along the Atlantic coast of South America. Mafic to felsic volcanism, plutonism, sedimentation, deformation, and hydrothermal mineralisation essentially took place during the Trans-Amazonian Orogeny, 2.26-2.03 Ga, the equivalent of the Eburnean Orogeny in the Birimian of West-Africa. Copper sulfides, especially chalcopyrite, bornite, chalcocite, and covellite, are minor but characteristic constituents of hydrothermal gold-bearing quartz veins, stringers, breccias, and disseminations in greenschist-facies metavolcanic rocks in Las Cristinas, Venezuela, Toroparu and Million Mountain in Guyana, and amphibolite-facies schists, BIFs, and carbonate rocks in the Tucano-Urucum gold deposit in Amapá, Brazil. The Montagne d'Or gold deposit in French Guiana contains four zones of stratabound sulphide disseminations, stockworks, and veinlets in which chalcopyrite is the main copper mineral. This deposit is informally considered an Au-rich VMS.

The second type comprises two unusual copper deposits without gold in high-grade metamorphic terrains in the interior of Suriname. In the UHT metamorphic Bakhuis Granulite Belt (2.09-2.03 Ga), copper and apatite mineralizations are associated with migmatitic granulites, calcsilicate granulite, metagabbros, and ultramafics. Bornite and chalcopyrite occur disseminated in the leucosome of the migmatitic rocks, and apatite beds are concentrated in calcsilicate granulites. An evaporitic origin for the phosphate deposits has been suggested.

The Weko Sula deposit in southeastern Suriname (2.06 Ga) forms a pluton-like body in the diatexitic southward extension of the greenstone belt and consists of plutonic-looking rocks composed of abundant large euhedral cordierite crystals with some plagioclase, and interstitial aggregates of coarse magnetite, apatite, biotite, and copper minerals, mainly bornite, chalcosite, and covellite. It is thought to result from anatexis of Kupferschiefer-like sediments, though the magnetite-apatite association also reminds one of Kiruna-type deposits.

Improving the Sustainability of Copper Mining Through XRF Sensing, Machine Learning, and Bulk Ore Sorting

Hannah Lang, Karim Halaseh, Giovanni Pamparana
MineSense Technologies Ltd., Vancouver, BC, Canada

MineSense Technology equips shovels and belts with XRF sensors to enable bulk ore sorting at the mine face and process optimization within the mill. Machine learning is applied to the XRF data to classify the material in every bucket during extraction and along the belt in the grinding circuit. Simple algorithms convert individual peaks from the XRF spectra into element grades to recover ore from being sent to the waste piles and tailings and reduce the amount of barren material the mill processes. This is achieved through direct integration with the fleet management system so that trucks can be automatically routed to the correct destination as soon as the truck is filled. More advanced machine learning models are applied to the entire XRF spectra to classify ore material by intrinsic rock properties such as mineral speciation and to predict rock response properties of the material during processing, such as grindability. These methods can also be applied to quantify the acid-generating potential of waste material. The high-resolution data sets generated by MineSense's ShovelSense© and BeltSense© technology allow for smart mining decisions to be made to optimize the entire mine-to-mill pipeline, ultimately allowing mines to reduce the environmental footprint per unit of metal produced. A calculation of pre- and post-adaption of MineSense allows for the environmental impact to be quantified by measuring the ore recovery and dilution reduction enabled by bulk ore sorting at the mine face and calculating reductions in GHG emission and water, energy, and reagent consumption achieved through improvements in recovery. MineSense is deployed globally in over a dozen copper mines throughout North and South America. Here we present real data use cases and their implications for sustainable mining practices applied to the copper mining industry.

Deciphering Fluid Origins in the Paleozoic Iron Oxide-Cu-Au (IOCG) Deposits, NW China: Constraints from Noble Gases and Halogens

Pei Liang, Yuling Xie, Lingling Xiao
University of Science and Technology Beijing, Beijing, China

Represented by the Laoshankou, Qiaoxiahala, and Heijianshan deposits, the northern margin of East Junggar and the Yamansu belt of East Tianshan are two important potential belts for Paleozoic iron oxide-Cu-Au (IOCG) mineralization in NW China. All IOCG deposits in these two regions formed in a basin inversion setting (dying back-arc basin) and show significant two-stage mineralization, i.e., Fe-oxide and chalcopyrite(-gold). In this study, we used halogen (Cl, Br, and I) and noble gases (Ar, Kr, and Xe) from fluid inclusions of quartz and epidote as reliable tracers to track the fluid sources and evolution of these deposits. The results (in mole ratios) obtained by a MAP 215-50 noble gas mass spectrometer through an UHV extraction line at the Noble Gas Laboratory at the Australian National University (ANU) and showed that three different fluid end members are mainly involved in the mineralization processes of these Paleozoic IOCG deposits: (1) a magmatic-hydrothermal fluid, with I/Cl, Br/Cl, and $^{40}Ar/^{36}Ar$ ratios of 16.3×10^{-6} to 18.0×10^{-6}, 1.03×10^{-3} to 1.06×10^{-3}, and 352 to 437, respectively; (2) surface-derived bittern brine derived from seawater evaporation, with highest Br/Cl ratios of 1.53×10^{-3} to 1.80×10^{-3} and I/Cl and $^{40}Ar/^{36}Ar$ ratios of 77.1×10^{-6} to 87.7×10^{-6} and 672 to 883, respectively; and (3) basin brine or formation water modified by organic-rich strata through water-rock reaction, with highest I/Cl ratios of 477×10^{-6} to $26{,}301\times10^{-6}$, low Br/Cl ratios of 0.39×10^{-3} to 1.28×10^{-3}, and $^{40}Ar/^{36}Ar$ ratios of 288 to 510, respectively. The obvious multi-stage mineralization and involvement of Ca-rich hypersaline non-magmatic brines with variable origins in the Paleozoic iron oxide-Cu-Au deposits in the Central Asian Orogenic Belt (CAOB) resemble the characteristics of other IOCG type deposits in the world. These findings further suggest that non-magmatic fluids dominate the formation of IOCG deposits in the basin inversion setting.

Implementation of a Tectonic and Magmatic GIS-Based Database for Porphyry Copper Deposits Across the Southwest USA

Roseanne Marshall[1], Thomas Lamont[1], Frances Cooper[1, 2], Robert Loucks[3], Adam Gorecki[4]
1. University of Bristol, Bristol, United Kingdom, 2. University College London, London, United Kingdom, 3. University of Western Australia, Perth, WA, Australia, 4. BHP, Melbourne, VIC, Australia

The Laramide belt in the southwestern USA is the world's second largest porphyry province in terms of economic copper. Despite this, there is limited understanding of the regional-scale tectonic and magmatic framework controlling the location, size, and grade of its porphyry copper deposits. Based on the notable presence of two high-grade hypogene deposits along the margin of the Colorado Plateau (Resolution and Bingham Canyon), we seek to understand whether high-grade porphyry copper formation could be linked to prolonged and focused compressional stress along the margins of rigid crustal blocks. To do this requires comparison of the spatial-temporal evolution of the regional stress field with the history of magmatism and mineralisation. We are therefore compiling an open-source geospatial database within a geographical information system (GIS) framework that integrates geochronological, geochemical, structural, magmatic, and mineralisation data across the southwest USA. Existing database query processes routinely used in mineral exploration use a multi-level interface that does not consider geospatial distribution of data. By contrast, our PostgreSQL database system is accessed within QGIS through the PostGIS and DB Manager plugins, allowing the user to easily query tectonic and magmatic data from a simple base map. Using these plugins, and the QGIS processing function "postgreSQL execute and load SQL" (PE&LS), a database query outputs a new layer onto the base map. The QGIS core process is to type the SQL query directly into the PE&LS function or DB manager, but to make it more user-friendly, we will add a dropdown-box-style interface, or similar function. The database also considers the best use of QGIS automated visual functions to allow for relevant data visualisation to occur in one software package, allowing for speedy and convenient report creation. Once completed, the database can be implemented in other areas of exploration where the tectonic-magmatic framework is important.

Understanding Climate and Weather Drivers of Tailings Risk in Chile: A Complex Multi-Hazard Challenge

Linda Jakschies[1], Blake O'Sullivan[1], John Coll[2], Stephen Edwards[2, 1], Katie McFall[1]
1. UCL, London, United Kingdom, 2. CGG Services (UK) Limited, Crawley, United Kingdom

Tailings are typically voluminous, reactive, and potentially toxic by-products of ore processing that must be safely and securely isolated from living organisms and the environment. Multiple tailings hazards involve water, which is why understanding the behaviour and transport of tailings and their derivatives when exposed to extreme weather conditions and shifts in climate is fundamental to reducing tailings risk. The aim of the research is to draw attention to these issues and better understand the interactions between tailings, weather, and climate by using Chile as a case study.

Chile's 757 active, inactive, and abandoned tailings deposits mostly derive from copper and associated metal production and reside almost entirely within the central and northern parts of the country, which are experiencing significant changes in weather and climate. From 1915 to 2010, seismic activity triggered 80% of tailings incidents in Chile, with the rest linked to hydrometeorological events. Since then, erosive flash floods from extreme precipitation have resulted in two of Chile's larger tailings incidents by volume. Moreover, windblown tailings dust is becoming recognised as a significant hazard.

Reanalysis of past hydrometeorological events that resulted in tailings erosion by wind and water has enabled better characterisation of preconditioning factors and interactions before and during these tailings hazard events. This has been used to identify certain tailings highly susceptibility to erosion. The tailings beach at Chañaral (Atacama Region) is one example of a hazard hotspot, as public health concerns have been raised over tailings dust in the town and several 100,000 m^3 of tailings were eroded during the March 2015 flash flood. Similarly, the May 2017 flash flood in the Marquesa valley (Elqui river catchment, Coquimbo Region) eroded about 100,000 m^3 of tailings upstream of urban centres and agricultural land.

Contrasting Tectonomagmatic Conditions for Porphyry Cu and Iron Oxide-Apatite Deposits in Phanerozoic Arcs

Xuyang Meng[1], Jingwen Mao[1], Adam Simon[2], Chao Duan[3], Guiqing Xie[1], Huimin Su[4], Tong Hou[1], Ke Shi[5], Nian Chen[1]
1. China University of Geosciences (Beijing), Beijing, China, 2. University of Michigan, Ann Arbor, MI, USA, 3. Institute of Mineral Resources, Chinese Academy of Geological Sciences (CAGS), Beijing, China, 4. China University of Geosciences (Wuhan), Wuhan, China, 5. MNR Innovative Center of Coverage Area Deep Resource Exploration Engineering Technology, Geological Survey of Anhui Province, Hefei, China

Porphyry Cu ± Mo ± Au and iron oxide (± apatite, IOA; ± copper-gold, IOCG) deposits commonly occur in close spatial and temporal proximity in Phanerozoic arc-related settings. However, the evolution of these deposit types in an evolving arc setting remains poorly understood. Here, we systematically evaluated the P-T-fO_2 conditions and S-Cl contents for the dioritic to granodioritic source magmas for representative magmatic-hydrothermal porphyry-skarn Cu ± Au (150–135 Ma) and IOA (~130 Ma) deposits that formed under transpressional and transtensional settings in the Middle-Lower Yangtze River Metallogenic Belt, China. Our estimates show that, compared to IOA deposits, the porphyry-skarn-related magmas were relatively felsic, cooler, more hydrous, and emplaced relatively deeper (264 ± 23 vs. 143 ± 17 MPa for IOA-related magmas; 2SE). Apatite data reveal that the porphyry-related and IOA-related magmas had comparable pre-degassed S concentrations (~0.13 ± 0.06 vs. ~0.16 ± 0.09 wt. % on average), but that IOA-related magmas had higher pre-degassed Cl/H_2O ratios (~0.10–0.15 vs. 0.02–0.08 wt. % for porphyry-related magmas) that decreased by a factor of two after magma degassing. Magmatic fO_2 estimated using zircon and amphibole, reported in log units relative to fayalite-magnetite-quartz (FMQ) redox buffer, gradually increased during cooling of the porphyry-skarn-related magmas (ΔFMQ +0.7 to +2.5) at 950° to 800°C and decreased to ΔFMQ +1 at 700°C. In contrast, the magmatic fO_2 for the IOA-related magmas varied significantly from ΔFMQ -1.5 to ΔFMQ +2.5 but generally showed an increasing trend with cooling from 970° to 700°C. We propose that the porphyry-skarn Cu deposits in the MLYRMB formed in a compressional setting in response to normal plate subduction that favors storage and evolution of S-rich hydrous ore-forming magmas at variable crustal levels, whereas the back-arc extensional setting that formed because of slab rollback led to rapid degassing of the Cl-rich magmas.

Isotopic Microanalysis of Calcite in the Permian Kupferschiefer System (Saale Subbasin, Eastern Germany)

Mohammed S. Mohammedyasin[1, 2], Joseph M. Magnall[1], Sarah A. Gleeson[1, 2], Hans-Martin Schulz[1], Maria R. Scicchitano[1], Michael Wiedenbeck[1]
1. GFZ German Research Centre for Geosciences, Potsdam, Germany, 2. Institute of Geological Sciences, Freie Universität Berlin, Berlin, Germany

The Kupferschiefer districts in Central Europe contain some of the world's highest-grade sediment-hosted Cu deposits. Stratiform to stratabound Cu, Zn, and Pb sulfides are hosted in calcareous, organic matter-rich shales and mudstones of the marine Kupferschiefer (T1), the underlying continental Rotliegend sandstone (S1), and the overlying Zechstein Limestone (Ca1). Detail petrography (cathodoluminescence, CL; scanning electron microscopy, SEM) and carbon and oxygen isotope microanalysis (secondary ion mass spectrometry, SIMS) have been combined to show the paragenetic relationships and origin of the calcite cement in samples (n = 47) from three drill holes (Wallendorf, Sangerhausen, Allstedt) in the Saale subbasin (Eastern Germany). Petrographic data show extensive in situ alteration of detrital clasts in the S1 unit. Primary porosity occluding early diagenetic calcite cement and subordinate dolomite and detrital clasts have been partially replaced by Cu- and Zn-Pb sulfides (bornite, sphalerite, galena) in the three units. Overlapping $d^{13}C$ (VPDB) and $d^{18}O$ (VSMOW) values in calcite cement in samples from the S1 and T1 in the Sangerhausen and Wallendorf drill cores show the calcite cement was derived from fluids of similar isotopic composition. The low $d^{13}C$ values of calcite cement in samples from the S1 (-15‰ to 1.3‰) and T1 (-15‰ to 0.7‰) in the two drill cores indicate sources of carbon and alkalinity from seawater-derived fluids and diagenetic pore fluids influenced by the oxidation of organic matter. The wide range in $d^{18}O$ values in calcite cement in the S1 (~18‰ to 31‰) and T1 (~ 22‰ to 31‰) often varies at microscales, suggesting pore fluid chemistry was influenced by early diagenetic alteration of volcanic rock fragments, perhaps with minor contributions from the influx of meteoric waters and evaporated seawater.

Lithostratigraphy, Mineralogy, and Structural Controls of the Late Ediacaran Tizi N'ousatour Cu-Ag Deposit, Northwestern Anti-Atlas, Morocco

Mehd Ousbih[1], Mohamed Charroud[3], Mourad Essalhi[2], Moha ikenne[1], Askkour Fatiha[1], Salih Amarir[5], Hassan Essaidi[4]

1. Ibn Zohr University, Agadir, Morocco, 2. Moulay Ismail University, Faculty of Sciences and Techniques, Meknes, Morocco, 3. Sidi Mohammed Ben Abdellah University, Faculty of Sciences and Techniques, Fes, Morocco, 4. National Office of Hydrocarbons and Mines (ONHYM), Rabat, Morocco, 5. Mohammed VI Polytechnic University, Benguerir, Morocco

The Imi N'Ozrou mining site, encompassed by various metallic indicators such as copper and silver, is situated in the Anti Atlas at the northwest boundary of the West African Craton. These metals are located in the volcano-sedimentary series of the late Ediacaran Ouarzazate Group deposited between 590 and 550 Ma. The volcanic facies at the site comprises porphyritic and aphyric rhyolites, vitroclastic porphyritic dacites and rhyodacites, as well as their pyroclastic equivalents. These formations are intersected by two systems of dikes, mainly oriented in the E-W and N-S directions. N-S dikes are the youngest and possibly Triassic, despite intersecting the lower Cambrian. The Precambrian terrains are overlain by a Paleozoic sedimentary series, topped by Neogene trachytes.

Three main fault systems have been identified, namely N-S, E-W, and NW-SE, which imply the occurrence of three deformation stages. Phase I is represented by N-S–striking strike-slip faults, carrying a suite of dacite and rhyodacite veins and developing at the basement. It corresponds to the post-collisional collapse that took place during the late Ediacaran, resulting in the formation of horsts and grabens. Phase II is marked by sub-meridian compression with dextral shear zones and medium E-W–striking reverse faults. This family is believed to be posterior to phase I. Phase III represents a compressional phase marked by the formation of dextral strike-slip faults, which is considered to have occurred after the previous two phases. The faults of this phase offset both the dikes of phase I and the faults of phase II. The E-W structures consist mainly of barite, calcite, quartz, Ag-Hg amalgam, argentite, polybasite, pearceite, pyrite, arsenopyrite, chalcopyrite, galena, malachite, azurite, oligist, and bornite associated with fractures and sub-parallel crushing zones in the E-W dolerite dike. The mineralized host rocks are characterized by stockwork mineralization and affected by hydrothermal metamorphism.

Application of Facies Analysis and Sequence Stratigraphy in the Zambian Copperbelt and Implications for the Tectonostratigraphic Evolution Katangan Basin

Martin Purkiss, Bruce K. Levell, Michael C. Daly
University of Oxford, Oxford, United Kingdom

The Neoproterozoic Katangan basin hosts the world's largest sediment-hosted copper province, the Central African Copperbelt. A key to understanding sediment-hosted mineral deposits and the ability to predict their presence in unexplored stratigraphy is an understanding of the dynamic stratigraphic context in which the sediments were deposited. Here we provide a sequence stratigraphic model for the Zambian Copperbelt based on a unified facies scheme, with significant implications for Katangan stratigraphy, hydrothermal pathways, mineral distribution, and tectonostratigraphic evolution.

The proposed facies scheme is based on detailed sedimentological logging of over 10 km of continuous diamond drill core. It provides the first complete facies framework for Katangan stratigraphy in the Zambian Copperbelt, which has allowed for the stratigraphic evolution of the basin to be constrained quantitatively. Clear sedimentological and stratigraphic evidence is provided for two phases of extension and subsequent thermal subsidence.

The proposed sequence stratigraphic model is based on an uncertainty-sensitive, rule-based workflow. This generates a relative chronological framework for a basin with a paucity of firm chronological constraints. It highlights the variety of viable stratigraphic frameworks for the Zambian Copperbelt by considering both accommodation space and sediment supply. The mineralised mudstone unit, locally termed "ore shale," appears to have been deposited in a variety of depositional environments, ranging from turbiditic to intertidal. It may not represent a single flooding surface, but rather local sediment starvation as a result of basin-wide aridification. Evidence is also indicative of the ore shale being highly diachronous, and possibly discontinuous along-strike.

These and other observations show our model provides a basis for increasing our understanding of the Katangan basin's facies distribution and stratigraphic architecture and their implications for mineralisation.

Integrated Re-Os, Ar/Ar, and U-Pb Geochronology Directly Dates the Timing of Mineralization at the Mina Justa and Marcona Deposits, Peru

Maria Alejandra Rodriguez Mustafa[1], Adam Simon[1], Robert Holder[1], Holly Stein[2], Andrew Kylander-Clark[3], Brian Jicha[4], Daniel Blakemore[1], Edson Machado[5]
1. University of Michigan, Ann Arbor, MI, USA, 2. AIRIE Program, Fort Collins, CO, USA, 3. University of California - Santa Barbara, Santa Barbara, CA, USA, 4. University of Wisconsin - Madison, Madison, WI, USA, 5. MARCOBRE, Lima, Peru

Iron oxide-copper-gold (IOCG) and iron oxide-apatite (IOA) deposits are major sources of Cu and Fe, respectively. They contain abundant Fe-oxides and may contain Au, Ag, Co, rare earth elements (REEs), U, Ni, and V as economically important by-products. In Peru, the Mina Justa IOCG deposit is located next to the giant Marcona IOA deposit. Constraining the timing of Fe and Cu mineralization at Mina Justa is fundamental to understanding the duration and type of processes that generated this mineral deposit, and ultimately to testing the genetic link with other deposits in the area. Previous authors used alteration minerals to indirectly date Cu mineralization at Mina Justa at around 100 Ma. We report Ar/Ar dates of actinolite, U-Pb dates of magnetite, apatite, and titanite collected by in situ laser ablation-multicollector-inductively coupled plasma-mass spectrometry, and Re-Os thermal-ionization dates of sulfides. Our results indicate that Cu mineralization at Mina Justa occurred at about 160 Ma and that Fe mineralization is older and coeval with the neighboring Marcona IOA deposit, consistent with Cu mineralization overprinting IOA-style mineralization at Mina Justa. This work also provides a new framework for exploration for these deposits in Peru.

3D Mineralogical Analysis Using X-Ray Attenuation Computed Tomography (CT) and XRF to Determine Geoenvironmental Reactivity of Future Non-Ore Material

Enrique Saez Salgado[1], Anita Parbhakar-Fox[1], Nathan Fox[1], Regina Baumgartner[2]
1. BRC - SMI. The University of Queensland, Brisbane, QLD, Australia, 2. Teck Metals Ltd, Trail, BC, Canada

Tailings from Cu-porphyry deposits typically contain pyrite (FeS_2), which may oxidise under surficial conditions to produce acid rock drainage (ARD). Evaluation of ARD potential is defined using conventional industry-standard laboratory-based static and kinetic chemical tests. These analyses do not often consider the mineralogical properties due to high analytical costs; therefore, statistics and representativity is impacted. Geometallurgical programs may sample between 0.5 and 1.0% of the ore while non-ore materials are poorly characterised in terms of mineralogy.

This research utilises computed tomography (CT) analysis of drill core samples from the Quebrada Blanca porphyry Cu-Mo deposit located in northern Chile. The Quebrada Blanca Phase 2 project (QB2) is currently under construction and is focusing on the development of the hypogene (sulphide-rich) zone. QB represents one of the world's most significant copper resources with reserves of 1.43 Bt @ 0.51% Cu.

CT is a non-destructive technique that uses high-power X-rays to image the internal 3D structure of various materials with broad applications in geoscience, particularly in the reconciliation of stereological errors generated by conventional 2D mineralogical analyses. The GeoCore X10 3D tomographic method combines X-ray attenuation and X-ray fluorescence (XRF) measurements of total or half-core samples. The sizeable 3D voxel resolution allows coarse textural analysis which, combined with elemental analyses by XRF, enables the detection of particularly high-attenuation minerals (e.g., sulphides, gold, silver). Finer sulphide textures that may drive reactivity will be analysed using SEM-based and micro-CT methods. The results of this research, combined with additional mineralogical and geochemical analyses, could generate geometallurgical indexes for ARD potential risk predictions of QB non-ore material. Robust non-ore-type domains will consequently be obtained at early stages of the life-of-mine cycle (LoM).

Insights into the Origin of Sedimentary-Hosted Copper Mineralization in Kazakhstan

Azam Soltani Dehnavi[1], Reimar Seltmann[2], Fereshteh Shabani-Sefiddashti[3]
1. Nazarbayev University, Astana, Kazakhstan, 2. Natural History Museum, London, London, United Kingdom, 3. N/A, Shahrekord, Iran

Kazakhstan has historically produced copper from the world-class Dzhezkazgan mining complex (2.0 Bt @ 1.1% Cu for 22 Mt Cu) located in the northern Chu-Sarysu basin, which is the third largest endowed sediment-hosted copper basin globally. Also, the Zhaman-Aibat deposits (160 km SE of Dzhezkazgan) and the Zhilandy group of deposits (located in northern Dzhezkazgan) bear significant copper mineralization in the Chu-Sarysu basin (Box et al., 2012). In addition, the Teniz basin located north of the Chu-Sarysu basin, separated via uplift from the basin, is the current prospect for copper. Mineralization is related to middle-late Carboniferous and Permian fluvial red-bed sequences. The Chu-Sarysu and Teniz basins are characterized by a continental-marine-continental depositional cycle from Devonian to Permian. Comparison of stratigraphic columns of both basins shows correlation at the base of the Dzhezkazgan and Vladimirov Formations (Chu-Sarysu and Teniz basins, respectively); marking several upward-fining clastic sequences of alluvial-lacustrine red beds (hosting most of the mineralized horizons) above Lower Carboniferous platform carbonates (lagoonal to marginal-marine salt-bearing strata) (Cossette et al., 2014). Occurrence of copper is structurally controlled adjacent to the intersection of F_2 anticlines (N-NW-trending) with the syn-depositional folding F_1 anticlines (E-NE-trending) within the zones of sandstone bleaching. The F_1 anticlines locally trapped petroleum deposits. These structures are the pathway of the flow of dense ore brines across the petroleum-bearing anticlines, resulting in ore sulfide deposition most likely via mixing of two fluids. The copper originates likely from Early to Middle Devonian intermediate volcanic and volcanoclastic basement rocks.

At regional scale, the formation of both basins can be the result of oroclinal bending of Permian Kazakhstan, generating a back-arc/rift-graben development. This is reflected in parallel strike-slip lineaments in eastern and western parts of the basins. Satellite images are interpreted to locate major structures within the basins, which can assist in further exploration.

Hydrothermal Apatite: A Distal Proxy for Potassic Alteration in Porphyry Cu Systems?

George Stonadge, Andrew Miles, Daniel Smith, Philip Bird
The University of Leicester, Leicester, United Kingdom

The chemistry of detrital apatite is a promising tool for identifying traces of porphyry copper deposits (PCDs) in distal settings. Apatite is particularly useful because it is abundant in PCD environments and has a large variety of trace element substitutions that re-equilibrate with hydrothermal fluids. However, the extent to which apatites from different alteration zones around PCDs have unique chemical identities has been understudied. This precludes the separation of ore-bearing potassic alteration from un-mineralized alteration in detrital sediments.

In this study, we investigated the chemistry of apatites in fresh and altered rocks from PCDs of the Baguio Mineral District, Philippines. Unaltered rocks contain fresh magmatic apatite with enriched Th, Sr, Mg, and LREEs (>3,500 ppm). By contrast, apatites from altered samples are depleted in LREEs, Th, XOH, and Sr and enriched in XF and HREEs. Specifically, apatites from central ore-associated potassic alteration were most enriched in XF, Mn (>1,500 ppm), and HREEs (e.g., Yb >10 ppm). This reflects changes in fluid temperature, pH, and composition with increasing proximity towards the deposit centre.

Using linear discriminant analysis, we created a model for predicting alteration assemblage based on apatite chemistry. The model was constructed using a global apatite database including metamorphic, magmatic, and PCD terranes to test the effectiveness of this tool in detrital mineral studies. An overall accuracy of 89% was achieved for this model, effectively identifying apatites from PCDs. Furthermore, apatites from potassic alteration were well delineated from all other affinities (95% accuracy, false positive rate <0.01).

Our results demonstrate that apatites from economically significant potassic alteration are markedly distinct from those in barren propylitic and non-porphyry terranes. This furthers the utility of detrital apatite not only for recognising PCD alteration, but also for identifying the presence of ore-bearing alteration from afar.

Detrital Zircon Indicators of Metallogenic Magma Fertility: A Potential Tool for Exploration Targeting for Porphyry Copper

Eliana Toro Paz[1], Frances Cooper[2], Byron Adams[2], Thomas Lamont[1], Anthony Kemp[3], Robert Loucks[3], Catherine Belgarde[4]
1. University of Bristol, Bristol, United Kingdom, 2. University College London, London, United Kingdom, 3. University of Western Australia, Perth, WA, Australia, 4. BHP, Tucson, AZ, USA

Magmatic systems associated with porphyry copper mineralisation typically contain zircons with distinctive trace element patterns indicative of metallogenic magmatic fertility (e.g., relative oxygen fugacity/sulphur-carrying capacity and hydration state). Combining this fertility signature with high-precision U-Pb geochronology provides an opportunity to examine spatial-temporal patterns of magma fertility, offering a powerful tool for reconnaissance-stage exploration. In many instances it is not possible or practical to reach every site of interest for bedrock sampling, so detrital sampling offers an alternative strategy.

Detrital studies can maximise exploration efficiency, minimise costs, and provide material derived from inaccessible areas. However, previous attempts to use detrital zircon fertility studies for exploration have focused on large-scale (<400-km^2) drainage basins, typically analysing ~1 zircon per km^2. While this approach covers a large area quickly, slow sample processing and ambiguous or false negative results have limited its effectiveness.

This project aims to establish a new protocol for detrital fertility analysis guided by a robust geomorphological analysis of the target area prior to sampling. This will ensure that the source region for each sample is well constrained and variations in sediment sourcing due to spatial changes in erosion rate are taken into consideration. To achieve this, we have targeted the porphyry-rich and well-constrained Copper Triangle in Arizona, USA. The primary goals are to constrain optimum basin sizes for sampling, determine the optimum number of zircons required for analysis per basin, and establish whether variability in erosion rate can lead to false negative results.

Samples have been collected from tributaries of the Salt and Gila rivers within the Copper Triangle and zircons separated for analysis. Paired U-Pb dating and trace-element analyses of individual grains via laser-ablation ICP-MS will be carried out to provide a time series representing the age span and variation in chemical fertility of igneous intrusions throughout the study area.

Mineralogical Characterization of Potential Source, Aquifer and Trap Rocks in the Central African Copperbelt Using the TESCAN TIMA® Automated SEM Mapping System

Veronica G. Trevisan[1], Jamie J. Wilkinson[1, 2], Martin Purkiss[3], Mike Daly[3]
1. Natural History Museum, London, United Kingdom, 2. Department of Earth Science and Engineering, Imperial College London, London, United Kingdom, 3. Department of Earth Sciences, University of Oxford, Oxford, United Kingdom

The Central African Copperbelt (CACB) is the world's largest repository of sediment-hosted copper and cobalt, which are critical metals for electricity transmission and the production of batteries needed to help decarbonise society. Key knowledge gaps regarding this mineral system include the scale and nature of alteration mineral assemblage zoning and alteration mineral major- and trace-element chemistry patterns. A detailed characterization of samples from the CACB can address these gaps and help guide exploration for technology metals across different sedimentary basins. Here, we report TESCAN TIMA® automated mineral mapping results on (meta)sedimentary rocks from contrasting geological domains in the CACB. Our initial findings confirm higher metamorphic grades and metamorphic textures in rocks from the Domes region and the Kansanshi Cu-Au mine and in Muva Group rocks from the Fishtie Cu deposit. Initial results also suggest a dominantly Mg-rich (±Na, ±Ca) metasomatism in the Domes region and Kansanshi mine, whereas in the Classical Zambian Copperbelt, a dominant K-rich (± Ca, ±Mg, ±Na) alteration is recognized. At Fishtie, alteration is principally characterized by Fe-Mg metasomatism. It is clear that the alteration-mineralisation characteristics in the different regions are sufficiently different for the regions to be considered separate mineral systems from a hydrothermal processes perspective. Future work will include conventional multielement whole-rock geochemistry; mineral chemistry of key alteration and accessory mineral phases using analytical SEM, microprobe, and LA-ICP-MS; and data interrogation and development of discrimination tools through supervised and un-supervised multivariate statistical techniques.

Alteration Assemblages in the Southern Congolese Copperbelt: A Comparison of Background Stratigraphy with Cu-Vein-Hosted Mineralisation and Salt Dissolution Alteration

Helen Twigg, Murrat W. Hitzman
iCRAG, University College Dublin, Dublin, Ireland

Integrated drill hole logging, multielement geochemistry, and petrographic studies have been utilised to develop alteration maps that provide potential vectors to vein-fracture-hosted Cu and Zn-Cu-Pb massive sulphides in the Southern Congolese Copperbelt—an underexplored region of the Central African Copperbelt.

Defining hydrothermal alteration halos in sedimentary sequences is commonly difficult due to the heterogeneous nature of the rocks and diagenetic alteration. The siliciclastic, carbonate, and evaporite rocks of the Neoproterozic-Cambrian Katangan Supergroup have undergone multiple alteration events starting with diagenesis and continuing through alteration due to movement of residual brines from evaporite deposition, multiple episodes of basinal brine migration, halokinesis and evaporite dissolution, and metamorphism. A number of these alteration events were associated with base metal mineralisation including those associated with potassic, magnesian, and sodic alteration. In this study, drill hole intercepts of background/least-altered lithologies were compared with Cu-mineralised rocks, Cu distal alteration halos, and examples of alteration along the margins of Roan breccia after diapiric salt bodies to map different alteration halos and footprints. The Zn-Cu-Pb massive sulphide deposit at Kipushi was associated with magnesian, potassic, and barium alteration while supra-salt Cu-vein mineralisation was associated with sodic and magnesian alteration. Alteration focused along former salt walls and diapirs in the area was also extensive and is potentially misleading in the exploration for copper deposits.

Rhenium – Potential as Proxy for Mineralization: Insights from the Altar Porphyry Cu-(Au-Mo) Deposit

Martin G. Vallejos[1], Laura Maydagán[1], David Selby[2], Kevin B. Heather[3], Mariano Poodts[3]
1. INGEOSUR-CONICET, Departamento de Geología, Universidad Nacional del Sur, San Juan 670, Bahía Blanca 8000, Argentina, 2. Department of Earth Sciences, Durham University, Durham DH1 3LE, United Kingdom, 3. Aldebaran Resources Inc., Santa Lucía, San Juan, Argentina

Rhenium is a critical metal ultimately sourced from molybdenite as a by-product of porphyry-type systems. At the middle to late Miocene Altar porphyry Cu-(Au-Mo) deposit (San Juan, Argentina), the rhenium abundance in molybdenite is up to ca. 3,500 ppm, suggesting that rhenium could have economic relevance for this project. Additionally, here we discuss how rhenium could have important applications in exploration and potential as a vector towards Cu-Au high-grade zones in porphyry-type deposits utilizing rhenium data from multi-element talus fines geochemical analyses and isotope dilution-negative-thermal ionisation mass spectrometry (ID-N-TIMS) rhenium-osmium molybdenite analyses. We identify two high-rhenium anomalies related to copper and gold high-grade sectors. The first is presented as superficial high-rhenium anomalies of about 2 km of lateral extent, located directly above the mineralized areas of the deposit, that open up the possibility of rhenium being considered key in exploration programs. The second is from an underground high-rhenium anomaly shown in monomineralic molybdenite separates related to molybdenite-bearing veinlets sampled over 1,000 vertical meters (Altar Central), coincident with copper- and gold-rich zones.

Future research in other sectors of the Altar porphyry mineral cluster (Altar Central, Altar Este, and Quebrada de la Mina-Radio areas) will further test the association of high-rhenium anomalies to vector Cu-Au mineralization and its potential as a recoverable by-product and, additionally, evaluate the duration of mineralization, which is currently estimated to be approx. 1.2 m.y. based on high-precision rhenium-osmium molybdenite dates.

Greenfield Porphyry Copper Prospects in Peru

César E. Vidal
Sociaty of Economic Geologists, Lima, Peru

Three porphyry copper prospects in Peru are presented at different erosion levels; from top to bottom, these are:

Arcopunco, Huancavelica

Junior operator, ALTITUDE, drilled a single 852-m hole in 2021 and intersected strong phyllic alteration with quartz stockwork and breccias, anomalous in Cu + Au sulfides, 400 to 500 m below a barren alunite-pyrophyllite lithocap. Ongoing studies include alteration logging and mapping, detailed geochemistry, ore mineralogy, and the design of a pertinent geophysical survey.

The metallogenetic setting of Arcopunco is the Miocene to Pliocene subaerial volcanic arc with peripheral, Ag-rich polymetallic veins of the Huachocolpa mining district. The blind Arcopunco porphyry center remains to be tested by additional 1- to 2-km-deep holes.

San Cristobal, Ancash

Located 50 km northeast of the ANTAMINA Cu + Zn skarn deposit of mid-Miocene age, San Cristobal consists of sericite-altered tonalite and monzodiorite plutons of the Cordillera Blanca batholith, cut by a central diatreme with hydrothermal breccias. Strong Mo (Au) anomalies characterize the leached oxide capping of a porphyry copper deposit that crops out emplaced in folded Cretaceous sandstones.

First-pass drilling was done in 1997-1998 by RTZ/CRA. Prospectivity remains at intermediate depths in the porphyry center and laterally into other targets.

Alto Quemado, Arequipa

The prospect is located 20 km east of TECK's Zafranal Cretaceous porphyries and 60 km northwest of FREEPORT's Paleocene to Eocene Cerro Verde porphyries. At Alto Quemado, Jurassic and Cretaceous plutons and high-level porphyrytic stocks of the Coastal Batholith intrude pre-Cambrian gneiss basement, overlain by Mesozoic sandstones and shales. Regional faults of the Sincha-Lluta-Guanaco systems dissect the Totorayoc porphyry copper leached capping and control ore shoots in associated quartz-chalcopyrite-gold veins.

Core logging of a 6,000-m drilling campaign done in 2020, surface alteration mapping, and IP/magnetometry surveys for district-scale target generation are in place.

Mineralization Timing and Characteristics of the Heavy Sulfide Zone at the Grasberg Porphyry Cu-Au Deposit, Papua, Indonesia

Kotaro Yonezu[1], Shunnosuke Omachi[1], Thomas Tindell[1], Adrian Boyce[2], Euis Yuningsih[3], Mega Rosana[3]
1. Kyushu University, Fukuoka, Japan, 2. Scottish Universities Environmental Research Centre, Glasgow, United Kingdom, 3. Padjadjaran University, Bandung, Indonesia

The Ertsberg-Grasberg district, Central Range of Papua, Indonesia, includes world-class porphyry and skarn Cu-Au deposits. A Cu-Au porphyry deposit hosted in the Grasberg Igneous Complex (GIC) was estimated at a total of 1,022 million tons of ore reserves at 1.01% Cu and 0.78 g/t Au. Porphyry mineralization is closely associated with the GIC, formed in the Pliocene. The GIC mainly consists of the Dalam (DI), Main Grasberg (MGI), and Kali Intrusions (KI). In addition, another mineralized zone, the Heavy Sulfide Zone (HSZ), is developed at a peripheral zone of porphyry Cu-Au mineralization; however, formation timing and characteristics of the HSZ have not been well examined. Therefore, we focus on the nature of mineralization characteristics at the HSZ to compare and understand mineralization conditions at the HSZ and GIC through petrography and trace elements in sulfide and sulfur isotope study.

Copper mineralization styles included in the HSZ are dissemination and veins. The dominant ore minerals in both types are chalcopyrite and bornite. Pyrrhotite, typically formed under low-sulfidation conditions, was observed with pyrite and chalcopyrite. Covellite, chalcocite, and digenite were precipitated along the fractures of pyrite, and native gold is present with covellite, suggesting that it was remobilized by cooling during mineralization (gold contents up to 73 ppm). The ore mineral assemblage observed in the HSZ was three stages. The first is porphyry copper mineralization formed at higher temperature and sulfur fugacity characterized by chalcopyrite and bornite. The second one is characterized by pyrrhotite and native gold under relatively low temperature and sulfur fugacity conditions. These two stages are considered to correspond to DI and MGI intrusion events. The third one is a retrograde cooling stage of mineralization by Cu-Fe sulfide and native gold. This stage is considered a later Cu mineralization stage (Kali intrusion-related) characterized by high Zn contents in sulfide.

A Trace Element and C-O Isotope Study of Sb-Au Vein-Hosting Carbonate-Talc Schists of the Murchison Greenstone Belt, South Africa

Andrea Agangi[1, 2], Axel Hofmann[2], Robert Bolhar[3]
1. Akita University, Akita, Japan, 2. University of Johannesburg, Johannesburg, South Africa, 3. University of the Witwatersrand, Johannesburg, South Africa

The Archaean greenstone belts of the Kaapvaal Craton of South Africa host numerous Au and Au-Sb lode deposits. The Murchison Greenstone Belt is composed of a ca. 3 Ga volcano-sedimentary succession metamorphosed to greenschist-amphibolite facies. Across the succession, the rocks are variably carbonatised as a result of sea-floor metasomatism. In the central part of the belt is a zone of sheared carbonate-talc rocks known as the Sb-Line that hosts Sb-Au deposits occurring as quartz-carbonate veins.

We analysed bulk rocks and carbonate minerals from mineralised veins and host-rock, in addition to C-O isotopes, to estimate the source of CO_2-bearing fluids and trace the origin of mineralisation. By comparing data of mineralised rocks from drill core and open pit, and regionally occurring unmineralised rocks, we test the hypotheses whether 1) carbonate of the Sb-Line was mobilised from the carbonatised volcanic rocks of the greenstone belt during deformation and metamorphism (internal derivation), or 2) carbonate formed from CO_2 originated outside the greenstone belt, either in magmas or the mantle (external origin).

Whole-rock analyses of talc-carbonate mineralised rocks and regionally distributed meta-mafic and meta-ultramafic rocks have similar, overall unfractionated chondrite-normalised trace element patterns. In situ analyses of carbonate-talc schist samples indicate coexisting magnesite and dolomite-ankerite with distinct trace element compositions. In particular, fine-grained dolomite-ankerite has overall relatively high REE contents compared with coarse-grained magnesite. In PAAS-normalised plots, dolomite-ankerite displays patterns with depleted light REEs over mid- and heavy REEs, near-flat mid-heavy REE trends, and positive Eu anomalies. The REE compositions of dolomite-ankerite are overall similar to whole-rock analyses. The $\delta^{13}C$ values of mineralised schists range between -6.0 and -3.1‰ and peak at -5‰, contrasting with regional carbonatised rocks with a range of -4.7 to 1.3‰ and a $\delta^{13}C$ peak at -2‰, suggesting contrasting sources of C and implying an external, deep derivation of CO_2.

Ambient Noise Seismic Interferometry Applied to Tailings Dam Monitoring in Minas Gerais State, Brazil

Amanda M. Bezerra
Federal University of Rio de Janeiro, Rio de Janeiro, United Kingdom

The latest accidents involving tailings dams in Brazil have shown the need to seek effective monitoring methods for these structures in order to prevent further accidents. Currently, promising results have been obtained through the continuous monitoring of these structures through microseismic events, through the approach of seismic ambient noise interferometry. This method identifies small variations in shear wave velocity (Vs) by pairing and cross-correlation of geophones. Small variations of Vs are directly linked to the stiffness modulus of the embankment or maximum shear modulus, associated with low soil deformations. Microseismic events occur as a result of the propagation of natural elastic waves or by anthropic action. Currently, continuous monitoring of the BR dam of this study uses 10 sensors of 4.5 Hz, 8 uniaxial and 2 triaxial, distributed along the crest, embankment, and abutments. The objective of this work is to prove the effectiveness of microseismic monitoring in real time associated with the increase or decrease of the structure's stiffness, correlating the variations of Vs with the rainfall index, WLM, blasting in the mining pit, and amplitude of peak ground velocity (PGV) and peak ground acceleration (PGA) in the structure during one year. The data are provided and processed by the IMS software. During this period the method was sensitive enough to record significant decreases in velocity variations across sensor pairings. The analysis of the presented data indicates its strong correlation with operational activities in the dam and its surroundings; however, other variables still need to be detailed. The research will consist of confirming the cause of these drops in velocity variation by observing the behavior of the sensors in isolation and correlating these data with geotechnical parameters.

Data Capture from Geoscience Documents Using Transformers

Amit Juneja, Henri Blondelle
Agile Data Decisions, Inc., Houston, TX, USA

When it comes to prospect or field evaluation in mature geological areas, accessing the key information from thousands or millions of reports and logs may be time challenging and the risk to miss a key observation may be very high. Artificial intelligence (AI) and particularly machine learning may be beneficial in capturing data and sentiment on the contents of reports or logs at scale and detecting the sentences supporting the key information. This paper will discuss use of computer vision and NLP (natural language processing) for the following:

- Detection of lithological descriptions and assigning them to appropriate depths in logs;
- Classifying the geological description on composite logs according to several criteria;
- Detecting some key positive or negative information in geological reports.

Text modelling has been done using the BERT Algorithm (Bidirectional Encoder Representations from Transformers), and visual object detection has been done using convolutional neural networks (CNNs) . For our case study, the inferences obtained from a set of exploration reports in using a BERT model have shown very interesting results with a minimum of training. We emphasize the importantceof "human-in-the-loop" in building successful AI models for capturing structured data from different modalities of documents—tables, images, sentences, and semi-structured layouts. We show how AI models continue to improve with feedback from geologists in these different modalities and how this feedback and subject matter expertise is critical in building useful AI models.

3D Deep Electrical Resistivity Tomography Study of the Calamita Distal Fe-Skarn Deposit, Elba Island (Italy): New Perspectives for Mineral Exploration

Damian Braize[1], Julien Sfalcin[1], Matteo Lupi[1], Kalin Kouzmanov[1], Andrea Dini[2], Gianfranco Morelli[3], Federico Fischanger[3]
1. Department of Earth Sciences, University of Geneva, Geneva, Switzerland, 2. Institute of Geosciences and Earth Resources, CNR, Pisa, Italy, 3. Geostudi Astier S.r.l., Livorno, Italy

The green transition has increased the demand for raw materials, making the search for new mineral deposits critical. Recent technological advances such as geoelectrical tools have allowed for the exploration of deeper orebodies and larger areas while reducing logistical challenges. In this study, we show the potential of deep electrical resistivity tomography (DERT) for mineral resource exploration. We use the Fullwaver technology to decipher the geological structure of the Calamita distal Fe-skarn deposit, Italy. Unlike conventional methods consisting in centralized acquisitions, this innovative, quasi-wireless system enables a flexible distribution of receivers, resulting in comprehensive 3D imaging of subsurface structures and orebodies' distribution at greater depth.

A 3D DERT survey was conducted in November 2022 to explore the Calamita deposit, involving 148 current injections on 48 receivers over an area of 2 km^2 to achieve exploration up to 700 m. Geophysical data were combined with shallow (200-m) exploration drill hole data.

The high-sensitivity 3D inverted resistivity and induced polarization models match well with the surface geology. The deposit consists of conductive massive magnetite-hematite orebodies hosted in skarnified marbles, overlying highly resistive micaschists of the Tuscan Units. Strong resistivity contrasts indicate the presence of sub-vertical conductive and chargeable pipes, connecting the various skarn bodies at depth. The pipes are interpreted as paleo-hydrothermal upflow zones, pointing towards the cupola of a potential causative magmatic intrusion. Additionally, high chargeable anomalies provide evidence for the presence of hidden massive orebodies and disseminated mineralization at depth. The model highlights the paleo-fluid circulation channels in this skarn deposit, bringing new insights into the complex genesis of hydrothermal Fe ores on Elba Island.

This study demonstrates the potential of DERT for fundamental research and mineral exploration purposes and highlights that new technologies may be a game changer for the exploration of ore deposits.

Petrography and Economical Potential of Banded Iron Formations of the Neoproterozoic-Age Buem Formation, Ghana

Ernest Brakohiapa
New Mexico Institute of Mining and Technology, Socorro, NM, USA

Historically, Ghana has relied mostly on gold, bauxite, and manganese as the major contributors to the domestic economy. Further investigation into other minerals to transform the mining sector in Ghana focused on detailed geologic field mapping and sampling. The Buem Formation of eastern Ghana, previously unrecognized as a potential source for ore deposits, was found to host banded iron formations.

The Buem Group comprises a complex series of metasedimentary, volcaniclastic, and volcanic rock units, including poorly defined banded iron formation occurrences. This study reports our initial petrographic observations and economic assessment of the Gyamurume-Wawase Range, one of the host regions for BIFs. Our petrographic studies show that stratigraphic contacts between dominantly steely hematitic ferruginous horizons and siliceous strata are distinctly sharp and abrupt. Most hematite comprises micron-scale, irregular grains interstitial to granular, undulatory-extinction quartz; crosscutting coarse quartz veinlets are barren of iron oxides. Although some BIF strata show former magnetite skeletal textures, even cm-scale ferruginous strata show only trace residual or relic magnetite. Initial studies of the mineralogical composition of Beum Formation BIFs show that steely hematite is abundant, with nominal or no magnetite. No other silicate minerals are observed. Our continued studies will assess the trace element contents of Beum Formation iron minerals, with emphasis on the P and Ti contents of hematite-dominant strata as a means of distinguishing low-contaminant, potentially economic horizons.

A Shift to Natural Gas, Nuclear, and Ocean Mining Could Initiate an Adequate and Resource-Viable Green Transition

Lawrence Cathles[1], Adam C. Simon[2]
1. Cornell University, Ithaca, NY, USA, 2. University of Michigan, Ann Arbor, MI, USA

M. King Hubbert's prediction in 1956 of peak U.S. hydrocarbon production and decline only departed from actuality when an entirely new petroleum resource was developed. The sources of conventional oil at the time were diverse and the demand/reward for new discoveries within the U.S. was strong, but Hubbert's predictions remained accurate until unconventional shale gas sources were brought online in the 2000s. This suggests that the predicted trajectory of metal production from mature methods and districts will not depart from expected trends unless and until entirely new sources are developed. The expected trends are very far from what would be required for the kind of green transition currently discussed. Ocean resources constitute a vast new resource that could change the metal production trajectory on a several-decade timeframe. A green future may depend on starting to develop this resource now. In the meantime, CO_2 emissions can be reduced at close to acceptable rates by transitioning from coal and oil to natural gas and accelerating nuclear power generation. Transitioning to natural gas will reduce metal demand by sustaining fossil-fuel–based transportation and heating. Reliable baseload electricity from nuclear, which has the highest capacity factor and lowest levelized carbon emissions footprint of any energy source, can support and encourage transition to an electrical economy at rates that can be resource-supported. This approach appears viable. Other approaches may also be viable. The resource profession needs to effectively communicate the resource challenge to the public and political class and identify resource-viable pathways to zero carbon.

A Multi-Approach Hyperspectral Analysis for the Mineralogical Characterization of Zn-Cu-Pb Vanadate Ores in the Otavi Mountainland (Namibia)

Francesca Corrado[1], Anna Sorrentino[1], Rita Chirico[1], Matteo Massironi[2], Sabrina Ferrari[2], Nicola Mondillo[1,]

1. Department of Earth, Environment and Resources Science, University of Naples "Federico II", Naples, Italy, 2. University of Padova, Padova, Italy, 3. Natural History Museum, London, United Kingdom

The Otavi Mountainland (OML) is located in the northernmost part of the Damara fold-thrust belt in northern Namibia. The area is characterized by base metals sulfide and nonsulfide deposits hosted by Neoproterozoic platform carbonates belonging to the Otavi Group. The Zn-Cu-Pb vanadate ores (i.e., descloizite, vanadinite, and mottramite; now mostly exhausted) of supergene origin occur in different styles, like veins or cements in karst breccias and cavities. Country rocks (limestones and dolomites), Fe-oxyhydroxides and V-ore minerals have diagnostic spectral absorptions in the visible-near to the shortwave infrared regions, allowing their identification and mapping through hyperspectral methods. However, the spectral response of V-bearing minerals is poorly described in literature so far.

The objective of this study was to characterize the OML vanadium ores in order to better define the mineralization styles, textures, and minerals' mutual relationships as well as expand the knowledge regarding the spectral behavior of V-bearing minerals by adopting a multi-technique approach based on the integration of laboratory-based hyperspectral bulk spectroscopy (through FieldSpec Pro3) and imaging (using HeadwallPhotonics Nano- and Micro-Hyperspectral cameras), both covering the 400- to 2,500-nm spectral range. The data processing was performed by applying feature-extraction band ratios and indexes. XRPD mineralogy was used for validating the results. The analyses allowed us to detect Fe-oxyhydroxides based on the Fe^{3+} absorption feature at 900 nm, to recognize vanadate minerals through the peak at 620 nm and double features at 1,990 and ~2400 nm, and to map and distinguish dolomite from calcite depending on the wavelength position of the CO_3 absorption band at 2,317–2,321 to 2,340 nm, respectively.

The use of multi-source high-resolution hyperspectral data represents a powerful tool for mineral exploration and can also be used for supporting satellite-scale hyperspectral remote sensing for mapping the outcropping host rocks and supergene alteration associated with orebodies at the regional scale.

How to Empower the Geologist's Expert Knowledge with Machine Learning in Drill Core Logging?

Christian Günther[1], Filip Siman[1], Nils Jansson[1], Foteini Simistira-Liwicki[1], Marcus Liwicki[1], Tobias Hermansson[2], Paul McDonnell[2]
1. Luleå University of Technology, Luleå, Sweden, 2. Boliden Mines, Boliden, Sweden

Drill core logging by geologists is essential in mineral exploration, which in turn is essential for securing the substantial amount of mineral resources needed to meet current societal goals, such as the net zero goal. However, when several geologists are involved in the core logging process, subjectivity commonly leads to inconsistent data sets, complicating further analysis and geological modelling. Additionally, core logging is time-consuming and cost-intensive. Machine learning (ML) could be a solution to more objective and efficient drill core logging. Still, current approaches in research are in the beginning and generally specific to their used context given by training and test data sets. Therefore, new innovative and flexible solutions are needed to aid core logging.

We present a solution where we integrate drill core logging with machine learning methods that, due to the complexity of the task, utilise a human-in-the-loop (HITL) approach to transform and empower the geologist's workflow. First, the geologist's visual analysis can be enhanced using ML labelling tools based on computer vision principles, allowing the geologist to label findings on drill core photos directly. The labels are then learned by existing ML models via transfer learning, so that these can predict labels themselves. Next, the geologist relabels incorrect ML labels to improve the ML model, with the potential to create a digital assistant in the analysis. With this approach, several ML models can be combined based on different labels and contexts (from joints to lithologies) for more complex analyses. Furthermore, since the analysis is based on digital drill core photos, a digital collaboration of several geologists is easy. This results again in several ML models that can be combined for better objectivity.

The current work focuses on the first results of the HITL approach using the open-source labelling software CVAT and ML Model Detecron2.

Exploration and Utilization Prospects of Coastal Placer Deposits

Min Liu
China University of Geosciences, Beijing, China

Coastal placer deposits are essential for strategic minerals such as tin, titanium, zirconium, gold, diamonds, and construction sand. More than 30 countries or regions are exploring and mining coastal placer mines and have established relatively mature commercial exploration and development models. Meanwhile, the impact of coastal placer mining on the environment attracts more attention. In terms of exploration, with the development of remote sensing technology, aerial and shipborne geophysical technology, underwater mobile detection technology, marine drilling technology, and big data fusion exploration technology, coastal placer mines have initially possessed the ability of satellite-air-sea-underwater integrated detection. For the mining aspect, large-scale mechanized and automated mining has been realized with the consideration of water and underwater collaborative development. In recent years, the newly developed mining and selection integrated ships have discovered electric, magnetic, and heavy joint selection, with high efficiency and significantly improved placer mining recovery rate. However, the mining activities also bring some unignored impacts on the surrounding environment, such as changes in terrain and landforms and the resulting biological effects. In the future, geophysical exploration methods may become dominant for exploring marine geological structures and energy resources and monitoring the environment, especially the seabed. Satellite, high-resolution aerial multi-parameter (magnetic, gravity, and deep electromagnetic) detection technology, and shipborne and underwater multidimensional (seismic, electromagnetic, magnetic, and gravity) detection technology will achieve high-precision detection and imaging of marine geological structures. Furthermore, intelligent detection platforms such as underwater submersibles adapted to deep-sea exploration will reach uninhabited detection. In terms of development, it is essential to establish low-disturbance and non-destructive mining modes, such as constructing a super-low ecological damage coastal mineral resource in situ development technology system, using high-precision sensors to achieve accurate identification of minerals and establishing onsite filling mining and selection technology and coastal wetland ecological restoration technology.

Determining the Lowest Quantity of Variables Required for Geometallurgical Machine Learning-Based Modelling: A Review

Ethan E. Mack, Bjorn von der Heyden, Margreth Tadie, Tobias Louw
Stellenbosch University, Stellenbosch, South Africa

Geometallurgy involves predicting the metallurgical performance of a mineral deposit, including its extraction and processing characteristics. A crucial component of geometallurgy is the prediction of extracted mineral content such as manganese, being a critical element in battery production. Machine learning has emerged as the leading industry standard for geometallurgical prediction, with random forest being one of the most universally utilised algorithms. However, using all available variables for prediction can lead to overfitting, poor generalisation performance, and increased computational cost. Therefore, determining the minimum quantity of variables required for accurate prediction is important. In this study, random forest machine learning was applied with SHAP (SHapley Additive exPlanations) analysis to identify the most important variables for manganese geometallurgy prediction modelling. SHAP analysis is an innovative technique that provides explanations for individual predictions made by machine learning models. It calculates the contribution of each variable to the prediction outcome, providing a ranking of feature importance. In this study, we used SHAP analysis to determine the minimum number of variables required for accurate prediction. Data was collected from Black Rock manganese mine located in South Africa. A random forest model was trained using all variables, followed by SHAP analysis to determine the feature importance ranking. Based on the SHAP analysis results, influential variables were retained to determine accuracy variance for the secondary model. The secondary model's results indicate that dimensionality reduction using random forest machine learning paired with SHAP analysis produces acceptable accuracy in the modelling of manganese geometallurgical processing while decreasing computational cost. In conclusion, this study highlights the usefulness of SHAP analysis in determining the most important variables for prediction modelling. By using this technique, reduction in the number of variables required for accurate prediction can be achieved, resulting in more efficient and effective manganese geometallurgy prediction models.

Deriving Plate Boundary Information from Deep-Time Palaeo-Reconstruction Models and Implications for Regional Mineral Exploration Targeting

Hoang Anh Tu Nguyen[1], Dène Tarkyth[1], Bruce Eglington[1], Jean-Christophe Wrobel-Daveau[2], Graeme Nicoll[2], David Evans[3], Drew Heasman[1], Samuel Butler[1]
1. University of Saskatchewan, Saskatoon, SK, Canada, 2. Halliburton, 97 Jubilee Avenue, Milton Park, Abingdon, Oxfordshire, United Kingdom, 3. Yale University, New Haven, CT, USA

Recent advances in palaeo-reconstruction models, such as the ability to perform deep-time Precambrian reconstructions, provide more opportunities for applications in regional mineral exploration. All palaeo-reconstruction models have similar foundations: a set of rotation records and a collection of polygon features (geodynamic units or GDUs) that describe their motion. Most Precambrian models tend to be snapshots (rather than kinematically continuous) and comprise a limited number of simplified GDUs without full present-day geographic coverage of the Earth. This precludes direct visualisation of all mappable features and limits data reconstruction and further geoprocessing workflows. More recent models provide interpreted plate boundaries, but other than for the last ~180 m.y. when preserved oceanic crust is available, these boundaries are manually inferred from the kinematics of the continental fragments, a time-consuming and potentially interpreter-biased approach. Models extending further back in time and incorporating thousands rather than tens of polygons need a more automated process to delineate plate boundaries. Geometrically constrained oceanic crust synthesised during this process provides an invaluable check on the geodynamic realism of any model. Coding is performed in Python and makes extensive use of the pyGPlates module and various GIS and scientific packages. Kinematics alone is insufficient for distinguishing upper and lower plate settings, so geochronological and mineral deposit data have been accessed from the DateView and StratDB public-domain databases to help identify geodynamic features like volcanic arcs.

We use two independent models to illustrate the quantitative identification of supercontinent cyclicity (SuperGDU's) back into the Palaeoproterozoic and its association with various styles of mineralisation; the delineation of different styles of magmatism related to SuperGDU break-up, accretion and collision; subduction of synthesised oceanic crust and the impact for mineralisation of predicted transitions from compression to extension in the upper plate in convergent margins.

Unconventional Mineral Resource Potential in a Low-Temperature Geothermal System

Valeria S. Nogales Herrera, Sean A. Fulcher, Karin E. Olson Hoal
Cornell University, Ithaca, NY, USA

New approaches to exploration and discovery of unconventional mineral resources are paramount to securing the materials needed for a green energy transition. One promising economic mineral target is weathered unconformities formed from low-temperature alteration by ancient fluid pathways. This study links unconventional mineral resources with geothermal energy using geomet concepts to aid mineral resource evaluation of a low-temperature geothermal system.

In 2022, Cornell University constructed a geothermal borehole observatory to a depth of 9,790.5' to characterize a potential low-temperature geothermal reservoir in upstate New York. The well intercepted Ordovician to Precambrian units and a ~500-m.y. unconformity between the Cambrian Potsdam sandstone and the Neoproterozoic Grenville basement. Moreover, the nearby Adirondacks host economic deposits of carbonate-hosted Zn, iron oxide-apatite (IOA), wollastonite, garnet, salt, and graphite. The observatory employs engineering techniques to assess the feasibility of geothermal energy, including target identification and underground analysis; however, those approaches lack mineral-specific data. Thus, this geomet work links mineral characterization for exploration under deep cover with geothermal development.

Wireline logs and samples showed unexpected variations in lithology, porosity, bedding style, and fracture density at and below the unconformity, indicating a deeply fractured basement and mineral alteration. Quantitative mineralogy of cuttings sampled every 5' between 9,330' and 9,575' using TIMA-X and hyperspectral scanning define the variability of mineralogy, texture, grain size distribution, and litho-fragment population with depth to identify potential economic indicators around the unconformity. There are notable occurrences of pyrite, chalcopyrite, and magnetite at the level of the unconformity zone, and thin-section analysis and XRD results document chlorite, hematite, phlogopite, muscovite, biotite, and graphite, indicating greenschist facies or hydrothermal alteration in an otherwise amphibolite-grade terrane. Future subsurface data from planned well pairs will have broader implications and applications elsewhere in the northeast US.

Differentiation Between Endoskarn and Exoskarn Using Portable X-Ray Fluorescence (pXRF) at the Antakori Cu-Au-Ag Project, Northern Perú

Luis Paredes Santa Cruz[1], David Portocarrero[2]
1. Universidad Nacional de Ingenieria, Lima, Peru, 2. University of Tasmania, Hobart, TAS, Australia

The Antakori project is located 50 km northwest of the city of Cajamarca, northern Peru.

The geology in the Antakori Project consists of Cretaceous quartzites, calcareous siltstones, and limestones. The Cretaceous basement is intruded by numerous porphyry dikes and stocks that have generated contact metamorphic haloes and metasomatic skarn alteration.

The skarn alteration is highly developed in the central and northern areas of the project, and it is a problem when you want to be precise when differentiating between exoskarn and endoskarn, since in some areas a "mottled" texture is generated, similar to that of a possible intrusive. The use of portable X-Ray Fluorescence (pXRF) has gained significant importance in mineral exploration. This technology has been able to collect geochemical data in a fast, accurate, and cheap way. In this study, (pXRF)-Olympus Delta has been used in geochemical mode on 130 core rock samples (limestone=20, marble=25, hornfels=10, exoskarn=40, unaltered intrusive=13, intrusive with moderate to strong alteration=20, samples with mottled texture [undifferentiated]=12) with three point measurements of 120s in each sample; finally, the three measurements were averaged to obtain a representative result for each sample. The data analysis was carried out with the ioGAS software, where bivariate diagrams of immobile elements were used. In addition, the most effective diagrams were Al vs. (Y, Nb, and Th), respectively. Undifferentiated samples with mottled texture correlated with the exoskarn field. The effectiveness of this method relies on the behavior of the elements used (Al vs. Y, Nb, Th), which during the environment and conditions of skarn formation, behave immobile; therefore, the geochemical signature of each wall rock (intrusive and sedimentary) is preserved and transmitted to its skarn equivalent (endoskarn and exoskarn). Geochemistry using pXRF has proven to be very useful to discriminate endoskarn from exoskarn and represents a powerful exploration tool.

TRIM4Post-Mining: Artificial Intelligence and Virtual Reality for Decision-Making in Remediation for Mining Companies

Luis Pizano[1], Jörg Bendorf[2], Andreas Knobloch[1], Diego Restrepo[2]
1. Beak Consultants GmbH, Freiberg, Germany, 2. TU Bergakademie Freiberg, Freiberg, Germany

On the forefront of current technologies driving the change in geoscientific data management is artificial intelligence. Countless challenges and questions can be approached now in efficient and holistic methods reducing subjective analysis while finding patterns within the data, allowing comprehensive spatiotemporal data analytics, feature extraction, and predictive modelling for the identification of potential hazards, and forecasting dynamics key in decision-making. This is the case in one of the big issues dredging the mining industry: closure and rehabilitation of affected areas, usually a final step ignored by the original inception of a mine. The showcase of artificial intelligence, and specifically machine learning, allows the identification of related information in the form of clusters, for instance, to recognize areas suitable for remediation activity and further planning. Therefore, this approach presents its role in support of decision-making and development during the transition from any commodity exploitation to a revitalized post-mining landscape, enabling alternative infrastructure development, such as for agricultural and industrial utilization, and contributing to recovery of landscapes originally used for mining dumps. Focusing on bringing the results into a more immersive environment and leveraging the derived information, an interactive visualization based on virtual reality and augmented reality technology permits the exploration of alternative and temporal scenarios with up-to-date data and models.

Quantitative Evidence of a Rift Origin for the Katangan Basin from Subsidence Analysis in the Zambian Copperbelt

Martin Purkiss, Michael C. Daly
University of Oxford, Oxford, United Kingdom

We present a quantified analysis of the basin-forming mechanism of the Katangan copper basin of Central Africa. A series of local "type logs" was constructed through detailed analysis of deep cores, which have been interpreted and form the basis of a model for basin extension.

Over 10 km of diamond drill core was logged from eight drill holes across the basin. A facies scheme was established to generate palaeobathymetric estimates. Local type logs were generated and decompacted. 1D backstripping was then performed assuming Airy isostasy. The McKenzie uniform stretching model was used to estimate stretching (β) factors that best fit the tectonic subsidence profiles generated from backstripping.

A clear trend is observed of tectonic subsidence increasing from the basin margin to the west and southwest, with indications of two distinctive early rift phases in the Tonian. The deepest parts of the basin experienced >50% crustal stretching, likely accommodating a syn-rift section in excess of 2 km and a full stratigraphic succession in excess of 8 km in thickness prior to basin inversion. The tectonic subsidence profiles of the backstripped cores and their good agreement with a uniform extension model are clear evidence of a rift origin for the Katangan Basin in Zambia. β-factors of 1.1 to 1.5 match those of known rift systems.

A rift origin was previously inferred based on regional geochemistry and lithostratigraphy. Our results support this, with evidence of lithospheric-scale stretching increasing from the east to a maximum along a deep Central Rift to the south of the Domes region. Basin inversion and high-grade metamorphism represent a data boundary to subsidence analysis. Establishing this quantified tectonic framework for basin subsidence has implications for stratigraphic architecture, early basin heat flow, development of hydrothermal systems, and ultimately control of the early mineral systems of the basin.

Maximizing the Value of Geoscience Databases: A Comprehensive Review of Advanced Data Processing Techniques

Mana Rahimi
Mineral Deposit Research Unit (MDRU) - The University of British Columbia (UBC), Vancouver, BC, Canada

As a part of recent mineral exploration, the mining and exploration industry is investing considerable money and time in having powerful exploration databases. While the availability of such data is undoubtedly a significant advancement for mineral exploration, the question remains: does everyone utilize the data to its full potential?

The value of data is not just based on the large volumes of data in the exploration database; the actual value of the database comes from the objective information that is extracted based on the insight and knowledge of data processing, analysis, and interpretation.

Exploration databases are expected to support exploration success through better decision-making, efficient exploration targeting, and enhanced productivity while reducing risks, times, and costs. They can streamline exploration workflows, ensuring that exploration efforts are targeted in the most promising areas.

This paper aims to provide an overview of the different geoscience data processing methods and discuss the advantages and limitations of each method. Three main types of data processing methods are used in mineral exploration: knowledge-driven methods rely on expert knowledge, data-driven methods use machine learning algorithms to analyze data sets and identify patterns and relationships that are not apparent to human analysts, and hybrid methods incorporate expert knowledge into the machine learning algorithms. With geoscience data growth, artificial intelligence (AI) and data mining techniques are used to automate data processing, feature detection, and modeling. These methods are beneficial for processing robust volumes of geoscience data quickly and efficiently. However, expert knowledge in all stages is crucial for accurate and relevant results.

This paper will discuss the methods and techniques used in mineral exploration through different case studies to provide mineral exploration professionals with a better understanding of the available tools and techniques.

A Novel Approach for Delineation and Mapping of K-Rich Igneous Rocks and Their K2O Content to Support Unconventional Potash Exploration

Otmane Raji[1], Muhammad Ouabid[1], Cheikh Elwali Malainine[1], Youssef Ahechach[1], Aaron Herve Mbissik[1], Hicham El Messbahi[2], Jean-Louis Bodinier[1]
1. Mohammed VI Polytechnic University, Ben Guerir, Morocco, 2. Polydisciplinary Faculty , Sidi Mohamed Ben Abdallah University, Taza, Morocco

Potassium(K)-rich igneous rocks encompass a range of alkaline and silica-undersaturated rocks, including syenites and their volcanic equivalents (trachytes). These silicate igneous rocks are characterized by high K_2O concentrations and have been the subject of numerous investigations, either for their geodynamic implications or for their association with critical mineral resources. More recently, researchers have dedicated further studies to assessing their potential as unconventional sources of potash for agricultural K fertilization. Their geological exploration has become a priority in many countries, particularly those with high agricultural productivity and dependence on imported potash fertilizers. In this study, we propose a new approach integrating remote sensing, GIS, and field data to efficiently identify K-rich igneous rocks over large regions and map the K_2O content with high spatial resolution within the selected potential areas. Our methodology involves the use of ASTER thermal data to rapidly delineate potential areas and select the most suitable for fieldwork and optimized sampling. To obtain a high-resolution geochemical map of K_2O content, we used a deep neural network method assisted by 10-m spatial resolution Sentinel VNIR data to interpolate a limited number of geochemical sample data. The proposed approach was successfully applied over a large-scale area of the Moroccan High-Atlas Mountains, where K-rich alkaline intrusions have been reported. It has proven to be relatively rapid and cost-effective. The results have allowed for the identification of the famous Tamazeght geological alkaline complex, which spans an area of 100 km² within an initial exploration area of more than 40,000 km² and contains alkaline igneous rocks of diverse natures. The obtained K_2O geochemical map for this geological massif—well-known to host the K-rich syenites—has shown a good coefficient of determination value of over 60%.

Automated Multi-Sensor Core Logging and Analysis for the Assessment of Sn-W Mineralization Pattern and Distribution – Hemerdon Deposit, Devon, UK

Reza Ramdhan[1], Gavin Hunt[2], James Shreeve[1], Briony Shreeve[1]
1. Geotek Ltd, Daventry, United Kingdom, 2. Spectra-Map Ltd, Daventry, United Kingdom

The Hemerdon deposit, located in Devon, southwest England, is the world's third largest tin-tungsten deposit, boasting an estimated resource of 347.7 Mt at 0.12% WO_3 and 0.03% Sn. The deposit is hosted in a sub-vertical granite dike that strikes NNE-SSW and consists of a series of steeply dipping veins that contain cassiterite and wolframite. The deposit displays complex geology and metallurgical characteristics due to its multi-phase alteration overprinting, ore-texture variations, and gangue mineral contamination. This study aims to re-evaluate the mineralization and alteration patterns and their relationship to the geological structures by using an integrated non-destructive core analysis. This approach includes a portable XRF, an ASD spectrometer, magnetic susceptibility meters, high-resolution linescan imaging, and a 3D laser profiler. The data from these analyses are then fed into machine learning algorithms that use K-Means clustering to generate litho-geochemical and alteration clusters. We also utilized a U-Net Convolutional Neural Network (CNN) algorithm to map the structural patterns and distribution of key ore minerals, using the linescan and 3D laser images as inputs. To test our methods, we examined three main drill holes with a total length of 760 m. Our analysis revealed a distinct pattern of alteration and mineralization, beginning with partly oxidized granite with low-moderate arsenopyrite alteration (low Ca, P, and Ni) at the top of the holes, followed by highly kaolinized granite with high arsenopyrite, and then the greisen with intense muscovite alteration (high P, Ca, Ni, Zn) at the bottom of the holes. We identified three primary structural trends, including the NNE-SSW steep-angle sub-parallel tourmaline veins, the NE-SW steeply dipping quartz and feldspar veins, and the shallow-angled E-W fractures. This study represents a significant step forward in our understanding of the deposit's characteristics and provides insights for future exploration and development efforts.

Prospects for Byproduct Critical Mineral Recovery from Mine Waste

Robert R. Seal, Nadine M. Piatak, Sarah Jane O. White, Sarah M. Hayes
US Geological Survey, Reston, VA, USA

Many mineral commodities identified as critical for maintaining supply chains occur as byproducts that typically end up in waste streams. Additional processing at both active and inactive mines could facilitate byproduct recovery. Some byproducts have strong associations with primary commodities such as Ge in sphalerite, whereas as others, such as Te in porphyry Cu ores, have variable deportment among ore-processing outputs. At inactive mines, critical minerals in wastes will vary based on deposit type and their deportment during ore processing. Depending on the age of the waste material, its geological characteristics, and climatic setting, weathering can influence reprocessing strategies for critical mineral recovery. For example, in the Tristate district (Oklahoma), weathering of sphalerite in mine waste results in the formation of the secondary Zn silicate hemimorphite, which sequesters higher amounts of Ge than the original sphalerite. Thus, any reprocessing strategy should consider the importance of hemimorphite, otherwise it will miss a substantial portion of the Ge.

Active mines with long life spans make return on investment for specialized circuits more feasible. For both active and inactive mines, the identification of nontraditional value may be needed for ultimate viability. For example, pyrite concentrates produced during ore processing (or reprocessing) can be leveraged to reduce long-term environmental management costs. In porphyry Cu deposits, pyrite can carry significant Au grades as either native Au or telluride inclusions. Oxidation of pyrite concentrates for Au recovery could also solubilize critical elements such as Te, Ni, or Co, facilitating their recovery. Other examples are gangue minerals, like olivine, pyroxene, and plagioclase in mafic/ultramafic Cu-Ni-Co-PGM deposits, which represent important targets for mineral carbonation—a source of carbon credits. Thus, the most tenable prospects to acquire critical minerals from mine waste will likely combine byproduct critical mineral recovery, additional primary commodity recovery, and improved environmental management.

Environmentally Friendly Low-Impact, Low-Carbon Footprint, Low-Power Electromagnetic Technique for Mineral Exploration Undercover

Gordon Stove
Adrok Ltd, Edinburgh, United Kingdom

Geophysicists in the 21st century now accept their commitment to being a sustainable and responsible surveyor of the Earth's subsurface. Reducing our carbon footprint on our route to Net Zero in the face of the challenges presented by the global pandemic and the subsequent return to work, fieldwork, and international air travel is not an easy task. Whatever our geophysicist tool of choice, we can take some basic steps to help, such as the following:

1. Protect the environment by lessening our impact through monitoring and reducing our carbon emissions;

2. Raise awareness without our own organisation of things that colleagues can do to reduce their personal carbon emissions as well as reducing carbon emissions within our organisations;

3. Engage colleagues in environmental issues and initiatives;

4. Identify environmental causes that we can support; and

5. Publicly hold ourselves accountable for our carbon footprint by publishing our carbon impact on our websites and public domain media channels.

Above all, we can develop and use our geophysics tools and technology in an environmentally friendly way. Adrok have been developing a low-carbon, low-impact geophysics offering for the past two decades. Over the past five years, we have been letting our clients know what our carbon footprint levels are for each geoscientific survey. Multiple surveys have been performed in Northeast England for geothermal and lithium brine exploration using novel electromagnetic (EM) technologies. This paper will explore this survey in more detail through a geophysical lens and an environmental lens.

Geometallurgy of a Complex Ore: Lappberget Zn-Pb-Ag-(Cu-Au) Deposit, Garpenberg Mine, Sweden

Glacialle Tiu[1], Christina Wanhainen[1], Nils Jansson[1], Yousef Ghorbani[1, 2]
1. Luleå University of Technology, Luleå, Sweden, 2. University of Lincoln, Lincoln, United Kingdom

Complex ores, which contain three or more metals along with additional precious and penalty elements, pose significant challenges for successful production forecasts due to the large number of variables that must be accounted for during the multi-stage separation process. In this study, geometallurgical characterization was conducted to understand the variability in the process performance of a complex and variable ore and to quantify the impact of geology on the metallurgical response and metal recovery processes. The approach is divided into five stages: (1) geological characterization, (2) ore classification, (3) grinding and flotation tests, (4) development of prediction models, and (5) geometallurgical domaining. Qualitative and quantitative mineralogical and textural investigations were performed on geological samples and flotation products using various analytical techniques, including scanning electron microscopy (SEM), automated mineralogy (i.e., QEMSCAN), electron microprobe analysis (EPMA), and laser ablation-inductively coupled plasma-mass spectrometry (LA-ICP-MS).

The study revealed the mineralogical, textural, and chemical variations of the sulfides within the deposit, reflecting the complex ore-forming conditions and tectono-metamorphic processes that have affected the deposit. Different generations of sulfide minerals (i.e., sphalerite, pyrite, galena, and chalcopyrite) were identified based on their distinct mineral chemistry, trace-element signatures, texture, and host rock associations. The study identified seven geometallurgical domains that were characterized by their specific mineralogical, chemical, and textural features, which were found to have a significant impact on the grinding and flotation performance. Mineral deportment of precious and penalty elements (Ag, Au, As, Bi, Hg, Mn, Mg) and potential by-products (Ga, Ge, In, Sb, Cd) were also investigated. A comprehensive and systematic approach for characterization was proposed to optimize data collection from both geological and process performance perspectives, towards a smart and future-oriented mining value chain.

Can Gold Mining Provide Critical Metals as By-Products Using Novel Solutions? A Case Study at Björkdal Gold Mine, Sweden

Lara Tritton[1], Gawen Jenkin[1], Dan Smith[1], Helena Moosberg-Bustnes[2]
1. University of Leicester, Leicester, United Kingdom, 2. Mandalay Resources, Skelleftea, Sweden

The transition away from hydrocarbons requires a huge increase in metal and mineral consumption for low-carbon technologies. However, mining and minerals processing is energy intensive and can be environmentally harmful, thus there is a need for novel solutions in extraction and processing. Te, W, and Bi are designated critical metals by both the BGS and USGS, and gold mining is coming under increased scrutiny due to its environmental footprint. Therefore, a method of recovering critical metals from existing gold mines would be ideal for meeting some of the demand for these metals without opening any new mines.

Gold deposits are often enriched in other rare metals; for example, the Björkdal mine produces gold concentrates and tailings variably rich in Te-, Bi-, and W-bearing minerals. Currently, there is no incentive to optimise for by-product recovery and few options by which to process and extract these additional metals for the mine operator. Deep eutectic solvents, which operate at low temperatures and are environmentally benign, are a promising development of recent years as an organic chemistry alternative to conventional metallurgy. Prior work has shown success with leaching of select metals and minerals including bismuth tellurides and gold, and there is some evidence that W-bearing minerals could also be amenable.

This work looks at tracing the flows of Te, Bi, and W through the Björkdal processing plant and mineralogically characterising feed, tailings, and concentrates. Bulk geochemistry of ore, concentrates, and tailings is compared to automated SEM analysis. Target metals deport almost entirely into bismuth telluride minerals (Bi and Te) and scheelite (W), which are well liberated in all concentrates and tailings. Characterisation enables assessment of the feasibility and impacts of inserting leaching stages with DESs, and/or altering the processing procedures, in order to maximise by-product recovery in an environmentally friendly and low-energy process.

The Green Transition Resource and Supply Chain Model: A Recipe for Global Energy Equity

Edith Newton Wilson
Rock Whisperer LLC, Tulsa, OK, USA

We live on the cusp of a profound transformation. Recent advances in technology have enabled more efficient harvesting of energy from primary sources, the sun's radiation and the Earth's heat, while bypassing the storage of energy in hydrocarbon molecules. For over a century, being an energy provider meant finding, extracting, refining, transporting, and using up vast quantities of a finite resource. The business was high risk, high return, and required deep pockets throughout a long investment cycle, which included the buildout of gathering and distribution infrastructure. In contrast, the distributed energy resources (DERs) available today are light on capital investment, low in risk, and lightning fast to deploy. Providing energy is changing from a commodity-based system in which value is sequestered among a few to a business that is focused on delivering the best service at the lowest cost to the most people. The resource model for converting our energy system to electricity generation and battery storage depends upon moderate quantities of recyclable metals rather than ever-increasing volumes of combustible hydrocarbons. Supply chains that support DERs can be localized and downscaled so that consumers have ownership and participate in the system. Metal ores, while limited in their global occurrence, can be readily transported to regional processing, manufacturing, installation, and recycling centers where value is created, and jobs are added. At the mine site, sustainable practices and secondary production reduce the resource footprint further. Like communication and banking, renewable energy is predicted to leapfrog traditional infrastructure to replace antiquated grids and pipelines with energy independence at the community scale. United Nations Sustainable Development Goal 7—ensuring access to reliable, affordable, sustainable, and modern energy for all—is within our grasp and offers potential for a dramatic improvement in global lifestyle driven by energy equity.

Geological Settings and Future Ore Potential of the Ordubad Region, Nakhchivan (Naxçıvan), Azerbaijan, Lesser Caucasus

Stephen J. Westhead[1], Robin N. Armstrong[2], Anar Valiyev[1], Rustam Abdullayev[1], Araz Orujov[1]
1. Azerbaijan International Mining Company (subsidiary of Anglo Asian Mining), Gadabay, Azerbaijan, 2. Natural History Museum, London, United Kingdom

The Ordubad district is located in southeast Nakhchivan and constitutes the Azerbaijan sector of the Zangezur-Ordubad metallogenic belt of the southernmost part of the Lesser Caucasus. The Ordubad contract area (OCA) is currently being explored for porphyry-style mineralisation by the Azerbaijan International Mining Company. The region contains several examples of porphyry- and epithermal-style mineralisation including the world-class Kadjaran deposit and the Agrak, Sungun, and Masjed Daghi porphyry-style deposits. The igneous rocks of the OCA were formed during the transition from Early Paleogene subduction to an Oligo-Miocene stage of collision to post-collision tectonics during the convergence of the African-Arabian and Eurasian plates. Previous work has demonstrated that these magmatic suites display strong "porphyry-fertile" signatures. The OCA is cut by the SE-NW Ordubad, Keleki, and Pazmara fault zones. East of the Ordubad fault constitutes the southwest part of the Mehri-Ordubad pluton (MOP) and hosts the porphyry Cu-Mo-Au centres of Misdag, Agyurt, Shalala, and Diakchay. The mineralisation at these centres presents as stockworks in NE-trending structural zones. The Pazmara fault juxtaposes a sequence of downthrown Eocene andesitic volcanic rocks against the Oligocene MOP. These Eocene units host the sub-epithermal Cu-Au vein systems of Shakardara and Piyazbashi, which are associated with extensive chlorite-epidote-carbonate alteration. Within the Eocene volcanic units west of the Keleki fault, an extensive area (12 km^2) of intermediate argillic to advanced argillic alteration is observed, which may be interpreted as a potential lithocap. This alteration exploits the more permeable pyroclastic/epiclastic volcanic units. At the margins of this alteration are located the Dyrnis malacite-dominated Cu-Ag occurrences, which are interpreted as supergene "washouts." The interplay of regional and local structures with favourable magmatic chemistry make the OCA a favourable district for both further discoveries and a natural laboratory for the study of the wider porphyry system model.

Trace Element Composition and Cathodoluminescence of Quartz in the Tsunheg Tungsten Deposit in Western Mongolia

Amitan Baatar[1], Jamsran Erdenebayar[2], Takeyuki Ogata[2], Takeyuki Ogata[2], Yondon Majigsuren[1]
1. MUST, Ulaanbaatar, Mongolia, 2. MiReSLab, Akita, Japan, 3. Mongolian University of Science and Technology, Ulaanbaatar, Mongolia

The Tsunkheg tungsten deposit is located in Nogoonnuur soum of Bayan-Ulgii province about 1,600 km from Ulaanbaatar, Mongolia. The Tsunkheg deposit is tectonically located in the Mongol Altai and Khovd deep fault zones and is characterized by the development of numerous faults to the northwest and northeast.
The Tsunkheg deposit consists of six large quartz-wolframite veins, 0.15 to 0.45 m wide, 400 to 600 m long, and one stock body with a diameter of 80 m in the central part. The bodies are mainly almost vertical and continuous to the 300-m depth. The quartz-tungsten veins contain scheelite, sulfide minerals: chalcopyrite, bornite, arsenopyrite, and molybdenite tetrahedrite; and carbonate minerals: siderite, malachite, azurite, fluorite, and beryllium. The tetrahedron has high silver content. The average contents of wolframite in the quartz veins are 2.5 to 5%, and in the stock body 0.24%. Cathodoluminescent (CL) texture and trace element signatures of quartz from the Thunkheg deposit have been investigated to discuss their genetic significance. Results show that the pre-ore quartz (QI) displays a bright and homogeneous luminescence texture, (QII) generally shows a bright and oscillatory luminescent texture, and the syn-ore quartz (QIII) generally has a dark and homogeneous (or slightly mottled) luminescent texture that may result from annealing of original CL textures; the post-ore quartz. EPMA analyses suggest Ti, Al, K, Fe, and Ca are the most abundant elements in various generations of quartz and mainly occur as solid solutions within the crystal lattice, in which Al can be incorporated into quartz by substituting Si in the crystal lattice, with additional cations (e.g., K, Na, Li, and Ca) to keep the charge balance. The low Ti concentration and oscillatory euhedral growth zones of CL textures suggest that syn-ore quartz (QII) may have precipitated from a low-temperature hydrothermal fluid.

An Insight into the Southern Extension of the Egyptian Golden Triangle and its Potentiality for Mineral Occurrences Related to Shear Activity

Mohamed Badawi
Miskolc Egyetem, Miskolc, Hungary

The Mubarak-Barramiya shear belt is a post-accretion nappe assemblage that extends in ENE-WSW direction alongside the southern boundary of the Egyptian Golden Triangle (EGT) that was announced in 2017 as an economic zone, amounting to 75% of reserves of Egypt's mineral resources—particularly of iron, gold, copper, and phosphate. This belt, which crosscut the main fabric in the Central Eastern desert, has been extensively intruded by multiphase magmatism, from syn-orogenic to post-orogenic granitoids, that is aligned with the NW-SE Um Nar shear zone and its prolonged shear activity. This regime, particularly sheared contacts of granitoids, has the potential to enrich precious base metals, i.e., orogenic gold occurrences and rare earth elements in the granitic pluton, or trigger hydrothermal mineral alteration to precipitate along quartz veins developed related to the master shear array. These veins form along extensional fractures as a result of post-amalgamation transtensional (i.e., the master shear array and related Riedel shears and tensional fractures) shearing, as a transition from transpressional (i.e., the master shear array and related P-shears, S-C fabric, and asymmetric folds) terrane accretion in response to Um Nar's shearing activity. Consequently, we applied an integrated approach to predict the potential mineral alteration localities that includes integrating detailed fieldwork (structural measurements and petrographic-oriented samples), remote sensing, and aeromagnetic data. Several data processing techniques, including false color composites (FCCs), band ratios (BRs), principal component analysis (PCA), minimum noise fraction (MNF), and decorrelation stretch (DS) are applied to ASTER, Sentinel 2, ALOS PRISM, and orientation entropy heat map (CET). The structural network of this fracture system greatly enhances the permeability within these Neoproterozoic rocks, allowing the precipitation of various potential mineral deposits as quartz veins due to the passing of hydrothermal fluids. These outcomes may be significant to broaden exploration further to the south of EGT.

Using Chemistry of REE-Bearing Phosphates for ISCG Exploration

Travis D. Batch[1, 2], Caroline Tiddy[1, 2], Adrienne Brotodewo[1, 2], David Giles[1, 2]
1. Future Industries Institute, University of South Australia, Mawson Lakes, SA, Australia, 2. Mineral Exploration Cooperative Research Centre, University of South Australia, Mawson Lakes, SA, Australia

The challenge of cover has forced exploration into deeper search spaces, thereby limiting the successful discovery of new ore deposits. This difficulty is reflected in decreasing rates of successful deposit discovery, which is impacting the global supply of minerals. Among these impacted minerals are critical minerals, which include copper and the rare earth elements (REEs). Copper and REEs are essential in the construction of renewable energy technologies and clean energy solutions. To secure supply of these minerals, more deposits that contain critical minerals need to be discovered. New techniques are being developed to be added to the exploration toolkit to aid exploration for such deposits. Among these developing techniques are geochemical targeting tools that use mineral chemistry to understand proximity to ore deposits.

Here we present the results from an investigation into using REE-bearing phosphate minerals (monazite, rhabdophane) in exploration for iron sulfide-copper-gold (ISCG) deposits. The Jericho and Kulthor ISCG deposits in the Cloncurry District, Queensland, Australia are used as case study areas. At Kulthor, hydrothermal REE-bearing phosphates associated with mineralisation are characterised by elevated light REE (except Ce) and S concentrations and depleted Th concentrations compared to samples with no association with mineralisation. Hydrothermal samples associated with mineralisation at Jericho also preserve elevated S and depleted Th, with the addition of elevated Ca compared to metamorphic monazite. The hydrothermal phosphates at Jericho yield low analytical totals (93–97%), which is attributed to the phase being rhabdophane, a low-temperature mineral that has a similar chemistry to monazite but contains structurally bound water. Rhabdophane is not apparent at Kulthor. Overall, the REE-bearing phosphate minerals show distinctive chemistries depending on their proximity to mineralisation, indicating their potential to be used as an indicator for proximity to mineralisation.

Assessing the Petrogenesis and Magmatic Sulfide Prospectivity of the Southwestern Laurentia Large Igneous Province (SWLLIP)

Thomas J. Boes, Simon M. Jowitt
UNLV, Las Vegas, NV, USA

The ~1.1 Ga magmatism of the Southwestern Laurentia Large Igneous Province (SWLLIP) generated a series of mafic-ultramafic dikes and sills emplaced within Mesoproterozoic sedimentary units and older crystalline basement rock throughout the southwest United States and northern Mexico. This LIP event remains enigmatic in terms of identifying the processes that generated this magmatism and any potential links to the contemporaneous magmatism of North America's mineralized Mid-Continent Rift LIP to the northwest of the SWLLIP. The contemporaneous 1.1 Ga Mid-Continent Rift LIP event hosts economically viable magmatic sulfide deposits in the Duluth Complex in Minnesota and elsewhere. This study aims to further our knowledge of the petrogenesis of the SWLLIP and the magmatic sulfide potential of this magmatic event.

Whole-rock lithogeochemical and Pt, Pd, and Au data have been obtained for 52 SWLLIP samples from California, Arizona, and New Mexico, allowing an initial assessment of the petrogenesis, magmatic sulfide fertility, sulfide saturation status, and crustal contamination of magmas of the SWLLIP. All these factors are critical aspects of assessing the potential for this LIP to host magmatic Ni-Cu-platinum group element (PGE) sulfide mineralization. Two suites have been identified within the SWLLIP, a potentially prospective mainly tholeiitic suite, and a seemingly unprospective alkaline suite, with the petrogenesis and magmatic sulfide potential currently being assessed in more detail.

The ongoing interpretation of these data will not only enable the determination of the petrogenesis and magmatic sulfide potential of this event but will place this LIP into the tectonic and global LIP record, potentially aiding in paleocontinental reconstruction. These data will also allow the further testing of the hypothesis that magmatism may be derived from a mantle plume that formed the magmatism of the Mid-Continent Rift of North America, as mentioned above.

Enhancing Exploration of Green Energy Transition Metals: Directly Dating Carbonate Vein-Hosted Cu-Au Mineralization

Sarah Bowie[1], Catherine Mottram[1], Dawn Kellett[2], James Darling[1]
1. University of Portsmouth, Portsmouth, United Kingdom, 2. Geological Survey of Canada, Halifax, NS, Canada

Developing new mineral deposits hosting copper, gold, and other base metals is of critical importance as countries transition to low-carbon energy technologies. Understanding when and why these metals become enriched in geological settings is central to locating and modelling mineralized systems and reducing the socio-environmental footprint of exploration efforts. In natural settings, many economic metals concentrate through magmatic-hydrothermal fluid processes. During collisional mountain-building events, faults and shear zones are known to act as planes of weakness facilitating the migration of magma and associated fluids up through the crust. Evidence of fluid flow is often preserved in the form of mineralized quartz-carbonate veins.

Establishing temporal controls on low-temperature vein formation has been difficult in the past, as classical isotopic dating methods are typically applied to high-temperature minerals such as zircon. Through recent developments in the application of U-Pb dating to carbonate minerals, it is now possible to date low-temperature fluid flow events in ore deposits for the first time.

This project is focused on ore deposits associated with faults and shear zones in the Canadian Cordillera, using the Llewellyn Fault in northern British Columbia as a case study. In this setting, the accretion of crustal fragments and volcanic arcs onto the western margin of North America over more than 200 million years has resulted in an array of mineralization styles, including Cu-Au porphyry, epithermal, carbonate replacement, and skarn.

Here we present preliminary results that combine field-based mapping of mineralized carbonate veins and breccias, novel geochemical and imaging techniques, and in situ U-Pb carbonate geochronology. The results provide timing constraints for mineralization and faulting, significantly improving our understanding of the timing, rates, and duration of fluid flow and metal (re)distribution in case-study mineral deposits. Ongoing work aims to link timing(s) of mineralization to the broader tectonic and metallogenic framework of this region.

The Santa Maria Pb-Zn-Cu-(Au, Ag) Deposit, Camaquã Basin, Brazil: Geology, Geochemistry and Geothermometry

Lucas F. Brito, Caetano Juliani
University of São Paulo, São Paulo, SP, Brazil

The Santa Maria Pb-Zn-Cu-(Au, Ag) deposit is hosted in the sedimentary clastic sequences of the Ediacaran Santa Bárbara Group, composed mainly of siltstones and arcosean sandstones crosscut by amygdaloidal andesites dikes. The mineralization is predominantly stratabound down to a depth of 200 m. The main ore is composed of sphalerite and galena with minor silver, chalcopyrite, bornite, and chalcocite in veins, veinlets, and lateral dissemination in sedimentary rocks. The Pb-Zn-(Ag) orebodies are enveloped by zones of intense pervasive leaching and bleaching of the sedimentary rocks with illite/sericite alteration, few pyrites, and local silicification. Below these zones, chalcopyrite and bornite occur associated with low-grade gold mineralization controlled by faults. Chalcocite prevails at the shallower levels. These copper ore minerals compose veinlet fracture infill and brecciated veins with quartz and carbonate. Sulfide-rich veins sometimes have platy calcite, indicating ore precipitation likely related to the boiling process. This deeper mineralization zone is associated with chlorite and sericite alteration halos. The sericitic alteration in this zone is more intense and has subordinated hydrothermal light green biotite. Electron microprobe analyses (EMPA) were conducted in hydrothermal chlorite and micas. Chlorite revealed composition varying between chamosite-clinochlore with Aliv from 1.2 to 2.5 a.p.f.u. and Mg/(Mg+Fe) ratio between 0.32 to 0.69, resulting in calculated temperatures from 208 to 417°C. Ripidolite commonly occurs closer to the ore. The white mica is usually fine-grained muscovite in all the samples, with Aliv between 2 and 3 and Si ranging from 3 to 4 a.p.f.u. The biotite has a siderophyllite to eastonite composition. The chlorite geothermometry indicates higher temperatures in the copper ore zones relative to the shallower sphalerite and galena mineralization zone. The deposit geometry, ore mineralogy, and hydrothermal zoning suggest a magmatic-hydrothermal genesis, with epithermal intermediate Zn-Pb-Ag mineralization grading to porphyry-like distal Cu-Au veinlets at depth.

Hydrothermal Alteration Zones Estimated by ASTER Satellite Image and ASD Terraspec on the Eastern Zone of Cerro de Pasco, Perú

Evelyn Caiza
CENTRAL UNIVERSITY OF ECUADOR, Quito, Ecuador

The research area is located in the Pasco Region of Peru, covering an area of 2,570 km² and geologically on the eastern zone of the polymetallic deposits associated with a Miocene volcanic diatremes belt. This investigation aims to target new hydrothermal alteration zones by interpretation of an ASTER satellite image.

The geological, tectonic, and structural setting of the studied area creates a suitable environment for formation of hydrothermal systems. With the presence of a mountain system with low vegetation cover, it is suitable to use an ASTER VNIR-SWIR multispectral image. The analysis and procedures on the satellite image were compared with SWIR spectral signatures obtained from the ASD TerraSpec in the mine sites of the investigated area; this allowed us to define areas of enhanced lithological and alteration signatures defining new exploration zones.

Four spectral techniques were used to identify hydrothermal alteration zones. The first technique involved mapping clay minerals and iron oxides using the combination of RGB bands. The second technique used band ratios to determine mineral alteration associations. The third technique used SWIR spectral indexes to corroborate the spectral results. Finally, the last technique was spectral mapping using the Spectral Angle Mapper (SAM) method, which needs a special procedure because it uses both field data obtained from hand samples using the ASD TerraSpec and spectral signatures from ASTER image.

The application of this last technique permitted us to corroborate the mineral alteration zones obtained with previous methods and also to refine the hydrothermal alteration halos based on mineralogical associations characteristic of mineral deposits. Taking into consideration the metallogenetic, geological, and structural environment, ten new exploration zones were defined: seven for high-sulfidation polymetallic systems with advanced argillic and intermediate argillic alteration, and three for porphyry systems with phyllic, intermediate argillic, advanced argillic, and propylitic alteration.

The Al:Ti Geochemical Classifier: Distinguishing Between Granites of Similar Characteristics Under Different Degrees of Alteration

Andrés Canales[1], Diego Guido[2], Julietta Zamora[1]

1. Facultad de Ingeniería, Universidad del Desarrollo, Santiago, Chile, 2. Astra Exploration, Vancouver, BC, Canada

The Pampa Paciencia vein system (PPVS) makes up the central zone of the Pampa Paciencia deposit (56-64Ma), located in the Paleocene mineral belt in northern Chile. In this zone, veins are hosted by the Sierra Gorda intrusive complex (Upper Cretaceous) and smaller Paleocene intrusives. Two units are identified as the PPVS' hosts. The microgranular quartz-feldspathic granite (GMQF) and the microgranular feldspathic biotitic granite (GMFB).

Similar mineralogy, texture affected by metasomatism, and the different degrees of alteration affecting the granites makes separating these two units quite a challenge.

To solve this, whole-rock geochemistry assay data (ICP-AES) from drill holes was used to geochemically characterize the hosts and alterations. To follow, representative samples were studied under hand sample/thin section and sampled by pXRF (averaging five measurements on flat surfaces) to relate their visual characteristics to their geochemistry and to test the diagrams created with the assay data.

First, feldspar Na-K GER diagrams revealed the hosts were mainly under two alterations (adularia and sericite, ±intermediate argillic). Al:Ti diagrams were created from the assays, with samples coloured by lithology. This showed two clusters, one richer in Ti dominated by GMFB and Al by GMQF, and samples were re-classified by them. After re-classification, the GMFB group had a linear regression with a slope of 13; the GMQF had a slope of 26. From thin section petrography, GMFB classified as a pyroxene bearing granodiorite-monzogranite (±biotite), and GMQF classified as a biotite-bearing syenogranite-monzogranite. The four samples classified petrographically as GMQF plotted (0.1-0.22% Ti; 5.5-6.5% Al) in the GMQF field in the Al:Ti diagram, while the 10 GMFB samples also did so in the GMFB field (0.2-0.35% Ti; 5.5-9% Al).

The Al:Ti plot is useful to separate similar granites with different degrees of alteration, not only with standard assay data, but also with lower cost alternatives like pXRF.

Study on Sedimentary Environments and Implications for Shale Reservoirs in the Permian Longtan Formation, Southeast Sichuan Basin

Yana Chen[1, 2]

1. PetroChina Research Institute of Petroleum Exploration and Development, Beijing, China, 2. PetroChina Hangzhou Geological Research Institute, Hangzhou, China

Marine-continental transitional shale is one of the most promising targets for shale gas exploration in the Lower Yangtze region. To investigate the sedimentary environments and the enrichment regularity of Longtan shale, multiple techniques, including core and thin section observations, geochemical and elemental analyses, X-ray diffraction, scanning electron microscopy (SEM) and low-pressure nitrogen adsorption (LPNA), were used to analyze the sedimentology, mineralogy, and pore structure of the Longtan shale. Core descriptions and thin section observations show Longtan shale was deposited in marine-delta transitional environments, including delta-front, shore swamp, mixed tidal flat, and shallow shelf facies. The Sr/Cu, V/Cr, CIA, EF (Mo), EF (U) and other major and trace elements results indicated warm and moist climates and water in reducing conditions in the Longtan period. Both the climate and water conditions are favorable for organic matter production and preservation. Additionally, the paleoenvironment and depositional model for the Longtan Formation was reconstructed. Geochemical results show that Longtan shale is in over-matured stage (Ro ranging from 2.4–3.57%) with II2-III kerogen type. The shale from shore swamp facies exhibits a high content of total organic carbon (TOC; avg 5.76%). The organic geochemistry of Longtan shale is mainly determined by sedimentary facies. The pore system of Longtan shale consists of inorganic pores with a small number of organic pores and microfractures. In the shore swamp and shallow shelf sedimentary system, pores are well developed. Porosity and specific surface area are mainly affected by TOC and clay minerals. An effective combination of brittle mineral particles and organic matter, clay minerals provide necessary conditions for pore preservation. The organic pores, clay mineral intergranular pores, and brittle mineral pores form the main network system for Longtan shale. In summary, the lithological combinations, organic geochemistry, and pore structure system were all affected by sedimentary environments.

Identification of Ore-Finding Targets Using Anomaly Components of the Ore-Forming Element Group Extracted by SVD and PCA in the Giant Gejiu Sn-Cu Ore Field, Southwestern China

Yongqing Chen, Aoyue Zheng
School of Earth Sciences and Resources, China University of Geosciences, Beijing, United Kingdom

The giant Gejiu Sn-Cu polymetallic ore deposit, located at the jointing site between the Yangtze and the Cathaysia Blocks, is one of the largest tin polymetallic ore deposits in the world. It is associated with a magmatic-hydrothermal ore-forming system triggered by the deeply buried geological structures and concealed granites. Multi-stage geological and ore-forming processes produce generally regional and local element concentration anomalies of ore-forming elements such as Sn and Cu. The principal component analysis (PCA) can be applied to establish various types of ore-forming element associations, which can be further divided into two different anomalous components with the help of the singular value decomposition (SVD). One can be defined as a regional geochemical anomaly generated by geological processes (i.e., magmatism); and the other can be regarded as a local geochemical anomaly created by ore-forming events. Three aspects are illustrated in this paper. (a) The ore-forming element combinations, including Pb-Ag-Sn-As-Mn-Cu and Be-W-F-Bi-Cu-B, are established by PCA. The ore-forming element group like Pb-Ag-Sn-As-Mn-Cu characterizes mesothermal-epithermal Sn-Cu-Pb-Ag polymetallic mineralization with alteration such as arsenopyrite and carbonate minerals; the Be-W-F-Bi-Cu-B ore-forming element group characterizes hypothermal-mesothermal W-Be-Cu polymetallic mineralization with fluorite and tourmaline alteration. (b) The regional geochemical anomaly established by magmatism and the local geochemical anomaly components created by hydrothermal ore-forming processes are respectively identified by the SVD from the ore-forming element groups. (c) The regional and local geochemical anomalies are delineated using the geochemical anomaly components corresponding to different eigenspaces. The PCA and SVD methods were applied to Sn-Cu polymetallic ore-prospecting target areas in the Gejiu ore field, southwestern China, by delineating their local geochemical anomalies. The results show that the SVD can extract not only the regional anomalies associated with magmatism, but also the local anomalies related to the hydrothermal ore-forming processes from multi-element groups established by PCA.

Carbonatite Melt as a Travel Agent for Magmatic Sulfide Liquid: Nature vs. Experiments

Maria Cherdantseva[1], Michael Anenburg[2], Marco Fiorentini[1], John Mavrogenes[2]
1. Centre of Exploration Targeting, School of Earth Sciences, The University of Western Australia, Crawley, WA, Australia, 2. Research School of Earth Sciences, Australian National University, Acton, ACT, Australia

The transport of dense sulfide liquid in mafic-ultramafic magma across the lithosphere has been a long-standing challenge in the studies of magmatic sulfide systems. Solving this issue is crucial for understanding the metal and sulfur cycles and for developing effective exploration strategies for valuable magmatic Ni-Cu-PGE deposits. Recently, it has been observed that sulfide mineralization is commonly found in intimate association with volatile-rich minerals, leading to a hypothesis that sulfide liquid transportation could be facilitated by water- and carbon dioxide-bearing fluids or gas bubbles coupled with sulfide droplets. However, this association with volatile-rich phases alone cannot explain the mineralogy, as minerals enriched in elements incompatible with mafic magma, such as Ti or Zr, are also present in close proximity to sulfide globules. The explanation of mineralogy by a pressure-derived infiltration of late, highly fractionated melt into the hollow spaces that left behind the bubbles presents similar problems, as some minerals cannot form in this manner. Additionally, this association has been found in both shallow and lower-crustal intrusions, where gas bubbles or fluid exsolution from the magma are unlikely due to high-pressure conditions. Therefore, a new model for sulfide liquid transport is needed, which we propose here based on mineralogy and textural variability. We suggest that the haloes enveloping sulfide globules are the product of crystallization of primary relatively buoyant and low-viscosity carbonatite liquid exsolved from the mafic magma, wherein the carbonatite globules will be immiscible with both sulfide and silicate melts. High-temperature, high-pressure experiments confirm that three immiscible liquids can coexist together, with carbonatite globules enveloping sulfide droplets in a coupled pair. The average density of these paired globules would be lower than that of the silicate melt, indicating that carbonatite liquid can be an effective transportation agent for sulfide globules in their journey from the mantle to the crust.

Investigating Metamorphosed VMS Mineralization from the Archean Yilgarn Craton, Western Australia: Preliminary Results from the King Zn Deposit

Cendi Dana[1], Steven Hollis[1], Darryl Podmore[2], Megan James[2], Ian Butler[1], Florian Fusseis[1], Adrian Boyce[3]
1. School of GeoSciences, The University of Edinburgh, Edinburgh, United Kingdom, 2. Black Raven Mining, Perth, WA, Australia, 3. Scottish Universities Environmental Research Centre, Glasgow, United Kingdom

Most ancient VMS deposits have been subjected to metamorphism, particularly those Archean in age. Although metamorphism may lead to protolith misinterpretation and increased complexity for exploration, it can also bring several benefits, including orebody thickening, coarsening of sulfides, and precious metal upgrading. Here we present the early results of a detailed mineralogical study of both the orebody and host rocks of the King Zn deposit as a precursor to better understanding the mechanisms of metal remobilization during metamorphism. In addition, representative lithogeochemical data of the footwall and hanging-wall stratigraphy were used to identify the protolith of the volcanic host rocks and associated alteration footprints. Mineralogical and geochemical data suggested that the footwall stratigraphy consists of several basaltic volcanic rocks that have been metamorphosed into garnet amphibolite and various types of schist (muscovite-biotite, quartz-chlorite-magnetite, chlorite-muscovite). Intermediate to felsic rocks stratigraphically overly these units and are capped by massive sulfide mineralization. The immediate hanging-wall stratigraphy consists of BIFs (likely meta-exhalite) and argillaceous metasedimentary rocks. The massive ore lens consists of a diverse suite of sulfides (i.e., pyrite, pyrrhotite, chalcopyrite, sphalerite, galena, with lesser marcasite, arsenopyrite, and rare gudmundite [FeSbS] and ullmannite [NiSbS]) and sulfosalts (i.e., tetrahedrite, boulangerite, rare canfieldite [Ag_8SnS_6]), as well as silver amalgam (Ag_9Hg). The underlying stringer sulfides include pyrite, chalcopyrite, pyrrhotite, sphalerite, and rare pentlandite. Textural evidence suggest that the formation of sulfosalts and silver amalgam is likely attributed to the sulfide recrystallization during amphibolite facies metamorphism. Magnetite and ilmenite are abundant and disseminated within footwall rocks, typically along the foliation, but minor in the massive ore lens. Titanite occurs late in the footwall stratigraphy, replacing earlier-formed ilmentite and rutile, both as coarse crystals and veins, indicative of extensive Ti remobilization during metamorphism.

São Félix do Xingu's Hydrothermal Gold and Base Metals Mineralization: Using Aerogeophysical Data to Find High Favourable Prospective Areas

Bruna de Freitas, Caetano Juliani
University of Sao Paulo, Sao Paulo, SP, Brazil

The São Félix do Xingu (SFX) region is located in the Amazonian Craton, Brazil. Geologically, the region is in the east part of the Central Amazonian Province, close to the west boundary of the Archean Carajás Mineral Province, and it is considered a promising area for epithermal and porphyry gold and base metals mineralization. The SFX region is mainly composed of andesites to rhyolites (ca. 2.0–1.88 Ga), usually undeformed and affected only by very low grade metamorphism. These units are considered part of the significant Uatumã magmatic event, constituted of several events of intermediate to felsic magmatism generated in continental magmatic arcs, with a final orogenic event (ca 1.87 Ga). These events include volcanic, volcanoclastic, and plutonic intrusive rocks and clastic sedimentary sequences. Several Au-Ag high-, intermediate-, and low-sulfidation and Au-Cu-(Mo) porphyry mineralizations have been recognized in these units along more than 600 km from the Tapajós Province to SFX region, encouraging, therefore, research of this important area.

However, the region has insufficient geological detail and a limited database for its high prospective potential. The use of aerogeophysical techniques for areas of difficult access, emphasizing mineral prospecting, has increased its application over time. Through predictive techniques, using gamma-spectrometric and magnetometric data, it was possible to develop favourability maps for exploratory targets. Applying ratios between channels and grid analysis CET techniques, with the integration of geological and structural data, regions of high favourability were highlighted.

The result shows that both methodologies effectively identified promising regions for mineral prospecting. Areas with potential mainly for Au, Cu, Fe, and Sn were defined based on the correlation between the magnetic data, where zones of significant discontinuities can host mineralization, and between the gamma-spectrometric data, characterizing regions of interest based on geostatistic and the study of regions with anomalous concentrations of potassium, uranium, and thorium.

The Epithermal-Skarn Transition at the La Colorada Deposit, Chalchihuites District, Zacatecas, Mexico

Joshua Ebner[1], Zhaoshan Chang[1], Pedro Francisco[1], Dante Juarez[2], Patricio Perez[2], Chris Emerson[2], Shiqiang Huang[1], Sidney Hemming[3], Richard Wendlandt[1]
1. Colorado School of Mines, Golden, CO, USA, 2. Pan American Silver, Chalchihuites, ZAC, Mexico, 3. Columbia University, New York, NY, USA

The La Colorada deposit in the Chalchihuites District of Zacatecas, Mexico, features intermediate sulfidation epithermal (IS), carbonate replacement (CRD), skarn, and porphyry styles of mineralization. IS veins continue from volcanic rocks into underlying carbonates, with combined measured and indicated resources of 5.3 Mt including 33.8 Moz Ag, 26.2 koz Au, 0.04 Mt Pb, and 0.08 Mt Zn (6/30/2022). Drilling since 2018 by Pan American Silver has discovered a Zn-Pb-Ag CRD-skarn with porphyry Cu-Mo mineralization below the eastern part of the IS veins. With 95.9 Mt of indicated resources containing 94.4 Moz Ag, 1.2 Mt Pb, and 2.7 Mt Zn (06/30/2022), the La Colorada skarn has become a major asset for Pan American Silver. This study seeks to identify the relationship between IS and CRD-skarn mineralization to assess if the distal expression of a skarn can be IS veins when volcanic rocks are encountered. If found to be transitional, this study will also identify vectors in the epithermal environment that can lead to the related CRD-skarn.

IS veins in volcanic rocks are found to transition downward into CRD veins in carbonates with the same quartz-calcite-sulfide assemblage, but with more calcite. K-Ar illite geochronology indicates a preliminary age of ~63 Ma for the IS veins, statistically indistinguishable from the U-Pb zircon/garnet and Re-Os molybdenite ages for the skarn-porphyry causative intrusion and mineralization. Illite crystallinities (IC) of altered dacitic volcanic rocks next to major IS veins range from 1.9 to 0.6. Homogenization temperatures of fluid inclusions in sphalerite and quartz from volcanic-hosted IS veins and carbonate-hosted CRD veins range from 325°C to 234°C (14 samples in this study and eight samples from literature). The two data sets spatially overlap, and together indicate two hot centers below the eastern part of the IS veins and a third hot center to the NW.

The Natural Enrichment of Phosphatic Sediments: Insights from the Moroccan High Atlas Phosphate Series

Radouan El Bamiki[1], Michel Séranne[2], Fleurice Parat[2], Jérémie Aubineau[2], El Hassane Chellai[3], Mohamed Marzoqi[3], Jean-Louis Bodinier[1, 2]
1. Mohammed VI Polytechnic University, Benguerir, Morocco, 2. University of Montpellier, Montpellier, France, 3. Cadi Ayyad University, Marrakesh, Morocco

Sedimentary phosphates are the major source of phosphorus and a potential source of critical metals such as rare earth elements (REEs). Sedimentary phosphates are non-renewable resources and witness an extremely increasing global demand. It is, therefore, critical to focus a lens on exploring new phosphate ores as part of many other proactive strategies to tackle a possible phosphate scarcity. In this study, we combine sedimentology, sequence stratigraphy, and geochemistry to explore the Upper Cretaceous-Paleogene sedimentary series of the High Atlas in Morocco. The sedimentological investigation revealed the occurrence of four genetically-linked lithofacies derived from a pristine phosphate by storm waves and bottom currents hydrodynamic differentiation, giving rise to winnowed phosphate, phosphatic lags, and karst-filling phosphate. The main phosphatic mineral in all lithofacies is carbonate fluorapatite (CFA), with 8.40 ± 0.37 wt % CO_3^{2-}. The sequence stratigraphic analysis indicated that the phosphate series of the High Atlas deposited during a second-order relative sea-level cycle, and phosphate layers preferentially accumulated at the base of the high stand systems tract and the base of the transgressive systems tract. Although some lithofacies have experienced a polyphase evolution, by winnowing, submarine reworking, transport, and subaerial weathering, CFA has resisted chemical changes and preserved its original geochemical signature, specifically an REE signal that reflects oxic seawater conditions. However, the differentiation processes triggered the oxidation of organic matter and sulfides, removing associated uranium and cadmium. Conclusively, the combination of winnowing/reworking and subaerial weathering is the most effective way of the natural beneficiation of phosphate, turning low-grade pristine phosphate (P_2O_5 <12 wt %) into a high-grade granular phosphate (up to 24 wt % P_2O_5) and reducing the content of chemical elements with environmental concern. Furthermore, points of sequence stratigraphic turnarounds are primary exploration targets of phosphate-rich sediments.

Geochemical and Mineralogical Characterisation of the Dulaan Khar Uul Pb-Zn-Ag Deposit in Western Mongolia

Jamsran Erdenebayar[1], Takeyuki Ogata[2], Batkhuu Baldorj[3]
1. MUST, Ulaanbaatar, Mongolia, 2. MiReSLab, Akita, Japan, 3. Chimchig Khan LtD, Ulaanbaatar, Mongolia

In this study, we focus on the Dulaan Khar VMS deposit in the Mongol-Altai belt. Four orebodies show lenses and veins and are stratabound by the early to middle Devonian Dulaankhar formation that consists of rhyolite tuff, flows, tuffaceous sandstone, rhyolite, and dacite porphyry subvolcanic bodies and andesite and diabase dikes. The $\delta^{34}S$ values of sulfide minerals of the Dulaan Khar Uul deposit range between +4.3‰ and +5.6‰.
Geochemical data of least-altered mafic and felsic volcanic rocks from the Dulaan Khar deposits show clear compositionally bimodal characteristics ranging from 43.7 to 52.3% SiO_2, and from 55.1 to 80.9% SiO_2. The Th/Yb vs. Nb/Yb diagram distinguishes oceanic arc from active continental magmas. All the samples plot within the active continental magma. Samples from the Dulaan Khar Uul deposit show a marked bimodal trend and consist of basalt and rhyodacite. In the Th/Yb vs. Nb/Yb diagram, samples of the Dulaan Khar Uul deposit are plotted outside the diagonal MORB-OIB array, on a vector at a steep angle but closer to E-MORB for all samples, and may be related to the variable fractionation processes and variable crustal contamination. All least-altered felsic rock samples are plotted on FII and FIIIa affinities field of Lesher et al., considered to have much higher mineralization potential.
The mineral composition of the deposit ore is characterized by mainly sphalerite and galenite with pyrite, a slight amount of chalcopyrite, and occasionally tennantite, marcasite, ilmenite, cerussite, smithsonite, hematite, malachite, arsenopyrite, siderite, gold, etc. Two types of sphalerite are identified: (Zn, Fe)S, marmatite with iron and native zin sulfide cleiophane.
The geochemical characteristics of whole rocks from the Dulaan Khar Uul deposit may be related to amalgamation being dominated by advancing rifting, which was favorable for the formation and/or preservation of VMS deposits.

Interpreting Machine Learning Neural Network Patterns of an Orogenic Gold Deposit, Yilgarn Craton, Western Australia

David M. First[1], Yush H. Rajcoomar[2], Daniel Mogilny[2], Ilia Sucholutsky[2], Farzi Yusufali[2]
1. Stratum AI, Brisbane, QLD, Australia, 2. Stratum AI, Toronto, ON, Canada

Machine learning is creating value within multiple facets of the mining industry. The authors provide a high-level introduction to AI, machine learning, and deep learning; widely recognized as one of the most powerful forms of machine learning. They will introduce deep learning tools like convolution neural networks, how they are already applied, and the economic considerations necessary for determining when deep learning may be the right solution to de-risking complex block modelling problems.

The authors present the results from preliminary mineral resource modelling of an orogenic gold deposit. They demonstrate how statistically significant non-linear correlated elements are used as direct inputs to the resource model to assist the target element's grade prediction at every block and how its results compare with classical geostatistical techniques such as kriging. This shows that 1) existing techniques for finding correlations between assayed elements do not reflect the complex geology of the asset; 2) non-linear correlations that are difficult to model as simple mathematical functions are representative of geological patterns in a deposit; and 3) non-linear correlated assayed data fed as inputs increase the performance of the resource model as reconciled through blind tests.

It is demonstrated that the deep learning element models provide an enhanced de-risking tool for interrogating the gold block model by improving confidence that high-grade (HG) gold clusters exist (e,g, V, Cu, Ta, Te, U) and as to whether unmined HG gold mineralisation remains (e.g., As, Cu, Ta, Te), due to element substitution in gangue (e.g., green micas) and ore (e.g., sulfides and tellurides) minerals.

The authors hypothesize that the patterns revealed by the block models are the result of geological processes that generated the mineral deposit; that is, the primary hydrothermal processes that deposited economic concentrations of metals are overprinted by secondary physicochemical processes, resulting in these complex patterns.

UV Fluorescence as an Exploration Tool in Carbonate Replacement Deposits

Spyridon Fragkou[1], Andrew Steiner[1], Shaun Barker[1], Tim Baker[2], Janja Mihajlovic[3]
1. The University of British Columbia, Vancouver, BC, Canada, 2. Eldorado Gold, Vancouver, BC, Canada, 3. Hellas Gold, Athens, Greece

Carbonate replacement deposits (CRDs) are challenging exploration targets because alteration haloes in carbonate rocks around the deposits are often difficult to identify (known as cryptic alteration). This study aims to characterize, quantify, and map the cryptic alteration footprint around the Au-rich Olympias polymetallic CRD, Greece (combined reserves [proven and probable] and resources [measured and indicated] of 5.2 Moz Au, 96.8 Moz Ag, 1.0 Moz Pb, and 1.3 Moz Zn as of September 2022). The Olympias deposit is located in the Cenozoic Western Tethyan Magmatic Belt of gold-rich magmatic-hydrothermal deposits. The replacement-style orebodies are hosted in marble, which is interlayered with feldspar-biotite gneiss within the Paleozoic Kerdylia Formation of the Serbo-Macedonian Massif. The orebodies generally follow the gently dipping marble horizons and steep normal faults, such as the Kassandra fault. Mineralization occurred during the Late Oligocene to Early Miocene, synchronous with local magmatism. The ore mineralogy of the Olympias deposit mainly consists of sphalerite, galena, and arsenopyrite, with Ag and Au strongly associated with galena and arsenopyrite, respectively. A portable XRF instrument and a short-wavelength UV lamp were used on drill core and rock samples from Olympias to assess how carbonate mineral chemistry in veins and wall rocks varies with distance to known mineralization. Haloes of Mn enrichment around mineralization (and interpreted fluid pathways) were recognized from portable XRF measurements. Manganese-enriched calcite veins and marble layers typically fluoresce pink-orange colour under UV light with variability controlled by Fe content, making UV fluorescence a rapid, inexpensive, field-deployable tool for visualizing cryptic carbonate alteration.

Unravelling the Genesis of the La Colorada Skarn Deposit, Chalchihuites District, Zacatecas, Mexico

Pedro Francisco[1], Zhaoshan Chang[1], Joshua Ebner[1], Dante Juarez[2], Patricio Perez[2], Chris Emerson[2], Shiqiang Huang[1]
1. Colorado School of Mines, Golden, CO, USA, 2. Pan American Silver, La Colorada, ZAC, Mexico

The La Colorada deposit in the Chalchihuites District of Zacatecas, Mexico, includes intermediate sulfidation epithermal (IS), carbonate replacement (CRD), skarn, and porphyry styles of mineralization. The current operation mines the IS veins that extend from the overlying volcanic cover into underlying carbonate-rich sedimentary rocks and transitions to CRD. Significant Zn-Pb-Ag skarn-CRD with porphyry Cu-Mo mineralization was discovered below the eastern part of the IS vein zone by Pan American Silver from extensive drilling in 2018. With 95.9 Mt of indicated resources containing 94.4 Moz Ag, 1.2 Mt Pb, and 2.7 Mt Zn, and 147.8 Mt inferred resources including 132.9 Moz Ag, 1.5 Mt Pb, and 3.4 Mt Zn (06/30/2022) the La Colorada deposit is ranked among the top five Zn-Pb skarns in the world. This research aims to identify the causative intrusion(s) and develop a genetic model of the La Colorada deposit to guide the ongoing exploration, by employing skarn vectoring tools.

An E-W cross section is built after extensive core logging, revealing skarn zoning patterns and locating potential causative intrusions. Garnet color changes from yellow-green to red and pyroxene color brown to light green from distal to proximal locations. Electron microprobe analysis of skarn samples showed Mn wt % variation in pyroxene (expressed as johannsenite mole percent) of 0.7-54.5 Jo% and in garnet (expressed as spessartine mole percent) of 0.2-4.4 Sp%. The results show Jo% and Sp% depletion trends that point to distinct "hot centers" at depth, serving as vectors towards potential causative intrusions. Porphyry molybdenite Re-Os and zircon U-Pb dating together with skarn garnet U-Pb dating results place the main porphyry-skarn event age at ca. 62 Ma. Current results suggest that multiple fertile intrusive phases produced multiple skarns. Additional core logging and geochronology is being conducted to refine the intrusive sequence and skarn events at La Colorada.

Lagoa Salgada VMS: Geochemistry Applied to Deposit Domaining, Geometallurgy, and Exploration

Sergio Gelcich[1], Vitor Arezes[2], Luis Lima[2], Joao Barros[1], Avelihno Pinheiro[2]
1. Ascendant Resources, Toronto, ON, Canada, 2. Redcorp Empreendimentos Mineiros, Lda, Braga, Portugal

The Lagoa Salgada polymetallic deposit (Zn, Pb, Cu, Au, Ag), located at northwest edge of the Iberian Pyrite Belt, comprises massive sulphide and stringer mineralization in the north sector and stacked fissural sulphide-bearing corridors in the south sector. Mineralization has been affected by structural transposition and the weathering process.

New collected downhole four-acid digestion multielement ICP coupled with ore-grade metals (oxidizing digestion) ICP has been extensively used for 1) refining the lithogeohemical character of the host volcanic rocks and their different domains and sub-domains; and 2) understanding metal zoning and mobility.

Lithogeochemically, the heterogeneous volcanic suite that hosts the different mineralized domains can be discerned by the dacitic tuffs that cap the mineralization based on differentiation trends (e.g., Eu anomaly). In terms of alteration, within the host volcanic packages, correlation between Alkalis/Al/Mg-Fe indexes (e.g., AI and CCPI) and mineralized trends exist. When properly normalized by host rocks, corridors of higher mineralization show a proportionally higher CCPI index in zones of higher strain. This qualitative feature works as a tool to delineate mineralized trends.

Metal ratios in the north sector reflect secondary processes where different metal solubility generates enrichment or depletions from a less variable primary ore. Gossan shows full depletion in Cu and Zn and residual enrichment of precious metals, V and Mo, and Transition shows enrichment of Cu and Zn depletion. RREE variations also support this classification scheme. The new Geochem-driven ore model has improved considerably the metal recoveries in current processing studies.

Metal distribution in the south sector is affected by deformation and contrasting metal solubility and sulphide rheology. In this high-strain environment, metal ratios define exploration vectors, particularly in terms of Cu, that form continuous elongated subdomains within some of the stacked fissural sulphide corridors.

On the Development of a Microstructural Signature During Alluvial Gold Transport

Juan Gomez-Barreiro[1], Santos Barrios Sánchez[1], Juan Morales Sánchez-Migallón[1], Jose Manuel Compaña Prieto[1], Kelvin Dos Santos Alves[1], Vanessa Leonor Escobar Duche[1], Inés Puente Orench[2]
1. University of Salamanca, Salamanca, Spain, 2. Institut Laue Langevin, Grenoble, France

The microstructure of crystalline solids is the result of the sum of multiple physicochemical processes. The understanding and discrimination of those processes has to rely on quantitative data. Due to the intrinsic material properties of gold, particles liberated in supergene environments can be subjected to dramatic chemical and mechanical changes. While some chemical processes are well documented (e.g., Ag de-alloying), microstructural evolution remains obscure. Deformation of gold particles in alluvial systems may result in recrystallization, grain-size reduction, and variation of crystallographic preferred orientation; those changes could have an effect on geochemical behavior of gold (e.g., diffusion).

We have conducted diffraction experiments for quantitative crystallographic preferred orientation (CPO) analysis of gold particles. The particles show different morphological indexes, tentatively related to a variety of transport distances in alluvial systems in Spain. Our results show a connection between particle shape and microstructure as an initial stage to discriminate transport/deformation models and source analysis.

Applications of Hyperspectral Analysis to VMS and Orogenic Gold Exploration in the Yilgarn Craton, Western Australia

Steven P. Hollis[1], Cendi D. Dana[1], Conal Rodgers[1], Jack Richardson[1], Darryl Podmore[2], Megan James[2], Allan Kneeshaw[2], Riquan Azri[2]
1. University of Edinburgh, Edinburgh, United Kingdom, 2. Black Raven Mining, Perth, WA, Australia

Despite countless advances in recent years, exploration for volcanogenic massive sulphide (VMS) and orogenic gold deposits remains challenging. This is particularly the case in the Yilgarn Craton of Western Australia, where there is a paucity of outcrop and weathering is deep and extensive. One of the most significant breakthroughs in recent years has been accessibility of low-cost hyperspectral data to industry. Here were present the results of two exploration projects in the southern Kurnalpi Terrane, where large short-wave infrared data sets have been acquired.

The King Zn-(Cu-Au) VMS deposit occurs 140 km east of Kalgoorlie as an overturned lens of massive sulfide mineralization in a basaltic to dacitic volcanic sequence. Chlorite and white-mica chemistry are strongly lithology dependent. Chlorite is dominated by mixed Fe-Mg compositions throughout the footwall stratigraphy, with localised zones of Fe- and Mg-rich compositions related to distinct lithologies and mineralization. Muscovitic white mica compositions predominantly occur in footwall basaltic rocks, with more paragonitic compositions in dacitic rocks directly underlying massive sulfides. An increase in white mica abundance occurs towards the ore lens. Hanging-wall white mica is phengitic.

At the EO04 exploration target, orogenic Au mineralization is associated with the margins of a dolerite dike cutting a mixed sequence of felsic to intermediate volcanic rocks. Anomalously high Te concentrations are restricted to west of the intrusion. Chlorite compositions are controlled by rock type, with Fe-rich chlorite in basaltic lithologies, and more mixed Fe-Mg compositions in andesitic and dacitic rocks. Mg-rich compositions are restricted to narrow zones within the latter, associated with anomalously high Te on the edge of current drilling. White mica compositions are dominantly muscovitic, with lesser phengite and occasional paragonite. Phengite occurs around zones of high-grade Au mineralization at depth within the central zone of drilling, within both the andesitic volcanic rocks and crosscutting dolerite intrusion.

Tourmaline: A Fertility Assessment and Vectoring Tool for Magmatic-Hydrothermal Systems

Wei Hong[1], Francisco Testa[1], David R. Cooke[1], Nathan Fox[2], Lejun Zhang[1], Michael Baker[1], Evan Orovan[1], Ayesha Ahmed[1], Pete Hollings[3], Ivan Belousov[1], Anthony Harris[4]

1. CODES/Earth Sciences, University of Tasmania, Hobart, TAS, Australia, 2. W.H. Bryan Mining & Geology Research Centre, The University of Queensland, Brisbane, QLD, Australia, 3. Department of Geology, Lakehead University, Thunder Bay, ON, Canada, 4. Newcrest Mining Ltd., Melbourne, VIC, Australia

Tourmaline's occurrence in a wide variety of ore-forming systems makes it a powerful tool for recording the temporal and spatial evolution of hydrothermal fluid compositions. It can provide key information regarding metal fertility and exploration vectoring within a system. We sampled tourmaline-bearing rocks from a diversity of magmatic-hydrothermal systems, including porphyry Cu-Mo and Cu-Au deposits and prospects, tourmaline-rich breccias, granite-related Sn mineralization, Archean Au lodes, metamorphic terrain, pegmatite, and granite batholiths. Analyses of >2,200 individual LA-ICP-MS spots revealed that tourmaline chemistry can discriminate different types of hydrothermal systems and geologic environments. Pegmatite-related tourmaline has extremely high Al, Li, and Pb and low Fe, Mg, and Ti contents, whereas metamorphic tourmaline is characterized by high Mn and Zn and low Y and Ba. High Sr, Ni, and Cr and low Ga and Mn/Zn characterize tourmaline from the Red Lake Archean gold deposit. Tourmaline from granite-related tin systems is characterized by high Al, Fe/Mg, and Sn and low Sr. Tourmalines from porphyry deposits and associated breccias typically have low Fe/Mg and high As, Sb, Sc, V, and Sr. Tourmaline's compositions vary systematically from deep to shallow regions in porphyry-related alteration zones and breccia pipes. Most transition elements in tourmaline commonly increase in concentration with increasing distance from mineralization centers. Arsenic, Sb, Pb, and Sr in tourmaline have their highest concentrations between 1 and 1.5 km from the deposit center, and define geochemical shoulders to the mineralization zones. Several trace elements in tourmaline can be used for fertility assessment to discriminate porphyry deposits from barren batholiths and to distinguish large porphyry deposits from smaller systems. Tourmaline is a new exploration tool that records hypogene low-level geochemical anomalies that assist in vectoring and fertility assessments for magmatic-hydrothermal systems, particularly those associated with porphyry Cu-Mo-Au mineralization.

Melt Inclusions and Chemistry of Detrital Titanomagnetites at Orokolo Bay, Papua New Guinea: Evidence for Genesis by Magmatic Immiscibility

Leeysmon Hulijeli, Yasushi Watanabe, Osamu Nishikawa, Carmela A. Tupaz
Akita University, Akita, Japan

Orokolo Bay in the Gulf of Papua (GoP) is known for its extensive placer deposits rich in titanomagnetite ores derived from igneous sources within the Central Range of Papua New Guinea (PNG). However, identifying the specific hinterland source of the detrital ore within the PNG Central Range is problematic owing to its complex history of orogenesis, bimodal arc magmatism, and varying subduction dynamics during the Cenozoic driven by successive arc-continent collisions. In this study, we analyzed grain textures, chemistry, and melt inclusions of 271 detrital titanomagnetites to investigate their source rocks and petrogenesis. The titanomagnetites exhibit rounded globular grain morphologies, and their textures indicate post-formation reequilibration processes. Titanomagnetites also host abundant apatite, zircon, pyrrhotite, and silica-rich glass (SG) melt inclusions. The SG melt inclusions have compositions resembling trachyandesite-rhyolite (55-73% SiO_2). Zircon inclusions are exclusive to high-Ti titanomagnetites (>10 wt% Ti), and pyrrhotite inclusions only occur in low-Ti titanomagnetites, while both apatite and SG inclusions are ubiquitous. Magnetite-ilmenite thermobarometry revealed two main stages of titanomagnetite formation, an initial stage below the NNO buffer at 1,054°C and a later stage commencing above NNO at 834°C. The results revealed liquid immiscibility processes, including separation of immiscible Si-Al-K-Na-rich and Fe-Ti-P-HFSE-S-rich melts, the latter hosting an Fe-S-(Ni-Cu-Co) residual melt. The varying titanomagnetite compositions, especially in terms of Ti, were interpreted to have been the result of immiscible Fe- and Si-rich melts under changing fO_2 and temperature conditions. Existing discriminant models classify the titanomagnetites as being derived from either an Fe-Ti-(V) or kiruna-type IOA deposit. The results of this study provide insight into the immiscible Fe- and Si-rich melts and discrimination of source rock type using detrital titanomagnetites. Furthermore, the study proposes the existence of a previously unknown IOA-type mineralization in the PNG Central Range as the primary source of detrital Fe-Ti oxides in the GoP.

Paragenesis of Alteration and Silver Mineralization of the Pomperia Veins Hosted in the Tacaza Formation, Southern Peru

Jhoan C. Humpiri Chavez, Milagros Apaza Aquino
Universidad Nacional del Altiplano Puno (UNAP), PUNO, Peru

Silver is one of the most important metals for modern industry. Around 57% of the world's silver production comes from the Americas. In 2020 Peru was the second largest producer with 110 million ounces of the 784.4 million ounces produced in the world. The central Andes hosts important Ag mineralization such as Antamina, Uchucchacua, and Inmaculada. In southern Peru there are several low- and intermediate-sulfidation deposits hosted in the paleo-neogene Tacaza Formation. In 2006 the Santa Ana and Corani world-class Ag deposits were discovered by Bear Creek in southern Peru. Santa Ana is hosted in the Tacaza Formation. Hydrothermal stages responsible for Ag mineralization have yet to be well defined in veins hosted in the Tacaza Formation. Our study was carried out on the Pomperia veins, 100 km north of Santa Ana near Puno, Peru. The aim is to study the paragenesis of alteration and mineralization and its relationship with silver geochemical variation.

Pomperia is a structurally controlled Ag-rich vein occurrence with mineralogical similarities to intermediate sulfidation epithermal deposits such as Santa Ana, but the quartz-poor character and the lack of adularia at Pomperia are distinctive. Veins cutting the Tacaza Formation lavas are composed of pyrargyrite, proustite, argentite, galena, sphalerite, hematite, chalcopyrite, and pyrite. Four paragenesis stages have been determined: I) fluidized breccias with a rock-flour matrix, which is believed to be the channel of the mineralizing fluids; II) barite, low-temperature quartz, specular hematite, and rutile; III) pyrargyrite, proustite, argentite, galena, and sphalerite (main mineralization); and IV) carbonates and amethyst quartz.

Mineralogical composition and identification of Ag mineralizing stages are significant in the mining cycle and can serve as a guide to exploration initiatives of the Tacaza Formation. Mineralogy of veining will be increasingly more important for geometallurgical evaluations as projects like Corani and Santa Ana continue to develop.

Stratiform Potash Mineral Systems Prospectivity Analysis in an Andean Setting

Christopher Lambert, Jean-Jacques Ferguson, Esme Whitehouse, Edward Bunker, Alexandra Scarlet, Jorge Gines
CGG, Crawley, United Kingdom

Mined potash contains various salts of potassium in water-soluble form, and over 95% of global production goes into fertilisers and other agricultural applications. Around 80% of the total global production within this US$23 billion industry (2022) is held by a small consortium of major companies in Canada, Russia, Belarus, and China. With the current geopolitical climate, potash is fast becoming a strategic mineral. A supply squeeze and a significant price increase in the last year has led to a focus on developing new long-term supplies for this vital commodity.

The application of mineral system analysis in mineral prospectivity modelling allows for the characterisation of a logical framework to define mineralisation potential. Using stratigraphical marker horizons coupled with improved geological mapping, it is possible to review the distribution of stratiform orebodies of bulk commodities within a given area. Using geospatial tools, we highlight areas of interest using a reducing gradational filter passed over the study area based on stratigraphical level and geographical distance away from target horizons. The approach aids the delineation of risk-reduced search spaces, which may include targets under cover.

As with most bulk commodities, environmental considerations and access to infrastructure are critical factors for the development of a project. In this study, we integrate environmental, infrastructural, and social data layers to define a holistic risk-mapping approach that transcends a typical mineral prospectivity model. This enables informed exploration decisions based on multiple non-geological data layers. The approach has promoted further research and development that includes workflow advancement and investigations into data-handling efficiency through automation.

3D Mineral Prospectivity Modeling for Targeting Concealed Mineralization Within the Chating Area, Middle-Lower Yangtze River Metallogenic Belt, China

Xiaohui Li[1,2], Feng Yuan[1,2], Chen Xue[1,2], Yuheng Chen[1,2]
1. School of Resources and Environmental Engineering, Hefei University of Technology, Hefei, China, 2. Anhui Province Engineering Research Center for Mineral Resources and Mine Environments, Hefei, China

The Chating area is situated within the Yangtze metallogenic belt in Anhui Province, China. Several concealed skarn and porphyry-type deposits have been discovered in this area, indicating high potential for hosting hydrothermal deposits. However, due to the complex geological structure, exploration risks significantly increase with increasing depth. To overcome this challenge, three-dimensional mineral prospectivity modeling (3DMPM) has begun to be widely applied for mapping the prospectivity of deep-seated and concealed mineralization. However, most previous studies on 3DMPM were based on shallow supervised machine learning models and dimensionality-reduced 3D predictive maps. Although these models have shown good results, they may lose spatial correlation within the 3D predictive maps and fail to explore nonlinear correlations between the 3D predictive maps and mineralization. Therefore, this paper initially builds and optimizes 3D geological models through implicit 3D geological modeling and a "total litho-inversion" approach. Subsequently, the 3D predictive maps are generated by employing various 3D methods, which are further integrated using a 3D convolutional neural network (3D CNN) model to identify highly prospective areas for mineralization. The results show that the highly prospective areas identified by the 3DCNN based 3DMPM include not only the training data but also other mineral deposits that have previously been discovered within the study area. In addition, the 3D CNN model performs better prediction capabilities due to its enhanced ability to capture the correlations between 3D predictive maps and multiple types of mineral deposits. It suggests that the 3DMPM based on the 3D CNN model has commendable predictive capabilities in identifying prospective mineralization areas, and some new highly prospective areas can be considered priority areas for future exploration of concealed mineralization within the Chating area.

What Lies Beneath? Sub-Seafloor Drilling of the Inactive Ultramafic-Hosted Semenov-1 Seafloor Massive Sulfide Deposit, 13°30'N on the Mid-Atlantic Ridge

Andrew J. Martin[1], Katie A. McFall[2], Bramley J. Murton[3], John W. Jamieson[4], Sven Petersen[5], Isobel Yeo[3], Christopher J. MacLeod[6], Anna Lichtschlag[3], Simon Allerton[6], Christian Bishop[3], Acer Figueroa[3], Szu-Ying Lai[3], Paul Lusty[7]
1. Department of Geoscience, University of Nevada, Las Vegas, Las Vegas, NV, USA, 2. Earth Sciences, University College London, London, United Kingdom, 3. National Oceanography Centre, Southampton, United Kingdom, 4. Earth Sciences, Memorial University of Newfoundland, St. John's, NL, Canada, 5. GEOMAR, Helmholtz Centre for Ocean Research Kiel, Kiel, Germany, 6. School of Earth and Environmental Sciences, Cardiff University, Cardiff, United Kingdom, 7. British Geological Survey, Keyworth, United Kingdom

Our understanding of seafloor massive sulfide (SMS) deposit formation is largely based on the study of ancient on-land volcanogenic massive sulfide (VMS) deposits and observations made at the seafloor, and these observations are heavily biased toward the sampling of active "black smoker" chimneys. Uncertainty remains when using seafloor observations to evaluate sub-seafloor characteristics of SMS deposits, such as deposit tonnage, how the composition of collected seafloor samples compares to sub-seafloor composition, and how metal enrichment processes vary between these two environments. However, internal core samples from seafloor drilling have only been recovered from a few SMS deposits, largely due to the expense and technical challenges associated with deep-sea drilling. Here we present data from the inactive Semenov-1 hydrothermal field located on the 13°30'N oceanic core complex on the Mid-Atlantic Ridge. In 2022 aboard the RRS *James Cook*, project ULTRA recovered cores to a depth of ~20 meters below the seafloor using the British Geological Survey's Rock Drill II. The aim of the project was to assess the resource potential of an ultramafic-hosted SMS deposit that has previously been reported to contain elevated Ni, Co, Au, and Te contents. Near the surface, below a layer of sediment, recovered core contains poorly cemented sulfide gravel and sand, with abundant chalcopyrite, pyrrhotite, and pyrite with native Au. Deeper in the section, the dominant sulfide mineral assemblage alternates between marcasite and pyrite-rich with elevated Ni and Zn, respectively, occurring as disseminations in clay-rich material that often exhibits sub-horizontal deformation fabrics. Our preliminary geochemical analyses indicate systematic trends in bulk chemistry, trace metal sulfide chemistry, and sulfur isotope composition of sulfide minerals with depth below the seafloor. These data provide a new perspective to understand the formation of ultramafic-hosted SMS deposits and to investigate metal enrichment processes below the seafloor in SMS deposits more widely.

Recognizing Intrusive Phases at the Red Chris Au-Cu Porphyry Deposit Using Integrated Chemical and Geological Drill Core Data Sets

Brian A. McNulty[1], Shaun Barker[1], Anthony Harris[2], Nicholas Fitzpatrick[2], Alan Wilson[3]
1. University of British Columbia, Vancouver, BC, Canada, 2. Newcrest Mining, Melbourne, VIC, Australia, 3. GeoAqua Consultants, Anguilla, United Kingdom

The Red Chris porphyry deposit in northern British Columbia, Canada, has a combined ore reserve and mineral resource estimate of 20 Moz gold and 5.7 Mt copper at an average grade of 0.30 g/t Au and 0.35% Cu (Newcrest Annual Report, 2022). The deposit is hosted in the Late Triassic (212-202 Ma) Red Stock which comprises a series of nested porphyry intrusions that display structurally reactivated contact relationships (Zhu et al., 2018). At least three porphyry families (P1, P2, and P3; Rees et al., 2015) have been identified based on subtle compositional and textural differences, truncated veins, and crosscutting relationships. Of the three porphyry families, the best copper and gold mineralization is focused within abundant quartz-sulfide stockwork veins in P2, leading to the interpretation that P2 is the causative intrusive phase (Riedell et al., 2020). The deposit is dissected by late- and post-mineral faults (Rees et al., 2015). However, with the exception of the P3 porphyries, the visual similarities of intrusive phases make reconciling the displacement along these structures challenging. The integration of systematic downhole 4-acid multi-element geochemistry with detailed logging, shortwave infrared, and X-ray fluorescence data sets provides an opportunity to refine the mineralogical and geochemical signatures of the porphyry intrusions at Red Chris. The improved recognition of the spatial and temporal relationships of the porphyry phases has significant exploration implications. Identifying least- and moderately altered P2 and understanding its footprint to mineralization could provide an additional mineral exploration vector. In addition, better understanding of the P1 phases, which are pre-mineral host rocks, combined with traditional magmatic-hydrothermal alteration mineralogical and geochemical vectors (e.g., Halley, 2020) could improve mineral prospectivity and lead to further discovery in the district.

Bridging the Gap: Rapid Geochemical and Mineralogical Characterization at the Hand-Specimen Scale

Andrew H. Menzies[1], Nigel M. Kelly[2], Alan R. Butcher[3], Falk Reinhardt[1], Roald Tagle[1]
1. Bruker Nano GmbH, Berlin, Germany, 2. Bruker Nano Analytics, Denver, CO, USA, 3. Geological Survey of Finland, Espoo, Finland

With increasing challenges to deposit discovery, there is a corresponding increased need for rock characterization across scales. These challenges will be best addressed through the successful integration of multiple analytical techniques, but with a view to balancing the desire for high-spatial-resolution analysis against rapid turnaround time for use in near-real-time decision-making—and cost. All analytical tools come with a cost. However, data that supports decision-making throughout the mining life cycle from discovery to remediation improves the value proposition. A key area that needs to be addressed is the transition from meso-scales (outcrop or drill core) to micro-scales (thin section, crushed sample), where detailed mineral-textural relationships are used to interpret mineralization processes and deposit viability, and decide on approaches to mineral processing, behavior of materials in waste piles, and much more.

Micro-XRF provides rapid geochemical and mineralogical data at the scale of microns but on hand-specimen-sized samples. With beam sizes of <20 μm and detection limits that allow measurement of trace elements, micro-XRF spectrometers yield fast results on samples that require no polishing or conductive coating, such as cut pieces of drill core. Data generated from the micro-XRF may be integrated back into broader-scale data sets (visual logging, hyperspectral scans) to provide ground truthing and assistance with ongoing exploration and provide context for down-sampling to higher-spatial-resolution and more costly techniques such as SEM.

This presentation will use case studies on Au-bearing, Co-bearing, Exotic-Cu– and REE-bearing deposits to demonstrate the full potential of data produced by the micro-XRF spectrometer. These case studies highlight the role micro-XRF data may play in brownfields and greenfields exploration projects in the initial early understanding of mineralization, through to its potential for early planning of mineral processing requirements and benefits throughout the life cycle of an economic deposit.

Cerro Manomó: Carbonatites and BIFs with Rare Earth and Radioactive Mineral Anomalies

Pablo D. Michel[1], German Uzquiano[2]
1. Universidad Mayor de San Andrés, La Paz, Bolivia, 2. Colegio de Geólogos de Bolivia, La Paz, Bolivia

Cerro Manomó is located in eastern Bolivia in the Bolivian Precambrian region, southwest of the Amazon craton in South America. It is a sub-Amazonian region with subtropical jungle and vast vegetation, savannah-type plains, a surface area of 26 km^2, and temperatures between 32° and 38°C. It is a conical and brecciated volcanic complex with extrusive rocks of fenitized carbonatitic magma, with lava flows and alkaline breccias. Various types of rocks containing rare earth elements, uranium, and thorium have been described. Some breccia dikes form annular domes on the top of the hill, with apatite lenses and the presence of BIF-type (non-marine) banded iron formations as well as pumice-type pyroclastic rolls. In the northern sector of the hill, concentrations rich in phosphates and uranium and thorium minerals have accumulated, where the BIFs with bands rich in silica and iron contain high anomalies in rare earth elements and radioactive minerals. The presence of iron carbonates, which are close to the alkaline annular cones, means that the complex is highly altered, with the presence of chalcedonic carbonatites with sharp edges, chalcedony and magnetite, and a record of 1500 cpm of radioactivity. The brecciated dikes have a composition of goethite, quartz-barite, and bastnaesite, which are interpreted as weathered carbonatites. Colluvium of carbonatites with goethite-type weathered iron oxide, in the upper middle part of the hill, has radioactivity of 3000 and 3600 cpm. On the slopes close to the volcanic breccias, there is presence of silver anomalies with little alteration. In the upper part of the hill, there are reddish clayey laterites, of high magnetic susceptibility, with high values of rare earths and with botryoidal goethite, goethite, hematite, jarosite, and magnetite.

Mineralogy and Petrology of the Tsagaantolgoi Fe-Pb-Zn Skarn Deposit in South Mongolia

Otgonbayar Munkhtsol[1], Jamsran Erdenebayar
1. Mongolian University of Science and Technology, Ulaanbaatar, Mongolia, 2. Department of Earth Science and Technology, Akita University, Akita, Japan

The deposit of Tsagaantolgoi skarn in South Mongolia has poly-metal mineralization of calcareous skarn type, formed at the boundary between limestone of the Neoproterozoic Oortzog Formation and granite of the Late Triassic-Early Jurassic Bor-Under complex. Petrographic and mineragraphic analyses of doubly polished samples were conducted at Akita University using the microscope Nikon eclipse LV100-POL (Japan) and scanning electron microscopy with energy dispersive spectroscopy (SEM-EDS).

The results of the petrographic and mineralographic study indicate that the samples collected from the study area /BO-1-BO-27/ are skarn. There are garnets, wollastonite, epidotes, vesuvianites, carbonates, and quartz, which show hypidioblast and microgranoblast textures and have a massive structure. In addition, they are subjected to silicification and carbonatization. There is an enriched concentration of ore minerals in some skarn samples ranging from 5 to 60%.

In the rock, there are ore minerals recognized in the study area including magnetite, pyrite, pyrrhotite, chalcopyrite, hematite, sphalerite, galena, and bornite and secondary minerals such as hematite, chalcocite, zincite goethite, limonite, and some Pb-Zn carbonate minerals. In addition, there are a few aggregations of pyrite and chalcopyrite that intruded sphalerite, and a tendency that big sphalerite and galena isometric grains and veins and veinlets can be noticed during the study. In the study area, there are two possible stages of ore mineralization:

1. Disseminated and patchy magnetites and pyrites formed in the host rock and mineralization range of magnetite-pyrite-chalcopyrite-bornite-sphalerite-galena with veinlets and disseminated structures. Ore minerals were replaced by secondary alteration minerals (hematite, goethite, and limonite).
2. The second stage of mineralization usually occurs along the chlorite, carbonate, and sericite alterations in the host rock, which are altered parts of the primary minerals.

Due to the fact that these alteration zones are also cut by late veins and veinlets, the mineralization stage is strongly correlated with these alteration zones.

Presence of Vanadium in Epithermal Systems of Patagonia, Argentina

Morena L. Pagola[1, 2], Sebastián M. Jovic[1, 2, 3], Diego S. Palma[1, 2], Daniel Hughes[4], Diego Guido[1, 2]
1. nstituto de Recursos Minerales, Facultad de Ciencias Naturales y Museo, Universidad Nacional de La Plata, La Plata, Argentina, 2. Consejo Nacional de Investigaciones Científicas y Técnicas (CONICET), La Plata, Argentina, 3. Departamento de Geología. Universidad de Buenos Aires, Buenos Aires, Argentina, 4. Trendix Mining S.A., Buenos Aires, Argentina

Vanadium has been considered a critical raw material by the European Commission since 2017 (Critical Raw Materials 2017 list) due to its economic importance and the high risk in its supply. The sustainable supply of this element is essential for the industry, as it is used in the steel industry (as an alloy for the manufacture of wind turbines, with aeronautical and space applications), and for its role in the transition towards clean energy (vanadium-based redox flow batteries). Vanadium is primarily extracted from V-Ti magnetite deposits as a co-product of iron. Also, it is recovered as a by-product in U-hosted sandstone deposits, or as a primary product in vanadate deposits and carbonaceous shales. The La Luz deposit, located in the Somoncurá massif, Patagonia Argentina, represents an epithermal deposit, composed of a 7-km NE and NW quartz-carbonates vein system, with widths up to 3 m. The main minerals identified are galena, sphalerite, malachite, chrysocolla, and native copper in quartz-calcite-rhodochrosite gangue. Surface samples analyzed (ICP on ore minerals) show high anomalies of gold, silver, lead, copper, and zinc and the presence of high values of vanadium (up to 1,455 ppm) associated with indium (up to 310 ppm) and gallium (up to 149 ppm). A positive correlation between Pb-V is observed, which could be related to the presence of some lead vanadate such as vanadinite. On the other hand, indium shows a good correlation with Cu. The determination of these anomalies in surface samples represents high potential at depth. The presence of these metals, mainly V, is relatively rare in this type of deposit and represents a discovery in the region. This assemblage increases the mining potential and must be considered during future exploration duties. More detailed studies are underway to determine the distribution and presence of these critical elements in this deposit.

The West Desert Zinc-Copper-Silver-Indium Deposit, Juab County, Utah, USA

Mark I. Pfau
Tellurian Exploration, Inc., Missoula, MT, USA

The West Desert zinc-copper-silver-indium deposit is in west-central Utah, USA, and is being developed by American West Metals Ltd. (ASX: AW1 and OTCQB: AWMLF). West Desert is a late Eocene age ferro-magnesium skarn-CRD replacement deposit, overprinted by zinc-copper-silver-indium hydrothermal mineralization and intruded by a porphyry molybdenum system at depth. Current resources are 33.7 Mt of inferred and indicated resource grading 3.83% Zn, 0.15% Cu, and 9.1 g/t Ag. The historic indium resource is being updated and stands at 869,000 kg grading 35 ppm indium. A combined open-pit and underground mine encompassing oxide and sulfide resources is being planned. West Desert is the largest viable base metal-indium deposit in North America. The Utah Geological Survey is working in conjunction with the U.S.G.S to further develop the project as a critical metal source. Gallium in the West Desert deposit is being evaluated as part of this process.

Indium at West Desert occurs in two forms: 1) a lattice substitution in sphalerite with lesser amounts in bornite and chalcopyrite and 2) the mineral roquesite ($CuInS_2$), which typically occurs in chalcopyrite-bearing sphalerite as replacement along crystal dislocations (chalcopyrite disease).Historical metallurgical testing indicates that about 94% of recoverable indium reports to a zinc concentrate and about 6% of indium to a copper concentrate. Further work is being conducted with QEMSCAN, TIMA, and column leach tests to refine oxide zinc and copper recovery and the indium deportment to each product. Critical to indium evaluation are the following: 1) type of mineral deposit being explored; 2) use of proper geochemical standards and sample digestion techniques for accuracy; 3) mineralogical distribution of indium within the various sulfide ore minerals. There are six exploration projects worldwide that may contribute to the near-term (5-20 year) supply of new indium sources. West Desert is currently the leading North American project.

Crosscutting Relations in Advanced Argillic Environment Revealed by Hyperspectral Core Imaging: Antakori Cu-Au-Ag Project Case Study

David G. Portocarrero
University of Tasmania, Hobart, TAS, Australia

The Antakori project is located 50 km northwest of the city of Cajamarca and is part of the Miocene cluster of Cu-Au-Ag deposits of northern Peru. An extensive advanced argillic alteration zone is developed in the shallower parts of the Antakori project, mainly above the contact between the cretaceous sedimentary rocks and the middle-Miocene volcanic rocks of the Calipuy Group, extending to the west to the Tantahuatay oxide mine and to the east near the Sinchao fault. The alteration develops open-space filling, vein selvages, and pervasive and patchy textures that destroy primary textures and give a powdery, white to grey appearance to the rocks. Epithermal high-sulphidation Cu-Au-Ag mineralization occurs in this environment as disseminations, veins, and semimassive subvertical structures and subhorizontal replacement bodies. The ore assemblage is dominated by enargite-luzonite-pyrite-tetrahedrite with lesser chalcocite, bornite, covellite, and Pb-sulphosalts. Three assemblages of advanced argillic alteration were recognized based on crosscutting relations: (I) pyrophyllite±diaspore, (II) kaolinite±dickite±white-mica±topaz, and (III) alunite±zunyite. There is a clear spatial and genetic association between high-sulphidation Cu-Au-Ag mineralization and assemblages (II) and (III), with increasing grades when zunyite and topaz are more abundant. The observations made in the present research led to the conclusion that the advanced argillic alteration found at Antakori is the product of two different hydrothermal fluids. An early assemblage of pyrophyllite±diaspore is developed at the base of an interpreted barren and extensive lithocap (stage I) formed at expenses of an underlying degassing magma. The early assemblage is then followed by an ore-related, structurally controlled, F- and Cl-rich hydrothermal fluid responsible for the formation of kaolinite±dickite±white-mica±topaz and alunite±zunyite assemblages (stage II) that may be related to a deeper mineralized porphyry system at depth. Thus, the occurrence of zunyite and topaz can be used to vector exploration for high-sulfidation mineralization in the lithocap environment.

A Stratigraphic Framework for the Zn-Pb Ba VMS Deposit, NW BC, Canada, Using New U-Pb Age Dates and Lithogeochemistry; Implications for Mineral Systems in the Golden Triangle

Kyle Powers, Shaun Barker, Kate Rubingh, Brian McNulty
University of British Columbia, Vancouver, BC, Canada

The BA Zn-Pb VMS system is located in the southern half of the well-endowed Stikine Terrane of Northwest British Columbia, Canada. It is hosted in the Jurassic (ca. 201-168 Ma) Hazelton Group; a bimodal island arc assemblage comprising submarine to subaerial coherent and volcano-sedimentary rock units that define multiple paleo-volcanic centers across a 150-km area. Several VMS deposits are hosted throughout the Hazelton Group stratigraphy including the silver-rich Dolly Varden deposit at the base and the gold-rich Eskay Creek deposit at the top. The Zn-Pb BA deposit is geographically between these two significant mineral deposits that lie 130 km apart in BC's Golden Triangle. However, the age and lithological provenance of the host stratigraphy remains ambiguous.

The volcanic facies at BA are assigned to both the Lower and Upper Hazelton Group using primarily textural classification with no proximal geochemical/geochronological constraints over the 25 Ma interval. Several unconformities have been mapped at BA and further complicate unraveling the stratigraphy in the study area and contextualizing its paleo-depositional setting within the Hazelton Group. In addition, the mineralization style at BA differs from the above precious metal-rich deposits, composed largely of the base metals Zn and Pb, which could be due to its location within the Hazelton Group.

Detailed field relationships and drill core logging are combined with new whole-rock major and trace element lithogeochemistry, shortwave infrared spectroscopy, and zircon U-Pb dates from key stratigraphic positions such as unconformities, mineralized horizons, and intrusive units. These results will establish the geological framework of the BA mineralizing system in both temporal and regional stratigraphic context in relation to the Dolly Varden and Eskay Creek VMS deposits.

Role of Hydrothermal Solution in Base Metal Enrichment in Alwar Basin, Rajasthan, India

Sajeev R[1], Dinesh Pandit[2], Mallickarjun Joshi[2], Sanjay Das[1]
1. Geological Survey of India, Bengaluru, KA, India, 2. Banarus Hindu University, Varanasi, UP, India

The Alwar basin forms part of the Meso-Proterozoic North Delhi Fold Belt (NDFB). The NDFB is known for its base metal occurrence for the last several decades. In the present work, the role of hydrothermal solution in base metal enrichment is assessed from the Amra Ka Bas area of Alwar Basin. Amara Kas Bas is situated 4 km NE of the Mundiyawas Khera (MK) deposit and is situated along the trends of the MK deposit with similar host rock and ore mineral assemblages. The main rock types exposed are a metavolcanic-sedimentary sequence, tremolite-bearing dolomitic marble, and carbonaceous phyllite of the Thanagazhi formation of the Ajabgarh Group. The characteristic hydrothermal alteration zones identified in the study area are phyllic, propylitic, and argillic. Sulphide mineralization is mainly associated with the phyllic alteration zones. Epigenetic sulphide mineralization is mainly associated with the propylitic alteration zone. The spatial association of the hydrothermal alteration zone with the weak planes indicates that they might have acted as the pathways for hydrothermal fluid migration and associated sulphide mineralization in the area. Ore petrographic studies reveal at least four phases of sulphide enrichment in the area. The first and second phases are manifested as fine dissemination of sulphide phases within felsic volcanics and are syngenetic phases. The third phase appears to be mainly responsible for scavenging, remobilizing, transporting, and depositing ore along favourable weak planes with the help of the hydrothermal solution. The fourth phase is a prominent, low-temperate phase marked by the development of colloform pyrite over the early phases. The weak planes available in the area have facilitated easy migration of the ore-bearing fluid and its deposition. The main key, therefore, for identifying a promising area for future exploration is the identification of hydrothermal alteration signatures in close association with prominent weak planes of the area.

Petrographic Study of Zinc Ore and Subordinate Veins and Veinlets in the North Sucuri Region, Vazante Mine, Minas Gerais – Brazil

Poliana Vidal Salgado [1, 2], Carolina Penteado Natividade Moreto, [2], Fernando Lucas dos Santos Peixoto de Villanova [1]
1. NEXA Resources, São Paulo, SP, Brazil, 2. State University of Campinas, Campinas, SP, Brazil

The Vazante district, located in the northwest of the Minas Gerais state, Brazil, holds one of the largest hypogene willemite deposits in the world, hosted in carbonate rocks of the Vazante Group. The unusual deposit has a total resource of 21 Mt 6.8% zinc, 0.34% lead, and 16.6 g/t silver and contains the North Extension and Vazante mines, both placed around 8 km from the city of Vazante (Minas Gerais). Its ore is structurally controlled by the Vazante fault, with orientation N50E/60NW. In the North Sucuri region, at the northeastern limit of the Vazante mine, were found sulphide zinc bodies imbricated or not associated with willemitic ore, which is uncommon to the deposit. Petrographic studies showed that the hydrothermally brecciated zone comprises the same lithotypes common to the deposit, such as iron-carbonated breccia, greenish to pinkish and sericitized metadolomites, metabasic rocks, and a carbonate bleached, greyish breccia that usually hosts the zinc sulphide ore unrelated to the willemite principal orebody. The mineral assemblage of the zinc ore is composed of the following: (i) principal ore body: willemite + dolomite + quartz + hematite ± barite ± apatite ± franklinite ± zincite ± magnetite; (ii) willemite ore with imbricated sulphide lenses: willemite + dolomite + hematite ± franklinite ± zincite associated with sphalerite and galena; (iii) sulphide isolated bodies: sphalerite + galena ± barite ± pyrite ± hematite ± zincite ± anhydrite. The sphalerites of types ii and iii are different in visual color and Fe content. The hydrothermally altered and brecciated zone also includes late venulation of oxidized iron-carbonate + hematite ± limonite, dolomite + quartz + barite ± hematite ± zincite, chlorite ± sphalerite ± galena, and dolomite + quartz + pyrite. Theses veins and veinlets can occur truncated, failed and overlapping each other.

Geological Modelling of the North Sucuri Region, Vazante Mine, Minas Gerais - Brazil

Poliana Vidal Salgado [1, 2], Carolina Penteado Natividade Moreto, [2], Fernando Lucas dos Santos Peixoto de Villanova [1]
1. NEXA Resources, São Paulo, Brazil, 2. State University of Campinas, Campinas, SP, Brazil

The northwest of the Minas Gerais state, Brazil, holds a world-class hypogene willemite deposit hosted in carbonate rocks of the Vazante Group, containing a total resource of 21 Mt 6.8% zinc, 0.34% lead, and 16.6 g/t silver. This deposit is divided by a severe karstication area that separates the North Extension and the Vazante mines, located 8 km from the city of Vazante (Minas Gerais state). The mines follow the northeast orientation of the Vazante fault, which structurally controls the willemite orebody. The North Sucuri region is placed on the northeastern limit of the Vazante mine, where the ore found with infill drilling is composed not only of the classic mineral assemblage of the willemite orebody (willemite+dolomite+quartz+hematite±barite±apatite±franklinite±zincite±magnetite and imbricated sulphide lenses) but also of sphalerite bodies composed of sphalerite+galena±barite±pyrite±hematite±zincite±anhydrite, unrelated to the principal ore. The infill drilling was unoriented, but geological mapping of the gallery and subadjacent areas in development prove that host rocks are the Lower Pamplona and Upper Morro do Pinheiro members of the Vazante Group, as expected, with N50E/20NW bedding orientation and N45W/50SW fractures. Macro-description of more than 9,000 meters of drill holes, allied to geological mapping data, resulted in the geological modelling of the host rocks and orebody with Leapfrog GEO (Seequent), where the carbonated/siliciclastic rocks of the Lower Pamplona and Upper Morro do Pinheiro member compose a hydrothermally altered and brecciated zone that hosts the ore. The modelling of the willemite and sulphide bodies shows that, despite the structural and composition complexity of the region, the orebodies remain following the Vazante fault and main body orientation. Moreover, the geological mapping of galleries and geological modelling results of the region can be useful for better comprehension of the geological setting and to the ore beneficiation process while identifying contaminating rocks.

Geochemistry of Apatite and Biotite in Tin-Mineralized Late Triassic Granites in Bangka Island, Indonesia

Enrico Gilrandy W. Suharjo[1], Ryohei Takahashi[1], Andrea Agangi[1], Syafrizal Syafrizal[2], Pearlyn Manalo[1]
1. Akita University, Akita, Japan, 2. Institut Teknologi Bandung, Bandung, Indonesia

Late Triassic granite in Bangka Island is classified into two types based on mineral associations, i.e., biotite granite and hornblende-biotite granite. The biotite granite is more enriched in tin (mean of 14.4 ppm Sn) compared to the hornblende-biotite granite (mean of 3.30 ppm Sn). Both of those lack magnetite, indicating reduced conditions of formation, which is considered to be suitable to tin mineralization.

FeO content of apatite in the biotite granite (mean of 0.53 wt %; range from 0.17 to 2.10 wt%) is higher than that in the hornblende-biotite granite (mean of 0.29 wt %; range from 0.04 to 2.53 wt %), indicating a higher degree of Fe^{2+}/Fe_{Tot} ratio for the former. This suggests that the biotite granite formed under more reduced conditions than the hornblende-biotite granite. Higher water content of apatite in the biotite granite compared to that in the hornblende-biotite granite would be due to elevated water contents in the melt. The $Fe^{2+}/(Fe^{2+}+Mg^{2+})$ ratios of biotite in the biotite granite and the hornblende-biotite granite are mean of 0.81 with a range from 0.75 to 0.83 and mean of 0.74 with a range from 0.71 to 0.75, respectively, indicating more reduced conditions for the former.

The results suggest that the biotite granite is strongly related to tin mineralization in Bangka Island, and the mineralization is attributed to water-rich magmas with reduced conditions. We consider that chemical compositions of apatite and biotite can be indicators of tin mineralization for granite in Bangla Island.

Generation of Paired Cu-Au Mineral Systems Within a Suprasubduction Zone Setting in Eastern Proterozoic Australia

Caroline Tiddy, David Giles
MinEx CRC, Future Industries Institute, University of South Australia, Adelaide, SA, Australia

Eastern Proterozoic Australia is well endowed with mineral deposits, including paired ca. 1.60-1.57 Ga IOCG and porphyry-epithermal Cu-Au mineralisation in the Gawler Craton, the generation of which is conjectural. We propose an alternative suprasubduction zone model for eastern Proterozoic Australia where the spatial and temporal distribution of mineral systems in the Gawler Craton was dictated by the evolution and chemistry of the fluids/melts derived from the subducting slab (Tiddy and Giles, 2020: Ore Geol. Rev. 122, 103483). The model draws upon modern geological analogues including the Late Cretaceous to Early Eocene Laramide Orogeny (western USA) and genesis of Cretaceous IOCG and Cu-porphyry deposits in the Chilean Coastal Cordillera.

A flat-slab subduction geometry prevailed from ca. 1.604-1.595 Ga. Anhydrous, F-rich, S-poor melts were produced and interacted with the subcontinental lithospheric mantle (SCLM). Progressive eclogitisation and densification of the frontal oceanic slab resulted in slab steepening from ca. 1.595 Ga, causing sudden upwelling of hot asthenosphere and further slab melting and SCLM metasomatism. Mafic melts were emplaced into the lower crust and underwent assimilation-fractional crystallisation processes, eventually forming the dominantly A-type Hiltaba Suite granites and Gawler Range Volcanics that were emplaced in the upper crust and are associated with IOCG mineralisation. Further slab steepening and slab retreat is evidenced by the present-day progressive westerly to southwesterly younging direction of the ca. 1.595-1.575 Ga Hiltaba Suite and saw development of an asthenospheric wedge and a return to a normal subduction geometry. Heat introduced from the asthenospheric wedge caused dehydration of the downgoing slab to produce hydrous, S-rich fluids that interacted with the asthenosphere to generate oxidised, S-rich partial melts related to porphyry/epithermal mineralisation. Therefore, within this suprasubduction zone model, mineralisation style is dictated by whether downgoing slab melts (anhydrous, F-rich, S-poor: IOCG) or slab fluids (hydrous, S-rich: porphyry/epithermal) interacted with the SCLM.

Anglesite-Bearing Polymetallic Hydrothermal-Shear Breccia in the Paja Formation, Eastern Cordillera of Colombia

Nicolás Tique-Ladino, María Alejandra González-Cepeda, Ariel Cadena-Sánchez, Juan Carlos Molano-Mendoza
Universidad Nacional de Colombia, Bogota, Colombia

Hydrothermal polymetallic occurrences in Colombia are commonly associated with volcanic-geothermal systems in collisional settings related to arc magmatism. Sediment-hosted and stratigraphically bounded mineralizations are sparsely distributed and fluid sources are still in debate due to lack of systematic data. Here we report an explosive breccia in a right-lateral shear system, bounded by two 20-cm levels extremely rich in organic matter in fissile black shales from the lower Albian from the Paja Formation, biostratigraphically dated with Dufrenoyia specimen samples from a facially correlated level 2 km from the main locality. The breccia is composed of angular fragments of carbonized organic matter, vitrinite, anglesite, malachite, sphalerite, Fe-oxides, and fuchsite. The paragenetic association, including Pb-Cu-Zn and Cr minerals with carbonate, sulfur, and carbonate anionic groups, indicates the interaction of a hydrothermal fluid with organic-rich levels in the shales.

Low-temperature associations give insights into a closed circulation convective system, cracking open shear faults crosscutting the strata and carrying fluids in between carbonized organic matter as mineralized breccias that crystallized in shallow levels and chemically do not require external fluid fertile sources. Field geology, thin section and charcoal maceral petrography, structural analysis, Raman spectroscopy, EPMA, and geochemical analysis let us conclude that organic-rich levels from black shales facies of the upper segment of the Paja Formation hold significantly metallic and REE elements in non-conventional geologic settings. Cr-rich shales can be used as a pathfinder and vectoring mechanisms for fluid transiting to upper stratigraphic levels of facially connected structures with favorable permeability settings, as reported near Zn-Pb vein-breccias. This study serves as a first report on polymetallic mineralizations on shale with proper stratigraphic, structural, and fluid properties description in order to explore for strategic resources in suitable settings from nearby rocks in the Eastern Cordillera.

Element Fluxes During Alteration in High-Sulfidation Epithermal Systems: A Case Study from the Pueblo Viejo Au-Ag(-Cu) Deposit, Dominican Republic

Ethan R. Tonks[1, 2], Jamie J. Wilkinson[1, 2], Robin N. Armstrong[1], Yannick Buret[1], Andrew T. Wurst[3]
1. Department of Earth Sciences, Natural History Museum, London, United Kingdom, 2. Imperial College, London, United Kingdom, 3. Barrick Gold Corporation, Toronto, ON, Canada

The alteration of rocks associated with hydrothermal ore deposits reflects a large-scale and complex migration of chemical elements, which may be added into the system by externally-derived fluids or mobilised from one alteration facies to another, following hydrothermal dissolution of minerals in the precursor rock and subsequent reprecipitation as new minerals elsewhere. Understanding chemical and mineralogical changes that occur during alteration and mineralisation provides an insight into the nature of ore-forming fluids and the changes in their physico-chemical properties during intense fluid-rock interaction.

To investigate the behaviour of elements and key mineral breakdown and reprecipitation reactions that occurred during alteration at the Pueblo Viejo Au-Ag(-Cu) high-sulfidation epithermal deposit, Dominican Republic, a whole-rock elemental mass transfer analysis was integrated with automated mineralogy (SEM-EDX) and mineral chemistry (EPMA and LA-ICP-MS). To assess spatial variations in elemental behaviour, multiple samples of andesite belonging to three alteration facies (chlorite-carbonate-illite, quartz-pyrophyllite, and quartz-alunite) were considered.

Elements conserved during alteration—REEs for chlorite-carbonate-illite alteration and LREEs for quartz-pyrophyllite and quartz-alunite alteration—were used to remove the effect of bulk mass change for the mass transfer analysis. Although epidote, an important host of REEs in the protolith, was destroyed during chlorite-carbonate-illite alteration, REEs were locally reprecipitated in newly formed carbonates. During quartz-pyrophyllite and quartz-alunite alteration, LREEs were preferentially incorporated on a local scale into newly formed alunite and APS minerals, following the dissolution of REE-bearing minerals (apatite, carbonates, epidote, and titanite).

Based on their behaviour both within and between each alteration facies, the 64 elemental components considered can be divided into seven groups. Elements such as Ti and Zr, typically considered to be immobile in hydrothermal systems, were mobilised during intense acid alteration. The changes in several chemical species (including Fe, Ca, C, S, Li, Zn, Se, and Te) were spatially systematic, which has important implications for geochemical exploration.

IOCG Systems: Are They Just a Problem of Ore Traps?

Fernando Tornos[1], Tobias Bauer[2], John M. Hanchar[3]
1. Consejo Superior de Investigaciones Cientificas, Madrid, Spain, 2. Lulea Technological University, Lulea, Sweden, 3. Memorial University of Newfoundland, Saint John's, NL, Canada

IOCG mineralization encompasses a range of hydrothermal deposits that share the presence of magnetite and hematite along with variable amounts, if present, of chalcopyrite and gold. Hydrothermal alteration, relationships with igneous rocks and structures, and origin of fluids varies between districts. A systematic study of the major IOCG belts suggests that the so-called IOCG deposits embrace a wide range of types of mineralization with no genetic connection; most deposits are formed by fluid-rock interactions overprinting already emplaced rocks during a metasomatic process geochemically similar to that of skarns. The mineralization is often formed by (a) reaction with magnetite-bearing shear zones as well as previous ironstones and magnetite-(apatite) rocks along structures; (b) skarn-like stratabound replacements of Si-Al-Ca–rich rocks (usually volcanic) with a zonation from feldspar+actinolite±biotite to actinolite/biotite rocks when in mafic rocks and feldspar-tourmaline when in felsic rocks; (c) shear zone-related alteration of equivalent rocks.

The hydrothermal fluids are very likely high temperature (ca. >400°C) and alkaline, equilibrated with Si-Al-rich intermediate rocks, and form the IOCG system at rather high rock/fluid ratios. Reactions with a previous ironstone destabilizes the Cu-Au-bearing complexes and can produce the Cu-(Au) mineralization. Reaction with intermediate-felsic volcanic rocks alone is not able to produce a significant change in fO_2 and pH and we suggest that fluid-rock interaction with reactive rock traps or fluid mixing with basin-derived water is the main mechanism of ore formation in the stratabound IOCG mineralization. Further cooling and oxidation when circulating along shear zones likely leads to the precipitation of low-temperature assemblages with hematite, white mica, chlorite, and carbonates. When the sequence hosts calc-silicate hornfels or limestone, skarns rich in garnet can form.

Our hypothesis is that IOCG systems can form in relationship with any magmatic-hydrothermal system related to intermediate intrusives and do not need a unique geodynamic scenario for their formation.

Advances in Knowledge: High-Grade Cu-Au-Ag Mineralization Hosted in Tourmaline Breccias at Soledad, Peru

Victor J. Torres Pacheco[1, 2], David R. Cooke[1], Lejun Zhang[1], Julie Hunt[1], Jerry Blackwell[2], David L. Kelley[2]
1. University of Tasmania, Hobart, TAS, Australia, 2. Chakana Copper Corporation, Vancouver, BC, Canada

The Soledad project is located in the Cordillera Negra of west-central Peru, within the Miocene Metallogenic Belt. Initial exploration focused on tourmaline breccia pipes hosting sulfide mineralization with strong Cu-Au-Ag grades. In 2022, an initial inferred resource was published on seven breccia pipes from the surface to an average depth of 330 m, containing 6.7 Mt grading 0.88 g/t Au, 0.88% Cu, and 54.2 g/t Ag. This resource reflects only a small portion of the potential of the Soledad mineral system as the tourmaline breccias are just one of several mineralization styles related to a major intrusive center at Soledad that are subject to ongoing exploration.

The pipes cut Cretaceous volcanic and epiclastic basement unconformably overlain by Tertiary Calipuy Formation with subaerial andesite flows, lithic tuff, and lesser dacite tuff ranging in age from 52.34 ± 0.67 to 40.03 ± 0.25 Ma. The magmatic-hydrothermal system lifespan spanned ~1 m.y., beginning with granodiorite at 15.65 ± 0.18 Ma and ending with mineralized breccia at 14.71 ± 0.07 Ma. All the intrusions have a fertile porphyry copper prospectivity signature according to Sr/Y and V/Sc ratios. The structural intersection of the NE with N-NW faults is a favorable location for breccia pipes.

Mineralization was introduced immediately after breccia pipe formation, initially with gold-bearing pyrite and arsenopyrite; followed by chalcopyrite-tetrahedrite, then Sb-Pb-Cu-Bi-Ag sulfosalts; and finally, electrum. Surficial oxidation extends to ~35-m depth. Within a pipe the upper levels are higher in Au-Ag, phyllic alteration is notable, and textures are shingle to mosaic. At lower levels, chaotic textures dominate, Cu is higher, with stronger quartz-tourmaline replacement.

Geometallurgical studies are ongoing to 1) resolve precious and critical metal deportment in sulfide and sulfosalt minerals to optimize mineral processing options, and 2) identify deleterious element deportment in sulfide and sulfosalt minerals for effective waste management planning.

Dating and Trace-Element Characterisation of Sn-W Skarn-, Greisen- and Vein-Type Occurrences from East-Greenland

Nynke Keulen, Benjamin Heredia, Sebastian N. Malkki, Tonny B. Thomsen, Diogo R. Rosa, David Whitehead
GEUS - Geological Survey of Denmark & Greenland, Copenhagen, Denmark

Different deposit types, such as skarns, greisen, and vein-type each have their own unique mineral trace element compositions in ore minerals like scheelite or cassiterite. A good understanding of compositional trends in these minerals can serve to assess what deposit types are expected to be found upstream from analysed stream sediment or till samples, and thus assist in mineral exploration and in the unravelling of the geological history of orogens.

The traceability of Sn- and W-minerals to their deposit type (skarn, greisen or veins) was tested on rock samples and stream sediments from East Greenland. Scheelite and cassiterite grains from nine occurrences were analysed for their host rock mineral assemblage and geochemical fingerprint (trace elements and dating by laser-ablation inductively coupled plasma mass spectrometry). Rock samples from eight localities were collected from skarn, greisen or vein occurrences, while the ninth sample represent loose blocks transported by a glacier and thus has an unknown origin. The mineralogy of the samples was investigated with automated quantitative mineralogy (AQM) and the trace element compositions of minerals were compared to trace element values for the same deposit types, as published in the literature. Additionally, the U/Pb-dates for scheelite, cassiterite as well as zircon, apatite and rutile presumed to have formed during the same mineralising events were used to estimate the timing of the mineralisation. AQM observations on the thin sections match well with trace element data and reveal a variety of mineralisation styles within and between occurrences. In addition, new dates of the minerals reveal a more complex mineralisation history than previously known, seemingly comprising three events during the Caledonian orogeny.

The Genesis of Zhijin REE-Phosphorite Deposit in Guizhou Province, China

Yuling Xie, Yunwei Qu, Ziwei Li, Pei Liang
University of Science and Technology Beijing, Beijing, China

The HREEs (heavy rare earth elements) hosted in phosphorite have significant economic importance, but the origin of the P and the REEs of the phosphorite has been controversial for decades. As a typical marine sedimentary deposit, the P of the deposit has been believed to originate from continent detrial material or from the river. Volcanic material involvement has also been proposed, but the key evidence for that is still lacking. Based on micrography, CL (cathodoluminescence) imaging, SEM/EDS (scanning electron microscope/X-ray energy dispersion spectrum), and TEM (transmission electron microscope) results of phosphorite and the silicate volcanic interlayers in the footwall dolomite, this paper discusses the genesis of the Zhijin P-REE deposit.

Our results show that the phosphorus spherulite usually has a layered outer zone, a microcrystal middle zone, and a crystal inner zone. The outer zone shows isotropism character. The middle zone comprises microcrystal dolomite, apatite, and zeolite. The inner zone includes apatites, dolmite, and quartz crystals, with minor barite, zeolite, asphalt, and pyrite. The CL images indicate high-temperature quartz and apatite characteristics. Laser Raman results show increasing crystallinity (informed from FWHM for the 964 cm-1 peak) of apatite from the outer zone to the inner zone. Some round phosphate and sulfate glassy globes can be found in the quartz of the volcanic rock, implying an immiscibility between silicate, phosphate, and sulfate melts. Combining the ore texture and the mineralogical results of phosphorite spherulites and the discovery of the immiscible phosphate and sulfate glassy globe in volcanic interlayers, it can be concluded that the phosphor spherulites are chilled volcanic droplets of carbonatite-phosphate melt. The carbonatite-phosphate melt originates from the immiscibility between silicate and phosphate melt. The submarine volcanic origin of phosphorite is significant to the P-REE exploration and also provides new insight into global climate change of that time.

A New High-Precision SEM – LA-ICP-MS Sample Coordinate Integration Method

Sebastian N. Malkki, Benjamin Heredia, Nynke Keulen, Tonny B. Thomsen
Geological Survey of Denmark and Greenland (GEUS), Copenhagen, Denmark

For geological and non-geological samples, multi-instrumental analyses are often required to obtain a comprehensive dataset from a given sample. Until now, trying to re-localise the exact area or mineral grain to analyse after transferring the sample from one analytical instrument to another to the next analytical instrument is not just time consuming, but often unprecise. For non-transparent and/or samples with uneven surfaces, it is extremely difficult and often impossible to locate single grains used for laser-ablation inductively-coupled-plasma mass-spectrometry (LA-ICP-MS) systems. We present a new analyses points coordinating approach, connecting the scanning electron microscope (SEM) and LA-ICP-MS analytical systems using the transposition of coordinates of a randomised Euclidean space. This approach allows for fast and precise localisation of mineral grains. By marking samples with more than 4 laser points using a CO_2 laser engraver, the transfer of the coordinates from the mosaic montage created by ZEISS Mineralogic software into a GEUS developed MATLAB script is possible by matching the fiducial points. Specific features, e.g., zircons, can be selected by thresholding their assigned colour from the false-coloured Mineralogic mineral map and their positions transferred as a list of points from the MATLAB script. Ultimately any succeeding laboratory instrument can be guided to the exact same positions in that instrument's coordinate system.

Shear-Related Sn-W Mineralisations of the Martinamor Extensional Gneiss Dome (Salamanca, Spain)

Daniel Bermejo-López[1], Juan Gómez-Barreiro[2], Lorena Ortega Menor[1], Santos Barrios[2], Pedro Castiñeiras García[1], Elena Crespo Feo[1]

1. Complutense University of Madrid, Madrid, Spain, 2. University of Salamanca, Salamanca, Spain

The Morille-Martinamor tin and tungsten mining district (Salamanca, western Spain) is part of the Martinamor extensional gneiss dome. Gneissic domes formed due to the gravitational collapse of the orogen during the late Variscan. Ductile extensional shear zones accommodated this process and led to a tectonic denudation and thermal attenuation of the crust. Strain increases as we approach the shear zone, with the development of a penetrative subhorizontal tectonic foliation (SE_2), recumbent and sheath folds, and shear bands (SCC') showing a top-to-the SE sense of shear. Foot-wall exhumation results in an isothermal decompression and heat transfer through the detachment zone. A syn-kinematic low-pressure metamorphism, reaching melting conditions in the dome core, overlaps the previous metamorphic signature.

In this context, the development and type of Sn-W mineralisation depends on the structural level and is mainly concentrated across the extensional shear zone. In the lowermost structural levels, Sn-W deposits occur in veins with quartz, wolframite, scheelite and cassiterite. On the other hand, in shallower zones, W stratabound-type mineralisation is predominant, with disseminated scheelite in metasedimentary calc-silicate levels.

Strain evolution within the extensional detachment led to structurally controlled mineralisations. For example, progressive deformation of recumbent/sheath folds, formed in the most competent levels, form scheelite saddle reef-type deposits in hinge zones and boudinage-neck and en echelon veins deposits along the limbs, resulting in the creation of favourable sites for the scheelite deposition and possibly in the redistribution of the stratabound mineralisation. Structural criteria have been used to define exploration targets in the area.

Fluid Evolution of the Ostra Hydrothermal Cu-Zn-Pb-Barite Mineralisation (Eastern Carpathians, Romania)

Balázs B. Boglárka[1], Árpád Ádámcsik[1], István Márton[2], Szabolcs Orbán[3]
1. Eötvös Loránd University, Faculty of Sciences, Institute of Geography and Geosciences, Department of Mineralogy, Budapest, Hungary, 2. Dundee Precious Metals Inc, Toronto, ON, Canada, 3. Goldron Geoconsulting Ltd., Budapest, Hungary

The Eastern Carpathians are a part of the Carpathian Mountains in Central Europe. They host a series of volcanogenic massive sulphide (VMS) deposits. The Ostra polymetallic mineralisation is situated in the northern range, in Romania. It is one of the various polymetallic Pb-Zn-Cu-Au-Ag, Cu-pyrite, barite, Fe-Mn, and uranium mineralisations, which are associated to the low-grade methamorphic Tulgheș Lithogroup of the Crystalline-Mesozoic Zone. The geologically correlated, close-by VMS occurrences (e.g., Leșul Ursului and Mănăila) have a proven [Cu (Au-Pb-Zn-Ag-barite)] content, which could provide an innovative approach to the belt scaled exploration.

Ostra was actively mined for barite in the 1900s; in spite of this, detailed mineralogical and geochemical data are not available. The barite, together with the underlying massive sulphide bodies, is hosted by a porphyroid rock. The ore mineralisation consists of pyrite, chalcopyrite, sphalerite, galena, and fahlore; the gangue minerals are barite and rare quartz. Barite occurs in two textural and paragenetic relationships: it can either be associated with distal parts of the massive sulphide lenses, or it can form massive bodies (without sulphides) in the host porphyroid rock. The composition of the barite is similar in each type; Sr content up to 3.64 wt % is common. Primary fluid inclusions of barite associated with the distal sulphides were trapped from a homogenous parent fluid, which can be modelled in NaCl-H_2O system. The homogenisation temperatures (minimum formation temperature) are Th(LV-L)=145-180°C; salinities are 3.7-5.3 NaCl equiv. wt. %, showing slightly higher values compared to modern seawater. Low formation temperature (<200°C) of the distal ore is supported by the low Fe content of syngenetic sphalerite, while the proximal ore of the massive sulphide lens formed at higher temperature (270±10°C, based on pyrite and sphalerite thermometry). More intense mixing with ambient seawater in the distal parts could cause this observed temperature decrease.

Re-Assessing the Controls on Magmatic-Hydrothermal W-Sn-Cu-As-Zn-Pb Mineralisation and Zonation Associated with the Cornubian Batholith, SW England

Kiara C. Brooksby[1], Robin K. Shail[1], Jens C. Andersen[1], Jamie J. Wilkinson[2, 3]
1. Camborne School of Mines, University of Exeter, Penryn, United Kingdom, 2. Natural History Museum, London, United Kingdom, 3. Imperial College London, London, United Kingdom

SW England hosts a world-class W-Sn-Cu-As-Zn-Pb ore field associated with the Early Permian granites of the Cornubian Batholith [1]. There has been a substantial hiatus in research on the lode systems of the region following the cessation of deep tin mining in 1998. The resurgence of exploration drilling for W, Sn, Cu, and Li, and deep geothermal energy, has provided a timely opportunity to provide a re-assessment of mineralisation processes using modern analytical techniques.

The project integrates studies of the principal ore-stage parageneses using transmitted/reflected-light optical microscopy, SEM, QEMSCAN, and electron microprobe with LA-ICP-MS analysis of fluid and melt inclusions. This aids in constraining the compositional variability of primary exsolved magmatic-hydrothermal fluids and their potential relations to different granite types [2], sampled in deep geothermal sites to depths of 5 km. Additionally, work is completed to evaluate the effects of structurally related fluid migration and mixing with meteoric fluid and/or basinal brines on controlling precipitation and variation in mineral assemblages.

Current drill campaigns have provided samples from historic lode structures including "The Great Flat Lode" intercepted by Cornish Metals at Carn Brea. This provides significant insights into the controls on the heterogeneous distribution of metals around “emanative centres” that indicate a complex relationship between episodic magmatism, separation of magmatic volatile phases, faulting and fluid mixing. As part of this, we will be addressing magmatic-hydrothermal mineral zonation (W-Sn ± As proximal and Cu, Zn, and Pb distal) and the development of one of the best examples of a copper-rich tin system [3].

This project is in collaboration with the Natural History Museum, Cornish Lithium, Cornish Metals, Cornwall Resources, and Cornish Tin.

[1] Jackson, N.J. et al. (1989) Economic Geology 84:1101-1133

[2] Simons, B.J. et al. (2017) Lithos 278–281:491–512.

[3] Sillitoe, R.H. and Lehmann, B. (2022) Mineralium Deposita, 57:1-11

Discovery, Geology and Timing of the Nautanen North Cu-Au-Ag-(Mo) Deposit, Norrbotten, Sweden

David Drejing-Carroll[1, 2], Dave Coller[3], Murray Hitzman[1], Danny Hnatyshin[2, 4], Sean Johnson[1, 5], Franziska D. Wilke[6], Robert Creaser[7], Adrian Boyce[8]
1. Boliden Mines AB, Boliden, Sweden, 2. iCRAG, SFI Research Centre in Applied Geosciences, Dublin, Ireland, 3. Earth Tectonics Ltd, Castlesallagh, Ireland, 4. Natural Resources Canada, Ottawa, ON, Canada, 5. Rio Tinto Exploration, Reading, United Kingdom, 6. GFZ German Research Centre for Geosciences, Potsdam, Germany, 7. University of Alberta, Edmonton, AB, Canada, 8. Scottish Universities Environmental Research Centre, East Kilbride, United Kingdom

Nautanen North is a newly discovered iron oxide-copper-gold deposit located in the northern Norrbotten ore province of Sweden that contains a resource of 21 Mt at 1.46% Cu, 0.78 g/t Au, 6 g/t Ag, and 99 g/t Mo. The deposit was discovered in 2012 by Boliden Mines AB while targeting ground EM and magnetic anomalies within the regional-scale Nautanen Deformation Zone (NDZ). Alteration facies at the deposit include sodic and high-temperature calcic-iron (HT Ca-Fe), high-temperature potassic-iron (HT K-Fe), and low-temperature potassic-iron facies at the core of the deposit, both of which host the highest grades of copper. Chalcopyrite and magnetite dominate the mineralization assemblage with pyrite, pyrrhotite, and molybdenite. Sulfides and magnetite occur within discrete zones of breccia and veining as well as along shears and as disseminations adjacent to structures. Re-Os dating of sulfides from the various mineralization styles highlights a protracted formation history of the deposit, which can be related to regional deformation and magmatic events. Early HT Ca-Fe veining with sulfides occurred at 1825 ± 8 Ma (molybdenite) and correlates with regional metasomatic alteration and metamorphism within the NDZ. Stockwork-style sulfide mineralization hosted within HT K-Fe altered rocks is dated to 1785 ± 8 Ma (molybdenite), and high-grade breccia formation is dated to1747 to 1766 Ma (molybdenite/chalcopyrite-pyrite); these dates correspond to a period of regional metamorphism and deformation and emplacement of the large batholiths of the Lina granite within northern Norrbotten. Vein-hosted sulfide mineralization adjacent to the main ore zone is dated to 1720 ± 8 Ma (chalcopyrite-pyrite). Magnetite geothermometry suggests magnetite at Nautanen North precipitated under conditions typical of hydrothermal systems. Temperatures derived from magnetite display a good correlation with the temperature ranges proposed for host rock alteration assemblages within the hydrothermal iron oxide and alkali-calcic alteration systems (IOAA).

The Newly-Discovered Ikkari Gold Deposit: Geology and Initial Insights into the Complex Controls on a Major Orogenic Gold Deposit in the Central Lapland Belt, Finland

Craig Hartshorne, Charlotte Seabrook, Kalle-Pekka Kotiaho
Rupert Resources Ltd, Sodankyla, Finland

The Ikkari gold deposit, discovered by Rupert Resources in April 2020 in Lapland, Finland, hosts a combined Mineral Resource of 4.39 Moz. It is the most significant gold discovery in the Paleoproterozoic Central Lapland Belt (CLB) since the Kittilä (Suurikuusikko) deposit was discovered in 1986.

The deposit occurs within a 4-km-wide belt of Savukoski group sediments and volcanics, bound to the north by the mafic-dominated Kittilä suite allochthon and to the south by the Kumpu group; a younger foreland basin composed of felsic sediments.

At the deposit scale, quartzite and siltstone slivers of Kumpu group sediments, up to 20 m in width, are structurally interleaved within highly strained ultramafic komatiites, introducing both geochemical and rheological heterogeneity. Ultramafic units containing centimetre- to millimetre-scale felsic clasts represent early fault zones formed during intercalation. Together with tourmaline-matrix cataclastic breccias, these represent structures active during the pre-mineralisation phase. Gold mineralisation, as free gold and associated with pyrite, was introduced during subsequent deformation events and is preferentially localised along the margins of felsic slivers and within early cataclastic units.

Late-stage hematite-bearing carbonate breccias, with a characteristic milled texture, host high-grade native gold and record the latest phase of gold mineralisation. The presence of at least two distinct mineralisation phases, repeated hydrothermal events, and the structural pre-conditioning are interpreted to have contributed to the high grades and tonnage present in the deposit.

Alteration is characterised by a broad envelope of iron and potassic mesothermal alteration, the products of which are controlled by the host lithology. Gold mineralisation has a characteristic Mo-S-Te-U-W association, distinct from the Kittilä deposit but analogous to other sub-economic gold occurrences within the CLB. Overprinting hydrothermal fluid and alteration events, coupled with multiple deformation phases, have produced a complex orogenic gold system an order of magnitude larger than other deposits in the district.

Developing New Soil-Based Mineralogical and Geochemical Exploration Tools for Sn, W, and Li in SW England

Edward Hill, Ben Williamson, Benedikt Steiner
University of Exeter, Penryn, United Kingdom

The UK government's target of net zero greenhouse gas emissions by 2050 will require huge quantities of "critical" raw materials, which include Li, mainly used in electric vehicle batteries, W for high-strength alloys and cutting tools, and Sn, which is used in solder and for plating. The SW England Cornubian granite batholith and world-class ore field has long been mined for Sn and W and hosts one of the largest Li resources in Europe. There is no current production of these metals, but there are mines in development for Sn and W and active exploration for all three metals. Such exploration is hampered by poor inland exposure and sometimes thick overburden.

The first author's Ph.D. project, which began in October 2022, aims to develop improved soil- and stream sediment-based exploration methodologies for Li, Sn, and W. The focus so far has been on a review of soil data from Tellus SW and local exploration companies to develop pathfinder and proxy elements for mineralisation. Based on geochemical analysis, multiple target areas have been identified as best suited to further development work. Samples will be collected for whole-rock analysis (including for Li which is absent from the Tellus data set) as well as automated mineralogical analysis to determine combined geochemical and mineralogical exploration criteria for Sn, W, and Li mineralisation.

Geological Characteristics of the Özyurt Gold Deposit: The Discovery of a Shear-Hosted Orogenic Gold Deposit in the Niğde Massif, Central Türkiye

Tuna Kaskatı, Hakan Yalçıntaş, Burak Gürel, Eray Yurtseven
Esan Eczacıbaşı, Istanbul, Turkey

Özyurt is a gold deposit located in the Niğde massif, forming the southernmost tip of the Central Anatolian Crystalline Complex. The massif is an extensional metamorphic core complex composed of Palaeozoic-Mesozoic–aged marbles, gneisses, quartzites, and amphibolites cut by Cretaceous (80–76 Ma) aplitic-pegmatitic dikes. Three detachment planes developed between metamorphic units characterized by ductile to brittle features. The upper detachment plane hosting the main mineralisation is characterized by cataclastic breccias. The N-S–trending, 23° to 40° E-SE–dipping main detachment plane and related second-order low-angle shear zones brought on cataclasis and extensive brecciation. Along the first- and second-order shear planes; high-angle normal faulting controls karstification and widely developed karstic breccias. Gold mineralisation at Özyurt occurs as both sulphide and oxide ore. Low-grade (up to 1 ppm Au) sulphide ore is characterised by discontinuous sulphide lenses within cataclastic breccias and sulphide veinlets or disseminations within chlorite-illite-calcite-siderite–altered amphibolite schist interlayers above and below the detachment plane. Main sulphide phases are pyrite and arsenopyrite accompanied by cinnabar, stibnite, galena, and chalcopyrite. Oxide ore represents weathered and supergene-enriched domains of the same sulphide ore. High-grade gold (locally up to 180 ppm) mineralisation within the oxide ore is associated with goethite-hematite-jarosite-ankerite–rich cataclastic and karstic breccias. Oxidation and deformation are intense along cataclastic and karstic breccias, where these are intersected by high-angle faults resulting in very high gold grades. Özyurt is interpreted as a shear-hosted gold deposit showing many similarities to orogenic gold systems. The Niğde massif, however, is still underexplored in terms of orogenic gold mineralisation and bears potential for discovery of similar mineralisation styles. In fact, Özyurt is the first economic deposit of this kind identified in the Niğde massif and puts forward a broad district for detachment fault-style orogenic gold mineralisation in greenschist facies rocks of the massif.

Exploration for REEs in Historic Iron Mining Districts – a Focus on Sweden

Aiden J. Lavelle[2]
1. Institute of Geologists of Ireland (IGI) (PGeo, EurGeol), Westport, Ireland, 2. Consultant Geologist, Westport, Ireland

The Olserum-Djupedal project in southern Sweden comprises a significant rare earth element (REE) resource with potential for near-term production. REEs occur in monazite, xenotime, and apatite hosted in intense biotite-magnetite alteration and veining located within gneisses of the Västervik formation and the Småland granite. The Olserum deposit was previously considered to be a high-temperature contact metamorphic-hydrothermal mineralising system (Andersson et al., 2018), with REEs and P probably derived from the Olserum-Djupedal granite by fluid interaction with the crystallising magma and hosted in metasediments near the granite contact. Recent re-logging of the core and mapping indicates that mineralization is shear zone hosted with characteristics of a deep IOA-IOCG-REE system and district-scale potential. Mineralisation is hosted in altered shear zones crosscutting the Småland granite intrusion. Intense alteration and shearing of the host granite has obliterated original textures and replaced original mineralogy. Recent rock chip sampling and historic drilling in the former small-scale iron mining district has shown that many of the other magnetite occurrences are mineralised with potentially ore-grade REEs. Widespread shear-hosted magnetite alteration indicates REE mineralization could be widespread in this district.

The genetic model for the Olserum-Djupedal REE mineralisation needs to be reconsidered, as REE mineralisation is hosted in heavily altered shear zones in a district that fits the IOA/IOCG model. Several iron oxide-apatite (IOA) or Kiruna-type deposits globally are known to be associated with enrichments of REEs, e.g., Per Geijer in Kiruna, the Adirondack deposits of New York, USA, and the Pea Ridge magnetite orebody in Missouri, USA, which hosts four REE-rich breccia pipes and is located in an IOA province.

The Olserum district warrants further exploration to assess the potential for additional economically extractable REEs in order to secure a domestic European supply of these critical metals.

Genetic Study of the Sphalerite from Rudabánya

Kristóf Lipp[1], János Földessy[2], Gabriella B. Kiss[1], Attila Kasó[3]
1. Eötvös Loránd University, Department of Mineralogy, Budapest, Hungary, 2. University of Miskolc, Miskolc, Hungary, 3. Rotaqua Ltd., Kővágószőlős, Hungary

The Rudabánya Ore Complex (northeast Hungary) has a mining history dating back to the Middle Ages. At first, native copper was exploited, which was followed by silver and iron mining. The mining ended in the middle of the 20th century, though new exploration took place in the late 1990s and in the 2000-2010s. These projects drew attention to the Zn(-Pb) and Ba prospects.

The Complex is hosted by Lower Triassic carbonate and siliciclastic rocks. Although the deposit is best known for its siderite content, novel studies confirmed its polymetallic character. According to our present knowledge, the mineralization formed during several superimposing events: (1) early stage, stratiform Pb-Zn ores in Lower Triassic siltstone; (2) metasomatic siderite hosted by dolomite in higher stratigraphic level; (3) Cu-Pb-Ba-Ag mineralization bound to the contact of the dolomite and the black shale; (4) supergene alteration of the primary sulphides, and (5) low-temperature hydrothermal Hg-As-Sb-Ag enrichment. Recent studies suggest that there are two types of different Pb-Zn ores with identical mineral paragenesis (sphalerite, galenite, pyrite, barite, carbonates), but their genesis is still a question of debate. The minerals of the type (1) ore form massive layers or lenses, whilst the type (2) ores are vein filling.

Sphalerite of type (1) ore contains lower amounts of iron (up to 0.2 mol % FeS) compared to sphalerite of type (3) ore (commonly 3-8 mol % FeS). This suggests lower formation temperature for type (1) ore (<100°C) and higher formation temperature for type (3) ore (100°–250°C). Detailed EPMA analyses of sphalerite revealed differences in trace element content: sphalerite of type (1) ore contains higher amounts of Hg, Ag, and V and lower amounts of Cd compared to sphalerite of type (3) ore. Mineral chemical studies confirm the formerly presumed MVT s.l. (so-called Alpine-type Pb-Zn) deposit classification for type (3) ore.

Petrology and Geochemistry of c. 1.8 Ga Granites Linked to Skarn W-F-Mo ± Cu Mineralization, Bergslagen District, Sweden

Edward P. Lynch, Stefan Bergman
Geological Survey of Sweden, Uppsala, Sweden

The Bergslagen mining district in south-central Sweden contains numerous skarn-related tungsten prospects located mainly along its western margin. Mineralization typically comprises disseminated scheelite, fluorite, and molybdenite ± chalcopyrite in skarn-altered marble horizons that form part of a c. 1.9 Ga syn-orogenic volcanic-sedimentary basin. Based on previous work, a link between late-orogenic (c. 1.8 Ga) granites and W-F-Mo ± Cu mineralization has been proposed. Here, we present a new petrological and geochemical assessment of c. 1.8 Ga granites in western Bergslagen to constrain key petrogenetic factors that promoted skarn tungsten mineralization.

Granitic intrusions closest to skarn W prospects comprise weakly foliated to massive equigranular syenogranite to alkali feldspar granite that locally contains xenoliths, mafic enclaves, and/or minor aplitic-pegmatitic parts. Biotite is the main mafic mineral along with accessory molybdenite, magnetite, fluorite, pyrite, chalcopyrite, zircon, and xenotime. Skarn-distal granites are weakly foliated to massive equigranular to megacrystic biotite ± muscovite ± hornblende granites, also associated with aplite-pegmatite.

All the granites generally have high-silica, peraluminous, and ferroan characteristics, are relatively enriched in SiO_2, total alkalis (K + Na), U, Th, Ta, and Y, and depleted in Ca, Mg, Ti, Nb, Zr and total REEs. A regional geochemical trend of increasing incompatible lithophile element abundances from skarn-distal to skarn-proximal granites is evident, suggesting fractional crystallization was an important control on enrichment of late-stage magmas linked to W ± Mo ± F mineralization. Overall, the granites have overlapping fractionated I- and A2-type affinities and likely formed via anatexis of mainly c. 1.9 Ga metaigneous source rocks affected by older subduction and metasomatic events. Coeval arc-related magmatism at c. 1.8 Ga and a switch to a net extensional tectonic regime may have initiated crustal anatexis and melt migration leading to granite formation in the mid- to upper crust, with a concomitant exsolution of W-mineralizing hydrothermal fluids.

Geometallurgical Characterisation of Sediment-Hosted Cu-(Ag) Ore from the Spremberg-Graustein-Schleife Kupferschiefer Deposit, Germany

Zahra Nourizenouz, Bradley M. Guy, Robert Möckel, Doreen Ebert, Max Frenzel
Helmholtz Institute Freiberg for Resource Technology, Freiberg, Germany

The Spremberg-Graustein-Schleife Kupferschiefer deposit is a sediment-hosted stratabound copper (SSC) deposit located in eastern Germany. The Cu mineralization occurs at depths of 980 to 1,580 m below surface, predominantly in a Permian carbon-rich black shale unit, commonly known as the Kupferschiefer. Mineralization is not only restricted to this unit and can extend into the overlying Zechstein carbonates, as well as the underlying Redbed sandstones. Recent studies have confirmed 91.7 Mt of inferred mineral resource at an average grade of 1.5% Cu and 24.0 g/t Ag.

In addition to Cu and Ag, elevated levels of Co, Ni, and Re are also found in Kupferschiefer ores. This contribution provides the first detailed evaluation of the mineralogy of the complete mineralization interval of the Spremberg-Graustein-Schleife deposit, including the mineralogical deportment of Cu. Mineral chemistry work constraining the deportments of Ag and Re is planned.

Fifty-three individual samples were collected from three drill cores. The samples were crushed, milled, and composited into 19 composite samples. These were then subjected to multi-element analytical methods (XRF, ICP-OES, ICP-MS) for bulk-ore geochemistry, X-ray diffractometry (XRD), and mineral liberation analysis (MLA) for mineralogy. Electron probe micro-analysis (EPMA) and laser ablation ICP-MS (LA-ICP-MS) for trace element geochemistry are currently planned. Mineralogical results reveal that major Cu-bearing minerals vary both spatially between the three different drill cores and vertically between lithological units. For example, chalcocite and covellite (and to a lesser extent bornite) are the dominant carriers of Cu in the most mineralized core (80% of contained Cu), whereas chalcopyrite is the most prominent carrier of Cu in the Pb-Zn Kupferschiefer facies (75% of contained Cu). By combining different analytical results, we aim to develop a quantitative predictive model to describe the mineralogical distribution of the copper and potential by-products that will be usable for mineral processing and mine-planning purposes.

Prospecting for Rare Earth Elements in Plio-Pleistocene Sands of the Cenozoic Mondego Basin, Portugal

Michael A. Pointon, Colm S. Pierce, Michael J. Flowerdew
CASP, Cambridge, United Kingdom

Rare earth elements (REEs) are critical to a wide range of applications, including several key technologies for the generation of clean and sustainable energy. Sedimentary placer deposits are a proven commercially viable source of REEs. Existing placer exploration efforts have focused on monazite and xenotime, which are common REE-bearing minerals in such deposits, particularly where sediment has been sourced from crystalline basement. Consequently, understanding sediment provenance is an important part of characterising a prospective placer deposit play. Demand for REEs in Europe is growing. Geologically recent, poorly consolidated placer sands within Europe are energy efficient to extract, offer geopolitical security, and could, at least in part, help to meet demand; however, such deposits are presently under-researched. This project was conceived in an attempt to address this by developing a low-cost, field-driven workflow to characterise heavy mineral and REE concentrations using Pliocene to Pleistocene sands of the Cenozoic Mondego Basin as an example. Poorly consolidated Pliocene to Pleistocene sands in the Mondego Basin are up to 70 m thick, unconformably overlie Mesozoic strata of the Lusitanian Basin and the Iberian Massif, and span a range of depositional environments from shallow marine to alluvial fan. Field-based geophysical analysis (magnetic susceptibility and gamma radioactivity) was conducted to attempt to detect heavy mineral concentrations by exploiting the natural radioactivity of several common heavy minerals (e.g., apatite, monazite, zircon) and the strong magnetic signatures of others (e.g., ilmenite, magnetite). Sand samples were collected for laboratory-based analysis of their heavy mineral assemblages using scanning electron microscopy-energy dispersive spectroscopy, whole-rock geochemistry, and U-Th-Pb geochronology. The sands contain monazite and xenotime and were derived from the adjacent Iberian Massif. The data collected so far reveal striking variations in monazite content, which is likely linked to subtle changes in sediment provenance through time.

Isotope Geochemistry of Native Sulfur, a Tracer of Microbial Activity in Secondary Ore Deposits

Jesús J. Rey-Samper[1], César Menor-Salván[2], Fernando Tornos[1]
1. Consejo Superior de Investigaciones Científicas (CSIC), Madrid, Spain, 2. Universidad de Alcalá de Henares (UAH), Alcalá de Henares, Spain

The Las Cruces mine (Iberian Pyrite Belt) is a VMS deposit with an unusual secondary zone, displaying high Cu grades in the cementation blanket but with a gossan that has been reduced to uncommon mineral assemblages (Conde et al., 2007). This singularity is related with the flourishment of deep microbial communities from the Messinian (Tornos et al., 2017), when the former gossan was transformed into siderite- and sulfide-rich assemblages.

Microbial activity at Las Cruces has been traced via microbiological techniques (Tornos et al., 2019) and biomarkers (Rey-Samper et al., 2022). In this late study, we detected elemental sulfur related to organic remnants. Native sulfur is an unusual mineral in these systems but potentially can be formed by microbial sulfate reduction. Elemental sulphur was extracted from 160 g of grinded sample for 12 hours using a Soxhlet extraction apparatus with a mixture of dichloromethane and methanol (8:1 ratio). Samples were then concentrated by evaporation until elemental sulphur crystallized, which was recovered and analysed by means of MC-ICPMS.

$\delta^{34}S_{sulfur}$ (CDT) is close to +12.2‰ (CDT). This result could be interpreted in two different ways: 1) the elemental sulphur was deposited as a by-product of the reduction of aqueous sulfate in underground water when reaching the gossan—$\delta^{34}S$ of aquifer water range between -8.9 to +20.4 ‰ (Scheiber et al. 2015)—or 2) the elemental sulfur is a consequence of the microbial reduction of the gossan—reduced rocks have $\delta^{34}S$ values of +12.3 to +22.9‰ (Tornos et al., 2017). Either way, the microbial-driven sulfate reduction could explain the formation and preservation of elemental sulfur in the modified gossan. This study shows that the presence of elemental sulfur in secondary assemblages can track the presence of microbial processes that are otherwise undetectable.

Study funded by the MINLIFE project (RTI2018-094867-B-IOO).

New Insights to the Characterization of the Kiruna Deposit – Contributions from Deeper Exploration

Carlos Santana, Orsolya Katai, Liz Lobo
LKAB, Kiruna, Sweden

The Norbotten region, in northern Sweden, is well known for hosting several mineral deposits emplaced during the Paleoproterozoic (2000–1800 My) Svecokarelian orogen. The Kiruna IOA deposit hosts the world's largest underground iron ore mine and is located in the Kiruna-Naimakka deformation zone, which itself is in alignment with the NNE-strike magnetite mineralisation. Several exploration campaigns have been conducted since 2015 to identify the continuation of the mineralisation both northwards and at depth. Historically, two different ore types have been identified, a P-poor (B-type) and a P-rich (D-type) magnetite mineralisation, with the latter generally overprinting the former, although more complex relationships are present. Pervasive sodic alteration is widely observed, of which albite±hematite is the most representative assemblage. Potassic (biotite±magnetite) and sodic-calcic (actinolite±titanite±magnetite-albite) are the most prominent proximal hydrothermal alterations, where several overprinting stages might be encountered. Large structures at deposit scale have been modelled based on drill hole data, historical maps, and geophysical surveys. They are grouped in three faults systems: 1) NS thrust fault (reverse oblique to dip-slip); 2) NW-SE oblique-slip and 3) NE-SW oblique-slip. The pre-syn–mineral structures are closely related with early stages of the mineralisation whilst post-mineral structures may offset the mineralisation by up to 400 m in certain areas. Particularly, a special hydrothermal corridor has been identified in the northern part, for which the connection with the Luossavaara and Per Geijer deposits remains uncertain. Thanks to the LKAB exploration, it has been possible to better understand the main controls and multi-episodic character of the magnetite. Geological modelling has revealed that the mineralisation is present in three stratigraphic positions, which are in connection with faults and alteration domains that reinforce the published character of an IOCG district. These results and findings have contributed to refine the exploration targeting process and open a new window for further research.

Proximal Hyperspectral Sensing for the Characterization of High-Sulfidation Epithermal Deposits: The Allumiere Quarry Test Site (Central Italy)

Anna Sorrentino[1], Francesca Corrado[1], Rita Chirico[1], Matteo Massironi[2], Sabrina Ferrari[2], Stefano Castelli[2], Barbara Marchesini[3], Stefano Tavani[1], Eugenio Carminati[3], Nicola Mondillo[1, 4]
1. University of Naples Federico II, Naples, Italy, 2. Department of Geoscience, University of Padua, Padua, Italy, 3. Department of Earth, University of Rome "La Sapienza", Rome, Italy, 4. Natural History Museum, London, United Kingdom

The Allumiere quarry is located northwest of Rome (Italy) in the Pliocene-Pleistocene Tolfa volcanic district and is locally known for the occurrence of an alunite+kaolinite mineralization intensely exploited from the 15th century onward. This mineralization is hosted in a lava dome and is considered to be genetically related to post-volcanic hydrothermal activity, which has produced a widespread structurally controlled alteration spanning from vuggy silica and advanced argillic to intermediate argillic. The main minerals characterizing these zones are quartz, alunite- and kaolinite-group minerals, and smectite. The hypogene assemblage was lately affected by supergene alteration, characterized by Fe-oxyhydroxides, jarosite, and other accessory phases. All these minerals are optically active in the visible-near (VNIR) to short-wave infrared (SWIR) range. In this study, we report the preliminary results of the hyperspectral analyses at proximal scale carried out on the distinct alteration facies using the Headwall Photonics nano- and micro-hyperspectral cameras, which cover the spectral range between 400 and 2,500 nm, with 270 (VNIR) to 166 (SWIR) spectral bands. The hyperspectral data have been processed by applying feature extraction indexes for determining the relative abundances of Fe-oxyhydroxides (hematite and goethite), by using the 900 nm absorption feature and hydroxyl-bearing minerals (white mica, smectite, kaolin group minerals, sulphates) and their compositional variations. The high-resolution hyperspectral data allowed us to (1) discriminate among kaolinite polytypes (kaolinite, dickite, and halloysite) based on the position of the OH absorption feature in the region around 1,380 to 1,400 nm, (2) discriminate alunite from natroalunite due to the wavelength shift of the 1,480-nm absorption feature, and (3) detect smectite through the double absorption feature deeper at 1900 and 2200 nm.

This relatively fast and cost-effective method represents a cutting-edge tool for the identification and characterization of potential exploration targets and can be used either in the laboratory or in the field.

Trace Element Contents in Sulfides from Porphyry Cu-Au(±Mo) Deposits, Central Srednogorie Zone, Bulgaria

Elitsa S. Stefanova[1], Irena Peytcheva[1], Stoyan Georgiev[1], Atanas Hikov[1], Sylvina Georgieva[1], Milen Kadiyski[2], Ventsislav Stoilov[3], Desislav Ivanov[3]
1. Geological Institute of Bulgarian Academy of Sciences, Sofia, Bulgaria, 2. Aurubis Bulgaria AD, Pirdop, Bulgaria, 3. Assarel Medet JSC, Panagyurishte, Bulgaria

Porphyry copper systems are of great economic importance as main suppliers of Cu, Mo, Au, and Re but also because they contain metals like Ag, Pb, Zn, Pd, Bi, Te, Se, most of them referred to as critical metals. In the present study we have applied a combination of SEM-EDS and laser ablation-inductively coupled plasma-mass spectrometry (LA-ICP-MS) techniques to measure metal contents in sulfide minerals from the Elatsite and Assarel porphyry Cu-Au(±Mo) deposits (PCDs). These are two of the biggest operating open-pit mines in Bulgaria, situated in the Central Srednogorie Zone as part of the Late Cretaceous Apuseni-Banat-Timok magmatic and metallogenic belt.

Special attention was paid to the content of critical metals in pyrite, as this mineral is part of all mineral associations in both deposits, but concentrates mainly in waste products during flotation. Our data shows that pyrite is a major host of Au and many other elements including Se, Co, Ni, Ag, Cu, Pb, Sb, Te, and Tl. Elevated Se content is established in pyrite from both deposits up to 1,116 ppm (in Elatsite). In some zonal pyrite crystals from the late quart-pyrite veins from Elatsite deposit, cobalt varies from a few ppm up to 1.6%. The zones with elevated Co content also have higher Ni (up to 0.5%) and Se (up to 530 ppm) contents. Chalcopyrite from Elatsite and galena from Assarel also reveal high Se content, up to 412 ppm and 2 to 3% Se, respectively. Consequently, ore minerals that are considered waste products for both deposits are major repositories and carriers of some critical metals.

Acknowledgments: The study is a contribution to the KP-06-N44/4 project, financed by the Bulgarian National Science Fund and Horizon Europe AGEMERA project (*Ag*ile *E*xploration and Geo-*m*odelling for *E*uropean Critical *Ra*w Materials); grant No. 101058178.

Piani Resinelli District: Preliminary Data on Ore Precipitation and Multiphase Diagenetic Evolution

Luca Summino[1], Michele Giorno[2], Luca Barale[3], Carlo Bertok[1], Max Frenzel[2], Marta Gasparrini[4], Luca Martire[1]
1. University of Turin - Department of Earth Sciences, Turin, Italy, 2. Helmholtz Institute Freiberg for Resource Technology, Freiberg, Germany, 3. CNR-IGG Turin Unit, Turin, Italy, 4. University of Milan - Department of Earth Sciences Ardito Desio, Milan, Italy

The stratabound, carbonate-hosted Pb-Zn-Ag (± fluorite ± baryte) deposit of Piani Resinelli belongs to the Alpine metallogenic province located between Italy, Austria, and Slovenia. The mineralization is hosted in the Lower Carnian stratigraphic succession of the Lombardian basin (Southalpine Domain, north Italy), mainly composed of peritidal limestones. The origin of the deposit is still unknown since the previous studies are largely outdated. The aim of this study is to constrain the ore-forming conditions using petrographic, geochemical, and fluid inclusion studies on both ore minerals and associated diagenetic cements.

Preliminary results show that the mineralization was preceded by dolomitization, silicification, and brecciation events. The occurrence of saddle dolomite locally associated with micritic sediments in geopetal structures points to a shallow burial and thus a hydrothermal origin of dolomite.

Preliminary thermometric data on fluid inclusions indicate the involvement of moderately hot temperatures (between 115° and 150°C), high-salinity brines (up to ~20 eq wt % NaCl) in ore genesis. Moreover, in situ analyses (EPMA, LA-ICP-MS) of trace elements on sphalerite are presently in progress and will be presented here, including their implications for ore-forming conditions.

The origin of the Resinelli district was probably linked to an Upper Triassic hydrothermal system, in which metal-rich fluids flowed upward, possibly through faults and associated fractures. This circulation in the Upper Triassic succession caused major modifications in the host rocks such as multiphase dolomitization, silicification, and brecciation, as well as the precipitation of ore minerals in a shallow burial setting. The results of this preliminary characterization show several similarities between the Piani Resinelli and the nearby Gorno districts in terms of the diagenetic evolution of the host rocks, temperature and composition of the hydrothermal fluids, and timing of the fluids involved in ore precipitation, suggesting that both districts could have formed during the same regional mineralization event.

How Fluids Make or Break Critical Metal Deposits: The Ivigtut Cryolite Body, SW Greenland

Anna Szreter[1, 2], Adrian A. Finch[1], Henrik Friis[2], Xiao-Ming Liu[3], Finlay M. Stuart[4]
1. School of Earth & Environmental Sciences, University of St Andrews, St Andrews, United Kingdom, 2. Natural History Museum, University of Oslo, Oslo, Norway, 3. Department of Earth, Marine and Environmental Sciences, University of North Carolina, Chapel Hill, NC, USA, 4. Scottish Universities Environmental Research Centre (SUERC), East Kilbride, United Kingdom

In this time of the green energy transition, the attempt to reduce fossil fuel consumption is inseparable from increasing demand for critical metals, including REEs. Currently, most REEs are mined and processed solely in China, although some deposits exist elsewhere in the world, including the north Atlantic region.

A key target area is Greenland, where igneous complexes associated with mid-Proterozoic continental rifting (termed “the Gardar Province”) host several potential REE deposits. In these complexes, magmatic fluids redistribute the elements of interest, often decreasing concentration but in some cases focussing the resources into a restricted area.

In this research, the Ivigtut cryolite deposit in southwest Greenland is examined. Ivigtut was the world’s largest cryolite deposit (before its complete extraction in the 20th century), and its structure comprises an orebody on top of a small granite cupola. We will present novel stable isotope studies (δ^7Li, δ^{34}S) to provide new insight into the origin of the fluid (i.e., meteoric vs. magmatic) and its relationship to the petrogenesis of the granite. At the same time, the atomic-scale processes are recorded by spectroscopy of the minerals crystallised from the fluid, including cryolite (in three colour variations: black, purple, and white), other aluminofluorides, and sulphides.

Lifetime luminescence and UV-Vis spectra of cryolites vary significantly. A few sharp, narrow peaks can be interpreted as lanthanide substitution in the structure, while broad peaks derive from structural defects. Both possibilities, and most importantly their petrological significance, are explored by a combination of spectroscopic methods, geochemical analysis, and thermal treatment.

Sulfide Mineralogy and Cobalt Deportment at the Viscaria Cu-Fe Deposit, Kiruna District, Northern Sweden

Koray Tasbicen[1], Marcello Imaña[2], William X. Chávez, Jr.[1]
1. New Mexico Tech, Socorro, NM, USA, 2. Copperstone Resources AB, Kiruna, Sweden

The Viscaria Cu-Fe deposit is hosted within 2.2 Ga Paleoproterozoic-age rocks of the Norbotten region, Sweden; these rocks also host world-class IOA deposits as well as Cu + Fe ± Zn and Cu ± Fe ± Au occurrences. Copper mineralization at Viscaria comprises three stacked, subvertical ore lenses of semi-massive to massive magnetite + chalcopyrite + carbonate, with disseminations and veinlets of pyrite + low-Fe sphalerite or pyrrhotite + high-Fe sphalerite. Locally, anomalous Co and Ni values correlate positively with copper.

Trace element characterization of sulfide shows that cobalt enrichment occurs with paragenetically early pyrite-magnetite and early pyrrhotite and sphalerite; in these assemblages, chalcopyrite is not favoured. Varying abundances of sulfides within different ore lenses and variations of carbonaceous content in host rocks impact cobalt concentration in sulfides. In the stratigraphically uppermost A Zone, cobalt is partitioned between pyrite, pyrrhotite, and sphalerite, with average Co concentrations of 846, 596, and 635 ppm, respectively. Cobalt favours pyrrhotite and sphalerite in the B Zone, with average Co concentrations of 1,370 and 749 ppm. Pyrite is the primary host for cobalt in the stratigraphically lowest D Zone, with an average concentration of 2,423 ppm. Co/Ni ratios of pyrite at Viscaria within the D Zone display the most significant values, from 10 to 100, while ratios for the A and B Zones vary from 0.1 to 1. High Co/Ni ratios of the D Zone might be attributed to greater-temperature fluids that we suspect originated from a mafic intrusion source. Mineral assemblages of the D Zone suggest more oxidizing conditions than A and B Zones, with Co/Ni ratios similar to iron-oxide and copper-bearing manto/vein/skarn systems, including those not occurring within carbonaceous stratigraphic sequences worldwide. Mineralization and alteration assemblages and textural features at Viscaria suggest a similar epigenetic origin for A, B, and D Zone ores.

The Origin of Stratabound Scheelite: The Future Resource for Europe?

Iñigo Borrajo, Fernando Tornos
Consejo Superior de Investigaciones Cientificas, Madrid, Spain

Most currently mined tungsten deposits in SW Europe are high-grade, low-tonnage discordant vein systems but in recent years, the discovery of the high-tonnage San Juan and Valtreixal (10 Mt @ 0.28 WO_3) deposits opens the possibility that in the future, stratabound scheelite can become a major resource. In Iberia, there are previously quoted small stratabound scheelite prospects that have been considered metamorphosed exhalative deposits or metamorphic skarns. However, Valtreixal and San Juan seem to be a new type of mineralization rarely studied up to now.

Mineralization forms large stratabound sulfide-poor bodies enriched in clinozoisite-muscovite-quartz and interbedded in thick siliciclastic sequences of dark shale of Late Paleozoic age; cassiterite U-Pb and muscovite $^{40}Ar/^{39}Ar$ ages suggest that the mineralization formed at ca. 313 Ma, coeval with nearby discrete intrusions of mesozonal to epizonal syn-tectonic Variscan granitoids. The connection between tungsten mineralization and the source granite, which can be several kilometers away, are large quartz-rich veins with cassiterite. Sr-Nd isotope geochemistry suggests a magmatic source for metals while ore-forming fluids seem to be CO_2-poor, intermediate-density water below the two-phase surface.

Early fluid-rock reaction of high-temperature magmatic-hydrothermal fluids with the source granitoids and the dominant Fe-bearing shales was only able to saturate the system in cassiterite, but this mechanism was not effective enough to saturate the fluid in W, suggesting that the key process for W precipitation, even from very low amounts, was reaction of the magmatic-hydrothermal fluids with the Ca-bearing layers.

This largely ignored type of mineralization is perhaps equivalent to the distal Zn-rich skarns or mantos in porphyry systems but formed in relationship with collisional felsic granitoids.

Funded by EIT Raw Materials Project 18036.

Morphotextural and Geochemical Analyses of Gold Particles Related to the Pinalejo Epithermal System (Salamanca, Spain)

Kelvin d. Alves[1], Juan G. Barreiro[1], Santos B. Sánchez[1], Lorena O. Menor[2], José M. Marchena[1]
1. Universidad de Salamanca, Salamanca, Spain, 2. Universidad Complutense de Madrid, Madrid, Spain

Understanding the morphological and geochemical evolution of gold particles during their transport is essential to define prospecting models for primary and secondary gold deposits (alluvial). In this work, we studied primary and alluvial gold samples from remains of Roman mining. The primary Au is hosted in shear-related epithermal quartz dikes (fluid inclusions: Th= 188.19-256.79°C), where the gold fill intercrystalline spaces in the quartz, which condition their morphology. Compositionally, primary gold shows a varying bimetallic alloy (Au_{90-94}-Ag_{10-6}). Alluvial gold particles were collected at nine points distributed along 4.5 km of the Pinalejo-Tenebrilla creek. Mean values of CFI, CSF, and Shilo morphological indexes are 3.15, 0.38, and 2.16, respectively. These values are consistent with transport distances of less than 5 km. Morphology analysis points to a complex mixture of two populations along the creek. All alluvial particles show a core-rim texture. Core alloys show a compositional range (Au_{72-98}-Ag_{28-2}) and inclusions of galena, cobaltite, and chalcopyrite. The gold-rich rims (Au_{98-99}-Ag_{2-1}) are the result of supergene processes (Ag-leaching) during the transport of the particles. Taking into account the chemical and morphological variation of gold particles, it is proposed that the Pinalejo epithermal system is an extended orogenic vein-field with a vertical/lateral compositional zoning, where undiscovered primary sources may exist.

Microchemical Signature of Placer Gold Nuggets from the Central Iberian Zone (Spain)

Santos Barrios Sánchez[1], Rafael Pablo Lozano Fernández[2], Juan Gómez Barreiro[1], Raul Merinero Palomares[3], Mercedes Suárez Barrios[1], Kelvin dos Santos Alves[1]
1. University of Salamanca, Salamanca, Spain, 2. Instituto Geológico y Minero de España, Madrid, Spain, 3. The Complutense University of Madrid, Madrid, Spain

Gold placers are abundant and have been intensively surveyed in western Iberia since antiquity. Three Cenozoic gold placers covering Neoproterozoic-Lower Paleozoic basement have been revealed: Salvatierra de Tormes (ST), Santibáñez el Alto (SA), and Casas de Don Pedro (CSDP). We have combined microchemical and inclusion analysis and morphology of gold nuggets to define gold signatures and relationships with primary sources and infer mineralization style.
Allotriomorphic nuggets have been identified in the placers, and gold with crystal faces and gold-bearing quartz fragments have been found in CSDP. Four different gold types have been defined: core (T1), gold-rich rim (T2), fine-grained gold in Fe-oxy-hydroxides (T3), and "mustard" gold (Au+Sb-Pb-Fe-oxides) (T4). Nuggets from ST-SA are zoned Au:Ag alloys (T1: 8.7 Ag wt%; T2: 0.6 Ag wt%), nuggets from CSDP only show fine T1 (0.6 Ag wt%), and all nuggets present late T3 (<1.0 Ag wt%) filling gold embayments. Filonian quartz includes an Au:Ag:(Cu) T1 alloy (13.5 Ag wt%), Fe oxy-hydroxides with Sb-Pb altered minerals (oxyplumboromeite), and secondary fine-grained gold (T4: Au:Ag, 5.3 Ag wt%).
T1 texture and composition in nuggets from ST-SA and gold in quartz from CSDP suggest hydrothermal origin, high variability of Ag (<5.0-15.45 wt%), and sulphides/sulphosalts microinclusions. Secondary Ag-leaching led to gold refinement (T2) that together with dissolution-precipitation mechanisms and oxidation of Fe^{2+} induced precipitation of T3. Finally, T4 reflects an early alteration of the primary mineralization, including reprecipitation of fine-grained gold and alteration of Sb-Pb phases.
Old mining works in primary mineralization observed in CSDP indicate it is a proximal placer (Plio-Pleistocene glacis). ST-SA deposits have similar characteristics pointing to a stage of relative dispersion from the source, being considered fluvial "trunk" placers (Eocene-Oligocene/Upper Pleistocene). Gold fingerprint is complex but evolves coherently from primitive-proximal deposits (CSDP) to more transported ones (ST, SA) providing a valuable tool for exploration.

Geological Settings and Mineralisation of the Vejnaly Gold Deposit, Lesser Caucasus, East Zangezur, Azerbaijan

Stephen J. Westhead, Anar Valiyev, Vusal Amirkhanov, Atabay Nushravanov, Araz Orujov, Aydin Bayramov
Azerbaijan International Mining Company (subsidiary of Anglo Asian Mining), Gadabay, Azerbaijan

The Vejnaly deposit and contract area is situated in the Zangilan region of the Republic of Azerbaijan, near the village of Vejnaly. In terms of tectonics and metallogenic origin, the area is confined to volcano-plutonic structures of the Qafan zone, Lesser Caucasus metallogenic belt, well known for its ore deposits. The Qafan tectonic block is usually interpreted as a discontinuous Jurassic to Cretaceous subduction-related magmatic belt. The mineralised target area of 30 square kilometres hosts 25 gold- and two gold-copper–bearing zones that have been identified generally with a NE azimuth (10°–60°) dipping 50° to 85° to the NW. The host rocks of the mineralised zones are Tithonian-Valanginian volcanic rocks of diabase composition and limestones, with Heterivian and Barremian marls. The deposit is associated with felsic intrusive rocks related to the Mehri-Ordubad pluton.

Vejnaly deposit ores are of gold-sulphide-quartz type, hosted within veins that exhibit mineralogy dominated by pyrite with subordinate chalcopyrite. Veins are distinct, with frequent pinch and swell geometry. The variable thickness of the orebodies in pinches ranges from 0.1 to 0.2 m and in swells from 1.0 to 2.5 m, rarely up to 6 m (as zones). The length of the vein zones ranges from 400 to 1,200 m. Near-ore alteration is widespread but does not have extensive thickness. The distribution of gold is nuggety within the vein systems.

Vejnaly ores have been reported to contain bismuth, vanadium, yttrium, and ytterbium. The mineralisation typically includes pyrite, chalcopyrite, and native gold. Other secondary minerals include magnetite, galena, sphalerite, quartz, calcite, ankerite, aragonite, and others.

The Vejnaly contract area of the Azerbaijan International Mining Company (subsidiary of Anglo Asian Mining) has significant prospects for volcanic belt porphyry-, hydrothermal-, and skarn-type mineral deposits.

Gold Metallogeny During Subduction Initiation and a Source Rock Control on Proto-Arc VMS Endowment

Thomas M. Belgrano[1, 2], James A. Milton[2], Larryn W. Diamond[3], Robin C. Wolf[3], Damon A. Teagle[2]
1. School of Earth Sciences, University College Dublin, Dublin, Ireland, 2. School of Ocean and Earth Science, University of Southampton, Southampton, United Kingdom, 3. Institute of Geological Sciences, University of Bern, Bern, Switzerland

Intraoceanic subduction intiation marks the transition between the mature divergent and convergent stages of the Wilson cycle, each with its distinct and relatively well understood metallogeny. The metallogenic systematics of the transition, however, are less clear, with uncertainties regarding controls on Au endowment in proto-arc volcanogenic massive sulfide (VMS) deposits and the potential for magmatic-hydrothermal ore formation. The lavas of the Samail ophiolite (Oman) formed during Tethyan subduction initiation and host VMS deposits with Au/Cu ratios that increase up through the volcanostratigraphy. Our new analyses of volcanic glasses reveal this increase is closely paralleled by Au/Cu ratios in the footwall lavas, implying a footwall source compositional control on proto-arc VMS Au endowment.

Supporting trace element analyses and magmatic modelling indicate that Au enrichment of the boninites was enabled by sulfide-undersaturated fractionation. This occurred due to remelting of mantle that had already lost most of its sulfide during production of MORB-like melts earlier in the subduction initiation seqeuence. The early melting left a refractory residue with only small amounts of Au-enriched sulfide, which was completely extracted during the later remelting to produce Au-rich but S-poor boninites. Despite H_2O enrichment in the boninites, low S contents may have suppressed efficient magmatic-hydrothermal extraction and transport of Au and Cu, thereby limiting magmatic-hydrothermal input to the VMS deposits. Subduction maturation and the refertilization of the wedge with S and metals by mantle corner flow would then have been necessary to produce melts capable of forming magmatic-hydrothermal ore deposits. This stage was not reached prior to detachment of the Samail lithosphere, but was by many other well-endowed Tethyan and circum-Pacific intra-oceanic arcs. The Samail ophiolite thus provides a record of both evolving metallogeny during subduction initiation and the limitations on the onset of porphyry-type mineralization in nascent arcs.

Structural Control on Fluid Flow Pathways and Vein/Ore-Shoot Emplacement at the Au-Ag LSE Pampa Paciencia Deposit, Antofagasta, Chile

Andrés Canales[1], Juan Pablo Radic[2], Diego Guido[3]
1. Facultad de Ingeniería, Universidad del Desarrollo, Santiago, Chile, 2. Enap-Sipetrol S.A, Santiago, Chile, 3. Astra Exploration, Vancouver, BC, Canada

The Pampa Paciencia deposit (56-64 Ma) is part of the Paleocene mineral belt, 57 km SW from the city of Calama. It is divided into three zones: north, central and south, central being the most important one and the focus of this study as it hosts the Paciencia vein system (PPVS). The PPVS is made up of three sub-vertical, NW-oriented veins: Paciencia Oeste, Paciencia, and Paciencia Este.

A total of 68 structural measurements were taken at surface (faults and veins) and summarized in six stations as rose plots. Riedel diagrams were generated using a hypothetical σ_1 at the syn- and post-mineralization stage. An RTP magnetometry map was interpreted with the predicted kinematics from the riedel diagrams. Au, Ag, Bi, Zn, Cu, and Pb values were numerically modelled on the PPVS vein solids to define the ore shoots and fluid flow pathways.

Syn-mineralization NNW σ_1 would have nucleated the PDZ, R, and T fractures along the pre-existing NW anisotropies observed in RTP magnetometry, controlling the vein emplacement (banded vein and siliceous breccia). A post-mineral event with an E-W σ_1 would have deformed the vein along strike, generating the limonitic breccia, and also NE dextral strike-slip faults (R' faults) (mapped in station 2) that would displace the PPVS and rotate the system 10°. Two ore shoots were identified, one in the Paciencia vein (biggest) and the other in the Paciencia Oeste vein. The first one is interpreted to be generated by a fault bend mode II and the second one by a lithological contact. Geochemical modelling supported this, showing high concentrations of Zn, Cu, and Pb in the deep zones and Ag and Bi in the shallow zones of the shoots.

This study shows the structural evolution of the PPVS as well as the mechanics and distribution of the ore shoots.

Geochronology of the Tocota Pluton and Au-Cu-Ag Mineralization of the San Francisco de Los Andes Quartz-Tourmaline Breccia, SW San Juan, Argentina

Brenda Gaia Carvalho Feitosa[1, 2], Ticiano José Saraiva dos Santos[1], Robert Lima Muniz[1]
1. UNICAMP - State University of Campinas, Campinas, SP, Brazil, 2. Turmalina Metals, Vancouver, BC, Canada

The San Francisco de Los Andes (SFLA) magmatic-hydrothermal breccia, composed of tourmaline quartz, is located in San Juan Province, east of the Cordillera Frontal. SFLA hosts Bi-Cu-Au breccia deposits, with mineral assemblies of Bi-Cu-Au-As-Fe-Zn-Pb-Ag. The mineralized Tocota pluton intrudes the carboniferous sedimentary Agua Negra Formation. The Pluton belongs to the Las Piedritas Granodiorite unit that integrates the Colangüil Batolite.The activity of late, boron-rich volatiles from the pluton causes hydraulic fracturing, forming veins and breccias filled by tourmaline and associated mineralizations. We aim to define the crystallization age of the Tocota pluton (TJBG-D1A and TJBG-D1B samples) and the mineralization age (TJBG-BRX sample) by U-Pb zircon and Re-Os on molybdenite, respectively. In this way, it will be possible to contribute to the understanding of the evolution of hydrothermal alteration of the San Francisco Breccia. Zircon separation from the intrusive rock was carried out at the Institute of Geosciences of the State University of Campinas (IG-UNICAMP). Two porphyritic facies were individualized according to the predominant alteration: chlorite ± sericite (TJBG-D1A) and k-feldspar (TJBG-D1B). Cathodoluminescence, BSE, and SE images were obtained from the scanning electron microscope (SEM), and 30 spots from each sample were defined for U-Pb method dating by LA-ICP-MS from the Laboratory of Isotopic Geology of IG-UNICAMP. The tourmaline quartz breccia sample (TJBG-BRX) was sent to the Applied Isotope Research for Industry and Environment Program at Colorado State University for Re-Os dating. The zircon grains are prismatic euhedral shape and display continuous or discontinuous growth zoning textures, typical of a magmatic source. Together with the petrographic description and the obtained zircon and molybdenite ages, this is expected to contribute to the understanding of the mineral-magmatic system evolution of this portion of the Andes.

Mineralogical Characterization of Auriferous Occurrences in Stream Sediments in the Carangola Region, Minas Gerais, Brazil

Gabriela O. Carvalho, Alexandre N. Nogueira, Matheus R. Violante, Sofia G. d'Orsi, Gustavo Luiz C. Pires, Silvia R. Medeiros
Universidade Federal do Rio de Janeiro, Rio de Janeiro, RJ, Brazil

During the 1980s, artisanal gold mining was performed by local residents from riverbeds in the vicinity of Carangola/MG, SE Brazil. The area is located at the Neoproterozoic Ribeira and Araçuaí belts. The Archean/Paleoproterozoic basement is composed of migmatitic charnokitic orthogneisses from the Juiz de Fora Complex (JFC) followed by Neoproterozoic paragneisses from the Andrelândia Group (AG). Metagabbroic/dioritic bodies and granitoids cut the previous units. They occur as NNE-SSW tectonic slivers limited by regional post-peak metamorphic shear zones. Despite the gold discovery, there are no scientific works or descriptions about the geology of these occurrences. This work presents the mineralogical characterization of stream sediments collected at 20 points along three distinct drainage networks in order to investigate the detrital gold occurrences, their spatial distribution, and possible source rocks. Preliminary results show mineral association composed of magnetite, garnet, monazite, ilmenite, zircon, rutile, sillimanite, leucoxene, spinel, hypersthene, amphibole, and apatite, as well as quartz, muscovite, and biotite. Gold grains 0.2 to 0.4 mm in diameter were identified in four samples. Farther upstream, the grains have irregular contours, whereas downstream the grains are rounded, indicating greater transport. The gold occurrence is restricted to the courses that drain the rocks of the AG and JFC, cut by a high density of lineaments of the Manhuaçu Shear Zone, post-peak metamorphic and of NNE-SSW orientation. The presence of minerals poorly resistant to weathering and gold grains with irregular contours, concentrated near the head of one of the catchments, corroborates the proximity of the source rocks and primary mineralization. The results presented in this work allow for effective reduction of the areas with potential for the occurrence of auriferous mineralization in the region, besides providing prospective guides for the next stages of mineral research in the area in a conscious and responsible way.

Nature and Source of Ore-Forming Fluid at Gadag Gold Field, Dharwar Craton, India

Sakthi S. Chinnasamy[1], Debasis Pal[1], Dnyaneshwar Gaonkar[2], Manish M. John[3]
1. Indian Institute of Technology Bombay, Powai, Mumbai, MH, India, 2. Ramgad Minerals & Mining Limited, Gadag, KA, India, 3. National Centre of Excellence in Geological Research (PPOD), Geological Survey of India,, Bangalore, KA, India

We investigated the gold-forming processes at Gadag gold field (GGF) from the Gadag greenstone belt in India by integrating fluid inclusions, stable isotopes (S and C), and tourmaline chemistry. GGF has three auriferous lode systems with distinct lithology, namely western lode (WL): metabasalt-hosted; central lode (CL): argillite- and/or metabasalt-hosted; and eastern lode (EL): BIF-hosted. Three generations of carbonates were identified based on their individual textures and deformation fabrics in the mineralized zones. The third-generation vein carbonates are associated with progressive dextral shearing and major gold mineralization. Detailed fluid inclusion (FI) study of mineralized quartz-carbonate veins from both of the lodes reveals low- to medium-saline (WL: 1.8 to 9.6; CL: 0.04 to 7.8 wt% NaCl eq.) H_2O-NaCl-CO_2 ± CH_4 primary fluid, except for a few inclusions with relatively higher salinity in the western lode (20.1 to 21.8 wt% NaCl eq.), and fluid immiscibility. Estimated P-T conditions of alteration and mineralization (1.6-2.9 kbar; 296°-333°C), by combining the FI and chlorite and arsenopyrite thermometry, reflect greenschist facies conditions at GGF. Tourmalines, closely associated with sulfides in the alteration zones, belong to the dravite or oxy-dravite group with low to medium Na (avg: 0.68 apfu) and low Fe^{3+}/Fe^{2+} ratio. Ore fluid $\delta^{34}S$ values deduced from sulfides in the mineralized zone lie between 1.0 and 3.4‰. Overlapping $\delta^{34}S$ values from the mineralized (1.5 to 4.5‰) and host-rock (–1.0 to 7.5‰) sulfides indicate desulfidation and/or dissolution of the early pyrite during continuous fluid flux along the shear zone. However, relatively depleted $\delta^{13}C_{fluid}$ values (−4.2 to 1.9‰) reflect the insignificant role of fluid mixing and source of carbon in ore fluid corresponding to decarbonation of the organic sedimentary sequences during metamorphism of the Gadag greenstone belt. Thus, a metamorphic fluid source is postulated for the origin of the auriferous ore fluid at GGF.

Geological and Geochemical Characterization of the Mugomo Gold Prospect, Tete Province, Northern Mozambique

Euclesia P. Cossa[2]
1. Instituto Superior Politecnico de Tete (ISPT), Tete, Mozambique, 2. Akita University, Akita, Japan

The Mugomo gold prospect is one of several gold occurrences in the northwestern part of Tete province, central Mozambique. Characteristics of geology
and mineralization of the Mugomo prospect are poorly understood. The prospect is hosted in a terrane composed of metavolcanic rocks of the Proterozoic Mualádzi Group, part of the Southern Irumide Belt. This study aims to describe the gold mineralization and associated hydrothermal alteration in the Mugomo prospect based on petrography, whole-rock geochemistry, and sulfur isotope data.

The mineralization in the Mugomo gold prospect is controlled by NE-SW– and N-S–striking regional Pan-African shear-zone systems. Epidote, quartz, and chlorite veins and veinlets are hosted by metavolcanic rocks and quartzite of the Mualádzi Group.

The metavolcanic rocks consist mainly of plagioclase, biotite, muscovite, and quartz, with secondary epidote, chlorite, calcite, and sericite; the quartzite is composed of quartz and minor plagioclase. Gold and sulfide mineralization occurs as dissemination of the host rocks and epidote-quartz veins.

The ore minerals consist mainly of pyrite and chalcopyrite with minor magnetite, bornite, covellite, silver-telluride, bismuth-telluride, and native gold. Bulk chemical compositions of ores show up to 2 ppm Au, and native gold is associated with pyrite and chalcopyrite. We consider that gold deposition in the Mugomo prospect was intimately related to sulfidation. The $\delta^{34}S_{CDT}$ of pyrite varies narrowly between +2.5 and +4.0‰, suggesting a single source of sulfur, either magmatic or metamorphic in origin.

Based on the geological setting, mineralogy, alteration, and sulfur isotope signatures, the Mugomo prospect can be classified as orogenic.

Timing of the Orogenic Jouhineva Au-Cu-Co and Huhta Au Mineralization in the Pohjanmaa Belt, Finland

Andressa de Araujo Silva[1], Clifford G. Patten[1], Jochen Kolb[1], Simon Hector[1], Benjamin Walter[1], Aratz Beranoaguirre[1], Elisabeth Eiche[1, 2]
1. Chair of Geochemistry and Economic Geology (EGG), Institute of Applied Geosciences (AGW), Karlsruhe Institute of Technology (KIT), Karlsruhe, Germany, 2. Laboratory for Environmental and Raw Materials Analysis (LERA), EGG-AGW, Karlsruhe, Germany

Some orogenic Au deposits show variable Cu and Co content, such as the Jouhineva Au-Cu-Co deposit and Huhta Au occurrence from the Pohjanmaa Belt (PoB), Finland. The reason for Co-Cu enrichment is still unclear, although multi-stage hydrothermal mineralization is the main hypothesis. We aim to constrain the timing of hydrothermal Au ± Co, Cu mineralization in Jouhineva and Huhta by combining in situ titanite dating with detailed petrography to link mineralization stages with the tectonic evolution of the Fennoscandian Shield. Jouhineva and Huhta are controlled by the same first-order structure, hosted in the same rocks, and contain similar ore mineral assemblages. Jouhineva and Huhta consist of at least two mineralization stages hosted in quartz veins: (1) auriferous arsenides, Co-bearing arsenopyrite, and $chalcopyrite_1$ with biotite, actinolite and $sericite_1$ on vein selvedges; and (2) $chalcopyrite_2$ and free gold with $sericite_2$ and chlorite alteration halos. Three titanite generations are determined. In Jouhineva, $titanite_{1J}$ is intergrown with arsenopyrite and $chalcopyrite_1$. $Titanite_{2J}$ is associated with sericite alteration halos. $Titanite_{1H}$ is intergrown with arsenopyrite, and $titanite_{2H}$ with $chalcopyrite_2$. $Titanite_{1H}$ and $Titanite_{1J}$ are dated at 1854 ± 15 and 1816 ± 12 Ma, respectively. These ages indicate that the first mineralization stages in Jouhineva and Huhta are not coeval, explaining the Au-Cu-Co and Au-only metal assemblages, respectively. The 1760 ± 14 and 1744 ± 30 Ma ages of $titanite_2$ overlap within error, indicating similar hydrothermal evolution. The first stage in Huhta is coeval with large-scale regional tectonic rearrangement during the formation of the Bothnian Orocline (1870-1850 Ma). The first-order structure was reactivated during continent-continent collision at the margin of the Fennoscandian Shield (1840-1770 Ma). The hydrothermal activity terminated during orogenic collapse and final cooling of the orogen (1760-1744 Ma). The complex tectonic and hydrothermal evolution of the PoB explains the different metal endowment in the two deposits.

Scheelite and Zircon U-Pb Dating Reveals Two Discrete Gold Mineralization Events at the Zaozigou Deposit, West Qinling Orogen, China

Shi-Guang Du[1, 2], Jian-Wei Li[1, 2]
1. China University of Geosciences, Wuhan 430074, China, 2. State Key Laboratory of Geological Processes and Mineral Resources, Wuhan 430074, China

The Qinling Orogen hosts numerous gold deposits that were formed mostly in the period of 220 to 200 Ma, coincident with late collisional orogeny involving the South China and North China cratons. The Xiahe-Hezuo district in West Qinling is characterized by several world-class sediment-hosted gold deposits spatially related to intrusions. A few Cu-Au skarn deposits also developed in the district coevally with the sediment-hosted gold deposits. Available data revealed that districtwide gold mineralization spanned from ca. 246 to 211 Ma, but whether individual gold deposits formed from single or multiple mineralization events is unclear and lacks detailed geochronological studies. Here we present new dating results to show the Zaozigou deposit of the Xiahe-Hezuo district is a product of two separate gold deposition events.

The Zaozigou deposit (~110 t Au) is hosted in both Triassic clastic rocks and intermediate to felsic dikes, the latter emplaced from 250 to 230 Ma. The orebodies are controlled by well-developed faults. Several mineralization styles have been recognized, including gold-bearing arsenopyrite-pyrite disseminations in altered wall rocks, quartz-pyrrhotite-chalcopyrite veinlets, and quartz-stibnite-gold veins in sequence.

Two quartz diorite dikes, one containing dense arsenopyrite-pyrite disseminations and the other hosting inclusions of quartz-pyrrhotite-chalcopyrite veins, yield zircon U-Pb dates of 231.8 ± 1.0 and 232.28 ± 0.99 Ma, constraining the disseminated and veinlet mineralization at ~232 Ma. Scheelite grains from the quartz-stibnite veins have core-rim textures, with the core and rim showing contrasting colors in CL images and distinct compositions, with the rims containing REE contents and U/Pb ratios one order of magnitude higher than cores. The core and rim give rise to ICP-MS U-Pb ages of 230.1 ± 2.9 and 209.9 ± 2.7 Ma, respectively. The new zircon and scheelite U-Pb ages thus indicate two discrete mineralization events at ca. 232 and 210 Ma in the Zaozigou gold deposit.

Morphological and Compositional Analysis of Alluvial Gold: Los Ángeles Gold Placer (Cáceres, Spain)

Vanessa L. Escobar Duche, Santos Barrios Sánchez, Juan G. Barreiro, Kelvin dos Santos Alves
Universidad de Salamanca, Salamanca, Spain

The morphotextural evolution and geochemical composition have been investigated from alluvial gold grains recovered from the Los Ángeles Gold Placer in the Central Iberian Zone of the Iberian Massif to understand the origin and evaluate the type and distribution of unknown primary mineralization(s). The main results revealed that a mixture of populations of gold particles has been identified. The gold particles derived from their first sedimentary cycle are mixed with recycled gold sedimentary particles. The transport range, between 0 and 20 km, is compatible with the location of primary gold deposits in the west of study area, which are associated to breccias and quartz veins emplaced in metasediments of the Schists and Greywacke Complex. Chemical analyses show a binary gold alloy (Au:Ag) and three gold types can be differentiated: Au_1 (core of particles) is characterized by a range of 62.6-96.2 Au wt% with 3.8-37.4 Ag wt%; Au_2 (gold-rich rim) includes a range of 88.3-99.7 Au wt% with 0.3-11.7 Ag wt%; Au_3 (allotriomorphic gold micro-aggregates <10μm) is formed by 90.8-100.0 Au wt% and 0.0-9.2 Ag wt%. Au_1 shows trace elements in its composition (Cu ≤ 0.14 wt%, Te ≤ 0.08 wt%, Bi ≤ 0.09 wt%) and microinclusions (sulfides, sulfarsenides, and arsenides) that reflect the orogenic genesis of a primary hydrothermal gold deposit generated at temperatures between 350° and 500°C, with a multi-vein system related to vertical zoning and/or at least two hydrothermal mineralization pulses, associated to sulphidation stages with a probable contribution of magmatic and/or metamorphic fluids. The last stages of gold evolution include the alteration of primary mineralization, formation of a first paleoplacer, and then formation of the Los Angeles gold placer. Once formed, the primary gold or Au_1 was affected by supergene processes that modified its initial composition through Ag leaching and lead to formation of Au_2 and Au_3.

Geological Setting and Mineralisation Styles of the Zəfər (Zafar) Epithermal Deposit, the Lesser Caucasus, Azerbaijan; a Newly Discovered Deposit in Western Asia

Anar Valiyev, Stephen J. Westhead, Javid Ibrahimov, Aydin Bayramov, Sabuhi Mammadov, Ərzuman Rizvanoğlu
Azerbaijan International Mining Company (subsidiary of Anglo Asian Mining), Gadabay, Azerbaijan

The Zafar deposit is located in the mineral-rich Gedabek contract area in Azerbaijan. It is geologically located in association with the Gədəbəy-Bittibulaq local deep fault in the Lök-Qarabağ volcanic arc of the Lesser Caucasus. The study of the regional geology of Zafar by Azerbaijan International Mining Company (AIMC) geologists since 2018, utilising a multi-layered geological data interpretation, identified the mineral potential culminating in the Zafar deposit discovery in 2020. Spatially, the Zafar deposit is situated between two main epithermal systems hosting the Gedabek-Gadir and Ugur deposits, which were discovered by the AIMC geology team.

The igneous rocks are extrusive and mostly felsic to intermediate in composition; rhyolites, dacites, andesites, and their tuffs are distributed from the surface to depth. The host rock of main ore mineralisation is locally termed a metasomatic quartz-porphyry (rhyolite-porphyry), which starts at depths of 250 to 300 metres from the surface to more than 500 metres.

Mineral content is visual; however, further petrographic studies by polished thin section were carried out on samples taken from outcrop and drill core. Resulting from these studies, pyrite is the main mineral observed in all drill core and outcrop samples in crystal structures in the form of massive, disseminated, euhedral, anhedral, etc. Other primary sulphides include chalcopyrite, covellite, digenite, galena, and sphalerite. The genesis of covellite and digenite is open for discussion. Other copper sulfosalt minerals, tetrahedrite and tennantite, were identified by ore microscopy. Associated non-sulphide minerals including jasper, magnetite, barite, haematite, and limonite have also been defined. Quartz is the main gangue mineral in the matrix of the host rocks.

The mineral composition, textures and relationships, and chemical composition suggest a genesis of intermediate sulphidation type of epithermal system for the Zafar deposit. AIMC is the subsidiary operating company of Anglo Asian Mining

Characteristics of Geology, Mineralization, and Alteration of the Khan Altai Gold Deposit, Mongolia

Khaliunaa Iderbat[1], Khashgerel Bat-Erdene[2], Imants Kavalieris[2], Mandalbayar Ganbat[3]
1. Mongolian University of Science and Technology, Ulaanbaatar, Mongolia, 2. Plus Minerals LLC, Ulaanbaatar, Mongolia, 3. Khan Urnud Mining LLC, Ulaanbaatar, Mongolia

The Khan Altai deposit, which is the largest gold deposit in Mongolia with 85.4 tons of gold, is located in southwest Mongolia on the northern margin of the Neoproterozoic-Cambrian Lake terrain. The host rocks are potentially Cambrian in age and have been strongly deformed and metamorphosed to lower greenschist facies. The host rocks consist of rhyolitic volcaniclastics interbedded with laminated siltstone, overlain by basaltic andesite, and faulted against dolomite. Two types of gold mineralization were identified: (a) low-grade mineralization associated with disseminated pyrite (2-10% by volume), mainly in rhyolitic volcanic and laminated siltstone with quartz veins; and (b) high-grade mineralization (up to 183 ppm Au over 1 m) related to centimetre-wide quartz-native Au-pyrite veins. Additionally, the Au mineralized zone includes VMS mineralization (currently minor), characterized by massive pyrite-sphalerite lenses and quartz-chalcopyrite stringer zones. The main alteration is quartz-white mica-albite, and gold mineralization is correlated with zones of strong tectonic foliation and the formation of metamorphic mica (phengitic-illite and phengite), which exhibit an outward zonation to muscovite and paragonite. Pyrite is typically euhedral and exhibits concentric growth zones as well as quartz pressure shadows to the enclosing foliation, consistent with syngenetic growth during metamorphism and deformation. Other sulfides include arsenopyrite, sphalerite, and possibly marcasite. Preliminary laser ablation-ICPMS mapping shows that pyrite rims are enriched in Au, As, Co, Cu, Ni, Pb, Ag, Mo, and Se, whereas the pyrite core is enriched in Co, Bi, Te, and the host rock in K, Ba, V, and Tl. Whole-rock geochemistry of basaltic andesite to rhyolite shows N-MORB characteristics for basaltic andesite as well as a subduction signature for all rocks, and high MgO (_8%) and TiO2 (_1%) and low Ni, Cr content in basalt, compatible with a back-arc tectonic setting.

Polymetallic Au-Rich Melts in Hydrothermal Systems and Their Roles in Gold Enrichment

Wei Jian
China University of Geosciences (Beijing), Beijing, China

Low-temperature polymetallic melts can scavenge gold from aqueous fluids and may be instrumental in the formation of some hydrothermal lode gold deposits. However, it is likely that metallic melts and their roles in gold enrichment have gone unrecognized in many gold deposits. This is due to the difficulty in recognizing textures that can be related to metallic melts and a lack of awareness that hydrothermal ore mineral assemblages can melt at low temperatures.

We present evidence for precipitation of Au-Ag-Te-rich melt directly from hydrothermal fluids and subsequent entrapment as melt inclusions in quartz veins of the Xiaoqinling lode gold district, central China. Au-Ag-Te-rich inclusions start to melt at a temperatures as low as 135°C, much lower than the eutectic of the Au-Ag-Te system at 304°C; native gold in contact with Au-Ag tellurides starts to melt at a temperature as low as 220°C. We also present textural evidence left by migration of melt inclusions, which can be used to distinguish previously trapped polymetallic melt inclusions from mineral inclusions.

We propose that Au-Ag-Te–rich melts formed through adsorption-reduction of metals on pyrite and quartz surface. Although quartz is often regarded as a chemically inert mineral, surface defect sites on quartz have a high capacity to reduce metals. We propose that polymetallic melt inclusions trapped early in the crystallization history of the host mineral could migrate to the crystal surface and enter into contact with aqueous fluids. The same effect can also be achieved by fracturing of the host mineral. In this way, relatively low volumes of polymetallic melts can continue to scavenge gold from the multiple pulses of ore fluids that are commonly involved in the formation of large gold deposits, without a requirement for the ore fluid to be saturated with the gold.

Geology of the Golddigger Intrusion-Related Gold Deposit in British Columbia, Canada

Randall Karcher[1], Thomas Monecke[1], James Reynolds[1], Russell Ashton[2], Manuele Lazzarotto[2], Bill Chornobay[2]
1. Colorado School of Mines, Golden, CO, USA, 2. Goliath Resources, Toronto, ON, Canada

Goliath Resources' Golddigger exploration project is located at the southern tip of the Golden Triangle of British Columbia, which is a region known for its unusual precious metal endowment. Extensive drilling in 2021 and 2022 revealed the presence of several high-grade sulfide-bearing quartz veins. The Surebet Zone is the main ore occurrence, which occurs as a planar 1.6-km^2 orebody that is crosscutting folded sedimentary strata assigned to the Hazelton Group. Ongoing research will unravel the ore-forming processes, timing, and mineral associations to inform ongoing exploration. The research includes core logging, whole-rock geochemical analysis, transmitted- and reflected-light microscopy, cathodoluminescence microscopy, scanning electron microscopy, fluid inclusion petrography, sulfur isotope studies, and Re-Os isotope dating. Initial petrographic investigations revealed that quartz formation occurred at conditions close to the brittle-ductile transition. Early vein quartz has been affected by intense recrystallization and is interpreted to have formed at lithostatic conditions, whereas late vein quartz forms euhedral crystals that formed at hydrostatic conditions. Early fluid inclusion populations have been affected by reequilibration during the pressure change from lithostatic to hydrostatic conditions. High CO_2 contents in primary and secondary fluid inclusion assemblages suggest that the deposit formed at a significant depth, likely >5 to 6 km below the paleosurface. The sulfide minerals postdate both generations of quartz. This includes pyrrhotite, galena, and sphalerite with minor pyrite and chalcopyrite. High Au and Ag assay values correlate with the occurrence of native gold and silver-rich tetrahedrite. Current evidence suggests that the mineralization is genetically linked to a deep intrusion within the Alice Arm region. Classification of Golddigger as an intrusion-related gold deposit opens up new opportunities for exploration within the region.

Trace Element Compositions of Placer Gold Samples: Implication for Gold Exploration in Northern Cameroon

Yanick Blaise Ketchaya[1, 2, 3], Taofa Zhou[1, 2]
1. Ore Deposit and Exploration Centre, School of Resources and Environmental Engineering, Hefei University of Technology, Hefei 230009, China, Hefei, China, 2. Anhui Province Engineering Research Center for Mineral Resources and Mine Environments, Hefei 230009, China, Hefei, China, 3. Department of Earth Sciences, Faculty of Science, University of Yaoundé I, P.O. Box 812, Yaoundé, Cameroon, Yaoundé, Cameroon

The type of primary source of gold deposit can be explored by using the study of trace element analysis of placer gold, which is a valuable exploration tool. Au-bearing deposits are investigated through the placer gold, which is an important indicator mineral. The hydrothermal fluids interacting with diverse geological settings exert an important function on the chemical composition of gold. Consequently, alluvial gold particles from the placer deposits within the Gamba district in northern Cameroon were examined by electron probe microanalyzer (EPMA) to show discriminant chemical signatures. The gold grains from different localities show the same trace element composition, which appears to be in solid solution in Au. These trace element compositions contained in gold grains indicate a homogeneous source. The placer gold particles have significant chemical characteristics (low Ag content) consistent with a mesothermal source. The gold particle signatures in the Gamba district, with high Te and Bi contents, reflect the chemical characteristics of the felsic host rock superimposed on the chemical signature of the hydrothermal fluid.

Using Borehole Data to Discriminate Magmatic and Hydrothermal Processes for Orogenic Gold Exploration

Mosayeb Khademi[1, 2], David Giles[1, 2], Caroline Tiddy[1, 2], Mark Lindsay[1, 3], Andrej Bona[1, 4]
1. University of South Australia, Adelaide, SA, Australia, 2. Mineral Exploration Cooperative Research Centre (MinEx CRC), Perth, WA, Australia, 3. CSIRO, Perth, WA, Australia, 4. Curtin University, Perth, WA, Australia

This paper demonstrates how integrated drill hole geochemistry, petrophysics, and geological observations can enable identification of geological processes related to an orogenic gold mineral system. The Yilgarn craton in Western Australia, one of the most well-endowed Au terranes in the world, is used as a case study area. Data patterns discriminate magmatic differentiation of the protolith from alteration footprints. In particular, whole-rock geochemical data exhibit patterns that align with in situ magmatic fractionation in the predominantly mafic host rocks, whereby Ti, Fe, and P positively correlate and Mg negatively correlates with Zr. Titanium and Fe reach a "transition point" between ~90 to ~130 ppm Zr consistent with the crystallization of magnetite during the later fractionation stages and supported by these zones preserving high magnetic susceptibility (>30 × 10^{-3} SI). We interpret these zones as the upper parts of dolerite sills that fractionated in place during crystallization.

Multielement geochemistry clustering analysis shows a relationship between Au and S, As, Sb, Cu, K, and Te. This is interpreted to reflect hydrothermal processes related to Au mineralization, suggesting S, As, Sb, Cu, K, and Te are pathfinder elements that can be used to delimit Au system targets. Enrichment of Au pathfinder elements is seen within core logged as felsic magmatic rocks and corresponds with zones of visibly strong alteration. The logged felsic samples have densities of ~2.85 g/cm^3, which is more typical of mafic rocks rather than felsic volcanic rocks (~2.4 g/cm^3). Therefore, the logged felsic magmatic rocks are interpreted here to be altered mafic rocks.

The outcomes of this study demonstrate how integration of multiscale, multifaceted data sets may be used to provide insight into magmatic processes and alteration zones related to mineralization, and may therefore be used to inform exploration programs targeting orogenic gold in the Yilgarn craton, Western Australia.

Carlin-Type Mineralisation at Lookout Mountain in the Eureka District, Nevada, USA

Simon Kocher[1], Zaid Nadhim[1], Thomas Monecke[1], Lesli Wood[1], Katharina Pfaff[1], Jens Najorka[2], Patrick Highsmith[3], Mark Osterberg[3], Steve Osterberg[3]
1. Center to Advance the Science of Exploration to Reclamation in Mining, Department of Geology and Geological Engineering, Colorado School of Mines, Golden, CO, USA, 2. Department of Science, The Natural History Museum, London, United Kingdom, 3. Timberline Resources, Coeur d'Alene, ID, USA

The Carlin and neighbouring trends in Nevada represent some of richest accumulations of gold on Earth. The Eureka district is located on the southern end of the Battle Mountain-Eureka trend and contains a number of active mines and exploration projects. Recent drilling has identified Carlin-style gold mineralisation in the Lookout Mountain area, situated approximately 10 km SSW of Eureka. The district is underlain by a thick sequence of Cambrian through Devonian sedimentary units that were folded and faulted by the Late Devonian to Early Mississippian Antler Orogeny. Continued deformation into the Cretaceous led to the formation of thrust faults, which themselves were deformed into a series of north-trending folds. Lower Paleozoic calcareous units were complexly folded and faulted as part of a Paleozoic thrust system, placing Cambrian rocks on top of Ordovician and Silurian strata. Basin and range extension and normal faulting subsequently formed the present topography. Lithologies exposed in the South Eureka district include limestone, dolomite, and minor amounts of shale and quartzite with a total thickness of roughly 4.5 km. Tertiary volcanic rocks and Mesozoic and Tertiary intrusions occur locally within the Eureka district. Gold mineralisation at Lookout Mountain occurs along a 5-km, extensively silicified structural zone and is associated with replacement of carbonate and shale lithologies, causing the formation of flat-lying, multi-staged collapse breccias. Carbonate dissolution, iron addition, and the formation of arsenic sulphides and fine-grained pyrite commonly accompany high-grade gold. In this study we combine optical petrography and SEM-based automated mineralogy analysis with clay mineralogy, geochronology, and detailed analysis of breccia textures with the aim to reconstruct mineralising processes and aid the identification of additional lithologies favourable to mineralisation.

Fluid Evolution and Carbon-Gold Systematics at the Fairview Mine, Barberton Greenstone Belt, South Africa

Rutger La Cock[1], Bjorn von der Heyden[1], Long Li[2], Yifan Du[2], Matthew Steele-MacInnis[2]
1. Stellenbosch University, Cape Town, South Africa, 2. University of Alberta, Edmonton, AB, Canada

The Paleoarchaean Barberton Greenstone Belt (BGB) plays host to a number of gold deposits, with over 350 tons of Au being extracted since its discovery in 1882. Gold has a wide variety of applications, with its most prominent role being how it serves as a safe haven for investors, which was exemplified during the recent Covid-19 pandemic. Of the many deposits within the BGB, one of the most economically significant is the Fairview gold mine in the northwest portion of the belt. Recent work has focussed on the structural aspects controlling mineralisation; however, the chemical controls on mineralisation are still poorly understood. This research focuses on understanding the evolution of ore fluids and characterising graphite found in and around high-grade shear zones.

Underground samples were characterised using scanning electron microscopy and laser ablation-inductively coupled plasma-mass spectrometry and document how variation in carbonate species and sulphide trace element distribution can be used to track temporally separate fluid flow events. Carbon isotope analyses provide insights into the potential source of ore fluids, with black shales of the Fig Tree Group appearing as a likely candidate. Graphite Raman spectroscopy reveals that graphite within shear networks is of nanocrystalline size and has a high degree of structural disorder. Geothermometry calculations based off of Raman spectra of graphite suggest that shear zones were subjected to maximum temperatures of around 400°C. A better understanding of how carbon contributes to gold deposition at Fairview will help geologists discern whether or not graphite is a useful parameter to consider during exploration. Graphite is strongly conductive and is therefore useful during geophysical surveys, which are cheaper and more environmentally friendly than conventional drilling programs. The findings of this research also have significant implications for ore-forming models pertaining to Archaean greenstone belts.

World-Class Gold Deposits and Emerging Exploration Opportunities in the Loulo District, West Africa

Thomas Stapley[1], Paul Stenhouse[1], James Lambert-Smith[2], Andrew Allibone[1], David Lawrence[1], Dioumacor Senghor[1], Germain Crestin[1]
1. Barrick Gold Corporation, Toronto, ON, Canada, 2. School of Earth and Environmental Sciences, Cardiff University, Cardiff, United Kingdom

The Paleoproterozoic wider Loulo district in western Mali contains +30 Moz of gold, with three deposits, Yalea, Gara, and Gounkoto, contributing +20 Moz of this endowment.

The Loulo district can be subdivided into two domains based on differences in geology and mineralisation styles. The eastern Kofi domain, hosting Yalea, Gara and Gounkoto, is dominated by metasedimentary rocks and mineralisation is more typically orogenic. Gold is commonly focussed along shear-reactivated D_1 structures, with pre-mineralisation albite or tourmaline alteration enhancing competency contrasts. In contrast, the western Falémé domain, which currently only hosts <1 Moz deposits, contains voluminous, broadly syn-mineralisation, high-K, calc-alkaline intrusive rocks (i.e., the Falémé intrusives), and mineralisation appears similar to oxidised intrusive-related systems. Syn-mineralisation strain is distinctly lower, and mineralisation is spatially associated with large hydrothermal ± magmatic breccia complexes, hypabyssal intrusives, and primary carbonates.

These domains were previously interpreted to be separated by the Senegal-Mali Shear Zone. However, mapping indicates that this shear zone does not exist, with intrusive relationships observed locally between the Falémé intrusives and Kofi domain metasediments.

Mineralisation-related carbonate veins with low $d^{18}O_{SMOW}$ (13.3–14.8‰) and sulphides with low $\delta^{34}S_{CDT}$ (–4.6 to 3.9‰) support the involvement of magmatic fluids within and adjacent to the Falémé domain, as typified by the Kabe-West deposit. Atypical H_2O-CO_2-NaCl-$FeCl_2$ fluid inclusions also indicate a component of magmatic fluids at Gara. However, evidence for magmatic fluids generally decreases to the east.

The lack of a major structural break between the Kofi and Falémé domains, and evidence for mineralisation-related magmatic fluids in both domains, suggests that the Falémé intrusives were both an important heat source for the district-scale hydrothermal system and, to varying extents, a direct contributor of fluids. This reassessment of the importance of the Falémé intrusives has highlighted the prospectivity of the Falémé domain, which is underexplored compared to the Kofi domain.

Chromium- and Vanadium-Bearing Phyllosilicates at the Kerr-Addison Gold Deposit, Virginiatown, ON: Implications for the Sources and Mobilities of Cr and V in Gold Deposits

Derek D.V. Leung, Andrew M. McDonald
Harquail School of Earth Sciences, Laurentian University, Sudbury, ON, Canada

An empirical relationship exists between Cr- and V-bearing green micas (e.g., muscovite var. fuchsite/mariposite and roscoelite) and gold deposits worldwide, given that 10% of the 100 largest lode-gold deposits in the world contain green micas. To better understand this relationship, a study was undertaken to evaluate the mobility of Cr and V by determining how these elements are incorporated into phyllosilicates (muscovite and chlorite). Drill core samples from the two principal protoliths at the Kerr-Addison gold deposit (ultramafic and mafic volcanic rocks of the Larder Lake group), collected proximal and distal to ore zones, were observed using optical microscopy to determine the paragenesis of the phyllosilicates. Mineral-chemical analyses were conducted using energy- and wavelength-dispersive spectroscopy (EDS/WDS) in conjunction with laser ablation-inductively coupled plasma-mass spectrometry (LA-ICP-MS). Results show that the phyllosilicates are the major secondary mineralogical hosts for Cr and V in the altered rocks, with muscovite replacing chlorite in the ultramafic suite (chlorite islands enveloped by diffuse rims of pitted muscovite, associated with magnesite and rutile, suggesting replacement of chlorite by muscovite). Phyllosilicates in the altered ultramafic rocks contain Cr, whereas those in the mafic sequence contain V. In the ultramafic suite, Cr contents in muscovite are variable but consistent, whereas V contents in muscovite increase away from proximal zones. These results suggest (1) the ultramafic and mafic rocks were the sources of Cr and V, and the fluids were wall-rock buffered, given the protolith control on the phyllosilicate mineral chemistry; (2) Cr is immobile, whereas V may be more mineralogically mobile at Kerr-Addison by preferentially partitioning into hydrothermal rutile, which highlights their contrasting geochemical behaviours; and (3) chlorite is likely to be a stable intermediate host for Cr and V before these elements are incorporated into muscovite.

Ore-Forming Process of Basalt-Hosted Carlin-Type Gold Deposits, Western Guizhou Province, China

Junhai Li[1, 2], Jianzhong Liu[2, 3], Pan Wu[4], Zhuojun Xie[5], Zepeng Wang[1, 2], Shuangju Zhang[6], Chengfu Yang[1, 2], Qinping Tan[5], Jingke Liu[1], Bingqiang Zhang[1]
1. No.105 Geological Party, Guizhou Bureau of Geology and Mineral Exploration & Development, Guiyang550018, China, 2. Innovation Center of Ore Resources Exploration Technology in the Region of Bedrock, Ministry of Natural Resources of People's Republic of China, Guiyang 550081, China, 3. Guizhou Bureau of Geology and Mineral Exploration & Development, Guiyang 550004, China, 4. Guizhou University, Guiyang 550025, China, 5. State Key Laboratory of Ore Deposit Geochemistry, Institute of Geochemistry, Chinese Academy of Sciences, Guiyang 550081, China, 6. Guizhou Institute of Technology, Guiyang 550003, China

The basalt-hosted gold deposits in western Guizhou, China, are newly discovered Carlin-type gold deposits, and the typical deposits include the Jiadi large-size and Damaidi medium-size gold deposits. Gold mineralization mainly occurs in the Emeishan basalt and the unconformity (SBT) between the Emeishan basalt and the underlying Maokou Formation. Research indicated that the ore-forming fluids were released from the deep buried granitic magma. Ore fluids are enriched in ore-forming elements including As Au, As, Sb, Hg, and Tl and volatile components including CH_4 and CO_2, and have the characteristics of high to ultra-high pressure. During the Yanshanian, ore-forming fluids upwelled to the regional structural detachment surface between P_2m and $P_3\beta$ along the regional faults. Some ore-forming fluids migrated laterally and reacted with the rocks near the regional tectonic detachment surface by water-rock reaction to form the SBT. Some ore-forming fluids migrated upward along the fault to the tuff or basaltic breccia in the interlayer fracture zone of $P_3\beta$, and the ore-forming fluids kept migrating laterally due to the high porosity of the rock. When the ore fluids converged at the culmination, they reacted with the Fe-rich basaltic rocks. Orthopyroxene, clinopyroxene, plagioclase, magnetite, ilmenite, and other minerals of the basaltic rocks dissolved and released Fe^{2+}, etc. The released Fe^{2+} combined with S and As in the ore-forming fluid to form arsenian pyrite and arsenopyrite. The Au-HS complex was decomposed, and Au was incorporated into the arsenian pyrite and arsenopyrite in the form of invisible gold. Sulfurization results in the precipitation of gold-bearing arsenian pyrite and arsenopyrite, and ultimately the formation of gold orebodies hosted by the SBT and $P_3\beta$ basalt.

This research was financially supported by grants from the National Natural Science Foundation of China (U1812402) and the Guizhou Provincial Science and Technology Projects (Qiankehepingtairencai-CXTD[2021]007).

Gold Source and Ore-Forming Process of the Linglong Gold Deposit, Jiaodong Gold Province, China

Yayun Liang[1], Pingye Ma[2], Xuefei Liu[2], Lei Shu[3], Wenhao Xue[1], Chenxi Zhang[3], Yao Ma[2], Chonghui Shen[4]
1. University of Science and Technology Beijing, Beijing, China, 2. China University of Geoscience, Beijing, Beijing, China, 3. MNR Key Laboratory of Gold Mineralization Processes and Resources Utilization, Shandong Institute of Geological Sciences, Jinan, China, 4. National Gems & Jewelry Testing Co. Ltd, Beijing, China

The Linglong gold orefield, situated at the northwest of the Jiaodong gold province, China, comprises more than 10 gold mines with a total gold resource of ~1,000 t. Gold mineralization is in the form of gold-bearing quartz veins. In this study, pyrites from three main mineralization stages have been investigated for texures, in situ major and trace element concentrations, and sulfur isotope compositions. Pyrites in stage I (Py-I) show euhedral crystals with homogeneous brightness in the BSE, without mineral inclusions, porous microtextures, and oscillatory zones and have low contents of Au, Ag, As, Bi, Cu, Sb, Zn, Co, and Ni.. Pyrites in stage II (Py-II) are fine- to coarse-grained and euhedral-subhedral with little fracture, and have similar element contents but higher contents of As and Sb than those of Py-I. Pyrites in stage III (Py-III) were produced in the quartz-polymetallic sulfide veins and show subhedral to anhedral porous texture features. Py-III have significantly higher contents of metals elements, especially Au and As.The $\delta^{34}S$ values of Py-I, Py-II, and Py-III show 6.76~7.47‰, 6.11~7.27‰, and 6.07~8.33 ‰, respectively. The textures and geochemical compositions of pyrites have proved that invisible gold mainly occurred in pyrite in the form of solid solution (Au^+) with a few of nano-form of gold. Ore-forming materials were mainly derived from mantle-derived mafic magma. Decrease of fluid temperature, water-rock interactions, and fluid boiling would lead to the deposition of sulphide and gold. In the Late Mesozoic, lithospheric delamination of the eastern block of the NCC triggered decompression and melting of the upwelling asthenospheric mantle and partial melting of the lithospheric mantle. These lithospheric mantle-derived mafic magmas provided the ore-forming materials for the large-scale gold mineralization in the Jiaodong gold province.

Structural Controls and Timing of Orogenic Gold Mineralisation in the Baie Verte Peninsula of Newfoundland, Canada

Georgian Manuc[1], Taija Torvela[1], Rob Chapman[1], Nick Roberts[2], David Copeland[3], Simon Tapster[2]
1. University of Leeds, Ores and Mineralisation Group, Leeds, United Kingdom, 2. British Geological Survey, Keyworth, United Kingdom, 3. Signal Gold Inc., St. John's, NL, Canada

This study is the first to robustly tie paragenetic associations of gold in the Baie Verte Peninsula (BVP) of Newfoundland, Canada, with regional- and deposit-scale structural observations and in situ geochronology. The BVP is one of the richest auriferous regions within the Newfoundland Appalachians, hosting more than 100 economic and sub-economic gold occurrences. The geological setting of the BVP resulted from the accretion of several micro-terranes of Iapetus origin to the Laurentian margin during the Ordovician Taconic Orogeny. Subsequent deformation developed under a sinistral transpressional setting during the Silurian Salinic Orogeny, followed by a complex dextral transpressional to transtensional tectonic regime during the Devonian Acadian and Neo-Acadian orogenic events. The studied gold occurrences, Stog'er Tight, Argyle, Romeo & Juliet, Deer Cove, and Pine Cove, lie along splays of the Baie Verte Line, the suture between the Laurentian margin and the accreted island arcs. All are currently either mined or under exploration by Signal Gold Inc..

New structural, mineralogical, and geochronological data has been collected through geological mapping, optical and SEM/CL investigations of mineralized samples, and in situ LA-ICP-MS U-Th-Pb geochronology of monazite, xenotime, and titanite. The structural data show that at least three deformational events (D_2-D_4) influenced the formation and present geometry of the gold deposits. Mineralogical and geochronological data reveal complex paragenetic sequences in each deposit and at least two periods of gold precipitation.

We interpret the first episode of gold mineralization to have occurred between ~420 and 400 Ma, broadly synchronous with development of E-W–striking and south-verging thrusts formed during D_2 progressive deformation. A later fluid phase was introduced in the system between ~390 and 370 Ma, during the D_3 and/or D_4 deformational events when the deposits were subsequently folded and faulted. Currently, it is unclear whether this fluid augmented the existing gold endowment or remobilized pre-existing gold.

Genesis of Gold Placers of the Cordón Baquedano, Tierra del Fuego, Austral Andes (54°S)

Carlos Marquardt, Tomás Roquer, Felipe del Valle, José Joaquínara Jara
Departamento de Ingeniería de Minería, Pontificia Universidad Católica de Chile, Santiago, Chile

Gold placers are formed by weathering, transport, and depositional processes that affect various lithologies, including gold-bearing ores and gangues. This work conducts a geological recognition of gravel placer gold deposits and their rock sources using geomorphologic and sedimentologic techniques. The study area is part of a mining placer gold district at the Cordón Baquedano, a mountain range of 600 m of altitude and NW-SE direction located in the northern part of the Tierra del Fuego Island (54°S). The origin of this placer gold district has been related to the great ice ages of the Upper Pleistocene. It is believed that glaciers eroded rocks from the Darwin Cordillera, transporting sediments 40 to 50 km to the west or northwest into the foreland basin, liberating gold particles in a fluvio-glacial sedimentary environment. Our results indicate that the origin of gold placers is related to plio-pleistocene foreland fluvial events. Sedimentological and geological results confirm the formation of fluvial deposits that represent the last event in the Upper Cenozoic part of the Magallanes foreland basin, probably during the Pliocene to Lower Pleistocene. Rock sources may have been located at the Darwin Cordillera, where rock outcrops with equivalent textures and mineralogy to the clasts of the placer deposits were observed, as well as structurally controlled gold-bearing hydrothermal alteration (e.g., silica-sulphide veins). Morphologic results suggest that repeated local episodes of fluvio-alluvial incision and deposition cut the uplifted fluvial deposit (related to the development of a regional pediplane or alluvial plain-like landform) and re-concentrate gold particles in the youngest fluvial morphologies (e.g., terraces and channels), likely since the Upper Pleistocene. This work highlights the importance of conducting detailed geological studies to prepare the geometallurgical models necessary to propose sustainable mining strategies in gold placers at Tierra del Fuego and similar gold deposits worldwide.

Compositional Characterization of Placer and Lode Gold Grains from Eastern Interior Alaska, USA

Erin E. Marsh[1], Robert J. Chapman[2], Renee Pillers[1], Heather Lowers[1], Ryan D. Taylor[1]
1. U.S. Geological Survey, Denver, CO, USA, 2. University of Leeds, Leeds, United Kingdom

The geology of eastern interior Alaska, from east to west, consist of the intersection and juxtaposition of two tectonic assemblages: the arc and basinal assemblages of the allochthonous Yukon-Tanana terrane and the parautochthonous continental margin Yukon-Tanana Upland (YTU). The YTU is bordered to the northwest by tectonically interleaved rock formations of the Livengood Assemblage. The region hosts various reduced intrusion-related and orogenic gold deposits and occurrences, (i.e., Fort Knox, Golden Summit, and Pogo).

Compositional characterization of placer gold and placer-lode gold pairings from across interior Alaska is based largely on the elemental signatures of opaque mineral inclusions revealed in polished section. The inclusion suites reflect the mineralogy coeval with gold precipitation in lode occurrences and correlate with the tectonic setting of the locality. Mineral inclusions in gold from arc and basinal assemblages have Ag-Te +/- Bi, Pb, and Sb signatures, which transition to Pb-As-Sb–dominated inclusion suites in gold from the parautochthonous continental margin. Gold from the ultramafic rocks of the Livengood Assemblage yields a Pb-As-Ni-Co-Cu signature. The observed differences suggest interaction between ore fluid and host lithologies, culminating in strong partitioning of minor elements to the mineral inclusions. In equivalent studies of placer gold in the neighboring Yukon Territory, samples with an Ag-Te mineral inclusion signature are postulated to be associated with epizonal lode deposits, whereas those with a Pb-Sb-As signature are related to orogenic gold deposits. The alloy compositions of gold from throughout the region do not provide robust discriminants, except for elevated Hg (to >3 wt%) in gold from the Livengood and 40 Mile Districts. The approach sheds light on the influences of the host terrain on the mineralogy of gold in orogenic belts and permits characterization on that basis.

Epithermal and Porphyry Style Copper-Gold Mineralization at Arthur's Seat Project, Jamaica

William Miguel, Stephen Hughes
C3 METALS, Toronto, ON, Canada

The Arthur's Seat project is located in the central inliers of Jamaica, approximately 45 km west-northwest of the capital city of Kingston. The Arthur's Seat epithermal vein swarm is a true maiden discovery, with almost no modern-day mineral exploration.

The project area is overlain by Lower Cretaceous andesitic volcanic rocks of the Arthur's Seat Formation. The Arthur's Seat Formation lithologies comprise a thick sequence of conglomerates and volcanic breccias, with intercalated ash and crystal tuffs and andesitic to basaltic andesite lava flows. The volcanic package is intruded by a series of monzonite and andesite dikes.

Three major vein and breccia zones have been discovered to date, varying from centimeters to 2 meters in true width, and typically strike northeast, north-northwest, and east-west. The main quartz vein zone is mineralized for over 3 km in strike length and is up to 25 m wide. Veins and breccia zones are characterized by massive, cockade, comb (dog-tooth), and drusy quartz textures; some of them contain rare carbonates. Shortwave infrared analysis indicates smectite to illite with chlorite alteration and a distal halo of illite-chlorite ± epidote alteration; adjacent to the vein, the smectite-illite alteration zone is predominant.

The Arthur's Seat project is a true maiden discovery and is early on showing significant potential to host an epithermal gold-copper deposit. Rock chip grabs from outcrop veins and breccias assayed up to 58 g/t gold and 11.1% copper. Rock slabs show identical vein and breccia textures to the little-known Pennants gold deposit located 11 km to the west. Although the project is early in its exploration development, two large, discrete magnetic anomalies coincident with the vein system, porphyry B-veins hosted in monzonite, and secondary copper oxides suggest there is a nearby porphyry system.

Uncommon La Virgen High-Sulfidation Epithermal Deposit, Northern Peru

Raul Montesinos[1], Zhaoshan Chang[1], Juan Carlos Castelli[2]
1. Colorado School of Mines, Golden, CO, USA, 2. Mochica Resources Inc, Lima, Peru

Almost twenty percent of world gold production comes from epithermal deposits. The central Andes, especially northern Peru, host world-class high-sulfidation epithermal (HS) deposits such as Yanacocha, Lagunas Norte (LN), and Pierina. The giant Yanacocha and Pierina HS deposits are hosted in pyroclastic units, but the rare Lagunas Norte with above 14 million ounces of gold is hosted mainly in siliciclastic rocks of the lower cretaceous Chimu Formation. The key factors controlling the advanced argillic alteration (AA) and HS mineralization in siliciclastic environments have yet to be well-defined. Our study was carried out on the La Virgen HS deposit mainly hosted in quartz sandstone of the Chimu Formation, 17 km East of Lagunas Norte. The outcomes fill a knowledge gap, enrich our understanding of HS ore-forming processes in siliciclastic environments, and guide exploration in siliciclastic terranes in northern Peru and worldwide.

La Virgen is structurally and lithology-controlled with textural, alteration, and mineralogical similarities to LN. The cracked and brecciated sandstone of the Chimu Formation is affected by highly acidic fluid, which produced AA. The alunite fills the interstices of resistant quartz grains of the sandstones and also microfractures. The minor K-feldspar grains in the sandstone are also altered to alunite and other clays. The alteration is cryptic, with the altered sandstone appearing fresh. The gold mineralization occurs in enargite-pyrite-luzonite–dominant sulfide zones, with intense silicification, partly as black silica with fine-grained pyrite and partly as creamy silica with high rutile content. The silicification also fills interstitial open spaces between grains of the sandstone, fractures, and occurs as cement of breccias. The lesson is that SWIR (short-wavelength infrared) analysis should be widely applied in siliciclastic terrains to identify cryptic clay minerals in reconnaissance exploration, even if the rocks look fresh.

Gold Targeting at the Kalimva Deposit, Kibali Gold District: Hydrothermal Alteration Vectors Interpreted from Multi-Approach Analyses

Yann Mpaka Waku[1, 2, 3], Bjorn P. von der Heyden[1], Etienne Mwandale[2], Gary Hurst[2], David Lawrence[2], Trézène Bampata[2], Fillsmith Luzolo[2, 4], Prince Cimanuka[2]
1. Stellenbosch University, Stellenbosch, South Africa, 2. Barrick Gold Corporation, Toronto, ON, Canada, 3. University of Kinshasa, Kinshasa, Democratic Republic of the Congo (Kinshasa), 4. University of Cape Town, Cape Town, South Africa

Explorationists use hydrothermal wall-rock alteration haloes to vector gold-mineralised veins or structures because of their limited thickness. These alteration haloes may be orders of magnitude greater than narrow auriferous veins and may indicate major mineralogy changes or more subtle "cryptic" alterations to wall-rock mineral chemistry. Thus, wall-rock alteration halo investigations are becoming mineral exploration guides.

The Kalimva gold deposit is located in the northern part of the Kibali gold district in the Neo-Archean Moto greenstone belt in the Democratic Republic of Congo. The deposit comprises a complex ore system that poses substantial challenges for exploration targeting. The main ore-bearing structures are hosted in metamorphosed volcanic and volcaniclastic sequences. This work used multi-element analysis of 1,140 samples from eight drill holes and 2,183 shortwave infrared (SWIR) spectra to map lithologies and fluid–wall-rock alteration envelopes. The selected samples underwent detailed petrography and scanning electron microscopy analysis with energy dispersive spectrometry to corroborate the data sets.

The results indicate that the distal and medial zones of the alteration halo are characterised by MgFe-rich chlorite, phengite, epidote, and hornblende. In contrast, rocks proximal to ore mineralisation have been altered to Fe-rich chlorite with minor Mg and FeMg-chlorite, calcite, and muscovite. This is supported by the SEM chlorite chemistry, which shows low Fe/(Fe+Mg) ratios in medial locations to mineralisation and higher ratios for samples proximal to the ore zone. In these analyses, lithological continuity was well constrained using immobile trace element measurement from the whole-rock geochemistry data set. Assay results highlight a strong correlation between gold and Fe-enriched mineral parageneses. These insights suggest an addition of Fe into the wall rock during the main gold mineralisation event; a finding which holds potential value for gold exploration targeting. This study thus demonstrates the effectiveness of integrating SWIR spectroscopy, petrography, and geochemical analyses for Au mineralisation vectoring.

Pressure-Temperature-Time Constraints on Synvolcanic Intrusive Activity in the Gold-Rich Doyon-Bousquet-LaRonde Mining District, Abitibi Greenstone Belt, Québec

Kevin Neyedley[1], Jacob J. Hanley[1], Patrick Mercier-Langevin[2], James L. Crowley[3], Zoltan Zajacz[4]
1. Department of Geology, Saint Mary's University, Halifax, NS, Canada, 2. Geological Survey of Canada, Quebec, QC, Canada, 3. Isotope Geology Lab, Department of Geosciences, Boise State University, Boise, ID, USA, 4. Department of Earth Sciences, University of Geneva, Geneva, Switzerland

The Mooshla Intrusive Complex (MIC) is an Archean low-Al polyphase tonalite-trondhjemite-quartz diorite-gabbroic magmatic body located in the Doyon-Bousquet-LaRonde (DBL) mining camp of the Abitibi greenstone belt, Québec, Canada. The MIC is spatially and temporally associated with numerous Au-rich VMS, epizonal intrusion-related Au ± Cu, and shear zone-hosted orogenic Au and/or remobilized VMS mineralization deposit types. The MIC is grouped into two distinct stages, the tholeiitic to transitional Mouska-stage and the transitional to calc-alkaline Doyon stage. Through a combination of petrography, high-precision U-Pb geochronology, accessory mineral thermobarometry, and zircon trace element chemistry from the different phases that make up the MIC, this study provides new constraints on the P-T-time evolution of this potentially gold-causative magmatic system. The results demonstrate that individual phases crystallized relatively synchronously and likely in distinct magma chambers at deep levels within the crust (T=~700° to 800°C; P=~5 to 11 kbar, 18.5 to 40 km), consistent with conditions expected for TTG magma generation and evolution. The entire intrusive complex formed over a maximum of ~1.44 Ma, and the relatively synchronous formation of the tholeiitic-dominant Mouska stage and the calc-alkaline-dominant Doyon stage implies a rapid change in tectonic setting from a rift- to arc-dominated environment early in the magmatic evolution of the MIC magmas. Based on distinct zircon trace element ratios in each lithology, this supports the change in tectonic setting and that each magma phase of the Mouska stage evolved separately. The duration and depth of crystallization of the early magma stages of the MIC are comparable to magmatism in major Cu-Au porphyry systems. Thus, the role of deep-seated magmatic processes (and not exclusively shallow degassing) should be considered for supplying gold and other metals to VMS and other hydrothermal systems in the Archean sub-seafloor environment.

Structural Controls of Gold Mineralization in Southwest Ghana: The Iduapriem Gold Mine (Pit 5)

Newman T. Odonkor
University of Ghana, Accra, Ghana

This study investigates the controls of gold mineralisation in Pit/Block 5 of the AngloGold Ashanti Iduapriem mine in Ghana, mainly by field mapping and drill core logging. The pit contains various rocks, including quartzite, Banket conglomerates, basalts, and rhyolites. The stratabound gold mineralization in the pit is associated with the Banket conglomerates, particularly in sub-facies Reef A, B, and C, which are well sorted and texturally matured. The ore geometry is generally subvertical with a thickness of 20 to 100 m. Locally, four deformational events were observed in the pit and denoted, Di_n to Di_{n+1}. Di_o in the study area is marked by preserved bedding and trough cross-stratification observed in the quartzites and the conglomerates. Di_1 affected all the rocks in the study area and is marked by an N- to NNW-striking S_1 foliation with dips ranging from 25° to 42° to the Northeast and East. The Di_2 event, an E-W compression event, resulted in the formation of tight isoclinal folds with wavelengths ranging from 2 to 5 cm and axial planes inclined between high 70° to 85° to the East, which are more pronounced within the quartzites. The Di_3 event is marked by the development of small or meso-scale ripple fall (F_3) open folds with a wavelength of about 0.5 to 1 m. Di_4 is defined by two subparallel shear planes striking N30E and N70E. The N30E shear planes have shallow dips ranging from 30° to 35° and the N70E shear planes have steep dips ranging from 70° to 85°. The N70E shear planes usually truncate at the N30E shear planes.

Overall, the findings indicate that the gold mineralization in the mine is mainly hosted by the Banket conglomerate and structurally controlled by the primary beddings. This means that the mineralization is mainly associated with the primary sedimentary deposition.

The Formation of Auriferous Jasperoid Under the Control of Multi-Stage Tectonism at Tavsan, Kutahya, Western Turkey

Furkan Oguz[1, 2], Burak Mert[1], Berkin Ugurlu[1], Erdin Bozkurt[2], Ahmet Kerim Sener[3]
1. Galata Mineral Madencilik San. ve Tic. A.Ş., Ankara, Turkey, 2. Department of Geological Engineering at Middle East Technical University, Ankara, Turkey, 3. Ariana Resources plc, London, United Kingdom

The collision of the Anatolide-Tauride Block with the Pontides along the Izmir-Ankara Suture Zone following the closure of the northern Neo-Tethys, and the extensive obduction of ophiolites between the Late Palaeocene and Early Eocene, is associated with several periods of magmatism and related hydrothermal mineralisation in the region of the Tavsan deposit.

This region contains several mineral deposits with varying metal endowments closely linked to host lithology and tectonism. To determine the genesis of these deposits, an integrated field study coupled with other disciplines of mineral exploration are used at the Tavsan deposit, in which Au, Sb-Au, and Sb mineralisation is identified.

Field observations and microscopic analyses define the relationships between lithological units, structures, and alteration zones at Tavsan. The presence of macrostructures (particularly faults and folds) and combined stress analyses indicate significant structural controls on dispersion of hydrothermal fluids, marked by the presence of silicification, quartz veins, and hydrothermal breccias. Microscopic studies distinguish the relationship between alteration zones and the auriferous jasperoid, and their precursors associated with ultramafic rocks and limestone units. Electron probe micro-analyser studies show that gold is localised with disseminated pyrite minerals within jasperoid after hydrothermal silicification and de-carbonatisation of host rocks.

The Tavsan deposit is morphologically controlled by thrust faults and their partial re-activation under a later extensional regime, which followed peak compression during the Eocene. The spatial zonation of metallic associations and variability in sources of ore-forming fluids indicate that Sb-Au deposition occurred as a consequence of a hydrothermal system developed in a post-collisional setting. Genetic links between these deposits, regional magmatism, and extensional fault systems controlling the distribution of mineralisation suggest a Miocene age for formation of the main gold deposit.

This study emphasises that direct or indirect integration of diverse exploration methods can enhance understanding of relationships between tectonic processes and mineralisation.

The Relationship Between the Structural Orientation and Copper-Gold Mineralisation in the Alta Floresta Mineral Province, Brazil

Renata A. Paes[1], Maria J. Mesquita[1], Thomas Blenkinsop[2], Guilherme M. Primo[1]
1. University of Campinas, Campinas, SP, Brazil, 2. Cardiff University - School of Earth and Environmental Sciences, Cardiff, United Kingdom

The Alta Floresta Mineral Province is in the southern part of the Amazonas craton, Brazil, where studies are encouraged due to its productive artisanal gold mining history. However, the spatial relationships between the mineralisations and their structural framework still need to be better understood once the available research focuses mainly on the deposit scale. Therefore, data integration is necessary to construct robust exploratory models that make discovery and mining feasible, responsible, and dynamic. Considering the flat topography and the tropical weathering, the preliminary structural analysis was made by integrating field mapping, aerogeophysics, and SRTM data. With the map data geostatistical analysis, it's also possible to establish the relationship between the number of copper-gold mineral occurrences and their main controlling structures and interpret the possible principal channels and traps that the mineralisation fluid used to percolate and allow precipitation and mineralisation. As a result, it's possible to recognise WNW continuous linear features, in the literature related to first-order shear zones, in correlation with continuous NE structures and discontinuous features with NS and NE trends. The mineralisation's host rocks are 2.04–1.98 Ga granitoids, quartz-feldspathic gneiss, and mafic amphibolites, with at least two deformation episodes. The first developed an Sn foliation trending 0°–50° and dips 70°–90° WNW, defined by feldspar-quartz–rich and phyllosilicate-rich domains in orthogneiss, and a well-developed spaced foliation in metagranitoids and metamafic rocks. The second deformation episode developed several WNW first-order shear zones. They are transpressional to strike-slip, mostly with dextral movements, with mylonite foliations, N20°–40°W dipping 50°–70°SW. Parallel to this first-order shear zone occur high-grade and low-tonnage deposits such as the Serrinha de Guarantã, the thickest copper-gold quartz vein at the area. The most concentration of copper-gold-quartz veins is at subsidiary shear zones. The subsidiaries are N-trending, controlling the Paraíba mine, or E-trending Peteca deposit.

Temporal and Genetic Relationships Between Gold Mineralization, Magmatism, and Deformation: The Neoarchean Windfall Gold Deposit, Northern Abitibi, Canada

Jean-Luc Pilote[1], Sébastien Castonguay[1], Patrick Mercier-Langevin[1], Benoît Dubé[6], Brandon Choquette[2], Édouard Côté-Lavoie[2], Sandra L. Kamo[3], Natasha Wodicka[4], Danny Hnatyshin[4], Robert A. Creaser[5]
1. Geological Survey of Canada, Québec, QC, Canada, 2. Osisko Mining Inc., Montreal, QC, Canada, 3. Jack Satterly Geochronology Laboratory, University of Toronto, Toronto, ON, Canada, 4. Geological Survey of Canada, Ottawa, ON, Canada, 5. Department of Earth and Atmospheric Sciences, University of Alberta, Edmonton, AB, Canada, 6. 4283 rue du Curé-Drolet, Québec, QC, Canada

The Archean Abitibi greenstone belt (Superior Province) is well known for its gold endowment (>300 Moz Au), exemplified by a variety of well-documented styles of mineralization such as gold-rich volcanogenic sulfide lenses, intrusion-associated dissemination/veins, and quartz-carbonate orogenic veins. These styles are related to specific time intervals in the tectonic evolution of the belt and display distinct alteration footprints. For instance, the main shortening phase associated with thick-skinned deformation (2670-2640 Ma) has produced most of the structurally controlled quartz-carbonate orogenic veins. Whereas the southern Abitibi hosts >85% of the production, reserves, and resources, its less-studied, -exposed, and -explored northern counterpart has recently witnessed growing interest with the discovery of multiple >1-Moz Au deposits. Much like the southern Abitibi, mineralization is commonly peripheral to major E-W–oriented high-strain zones and/or along second-order structures. Mineralization that post-dates the bulk of volcanic cycles (<2700 Ma) and pre-dates the main shortening phase (>2670 Ma) possesses timing or genetic characteristics and complex relationships with structural events that require more detailed study. Here, we focus on the 7.4-Moz Windfall gold deposit in northeastern Abitibi, which is hosted in folded 2715 Ma tholeiitic bimodal volcanic sequences and gabbroic dikes along the structural footwall of a regional SE-dipping deformation zone. Units define a NE-plunging synform and are cut by ~2698 Ma axial-planar, calc-alkalic intermediate-felsic dikes. Two penetrative fabrics are documented: a NE-dipping foliation (S_n) crenulated and folded by a steeply dipping, NE-trending S_{n+1}. Gold is associated with pyrite-quartz-sericite stockworks and quartz-pyrite±carbonate veins that are oriented (sub-)parallel to S_{n+1}. Petrochronological and Re-Os sulfide dating suggest gold introduction penecontemporaneous to the emplacement of ~2698 Ma felsic dikes, whereas relative timing of deformation in relation to the gold-bearing veins suggests a possibly younger event. Resolving this apparent dichotomy is critical for developing robust metallogenic and exploration models for gold systems.

A Comparison of Analytical Methods Testing Soils for Gold Exploration in Pembrokeshire, West Wales

Maxwell L. Porter[1], Iain McDonald[1], Tristan English[2]
1. Cardiff University, Cardiff, United Kingdom, 2. Sarn Helen Gold, Cardiff, United Kingdom

West Wales has experienced a resurgence in gold exploration, with Sarn Helen Gold (SHG) seeking to find new gold resources across the historical mining region located near Dolaucothi Gold Mines. The Plumstone Mountain locality within Pembrokeshire has been an area of focus from prior exploration efforts by the British Geological Survey (BGS), revealing several prospective features through regional-scale geochemical and geophysical surveys. From July to October 2021, the author collaborated with SHG to complete a soil sampling, rock chip sampling, and gold panning programme to follow up on these initial findings by the BGS. The synthesis of completed data analysis addresses a series of aims: (1) assess the prospective nature of the Plumstone Mountain study area; and (2) evaluate the effectiveness of pXRF, fire assay, multi-element, and ionic leach analytical techniques.

An approach that integrates geochemistry and conceptual thinking is required to explore the mineral potential of Plumstone Mountain due to the lack of surface exposure. The pXRF has the ability to provide quick and more cost-effective data over alternative commercial lab techniques. These advantages are critical during an early-stage exploration programme despite the higher limits of detection provided (LOD), with continued pXRF use recommended for future exploration work undertaken within the region.

Plumstone Mountain contains intensely altered portions of the Roch Rhyolite Group, a rock package that has previously been revealed to contain enriched concentrations of base metals and trace gold mineralisation. A geochemical soil anomaly is present in the north of the study area, corresponding with high concentrations of pathfinder elements (As, Ba, Cu, Fe, Pb, Zn), alongside an Au anomaly identified using ionic leach data. When assessed and compared to other known deposits, the analysis indicates bimodal-felsic volcanogenic massive sulphide mineralisation and is supported by previous discoveries of Au associated with pyrite in quartz veins.

Evaluation of the Potential Intrusion Related Gold System (IRGS) of the Urcos-Sicuani-Ayaviri Structural Corridor System in Santiago de Pupuja, Puno, Peru

Mirian Puma
Universidad Nacional del Altiplano, Puno, Peru

The area of study is within the strip of Sn-Cu-W deposits related to Oligocene-Miocene intrusives and epithermal Ag-Pb-Zn (Au), part of the northern continuation of the Bolivian polymetallic strip where the Kori-Kollo and Tasna deposits are found, both deposits previously studied and classified as intrusion related gold systems (IRGS).

The IRGS is a type of deposit that has been little studied and will benefit the scientific community of geologists and companies involved in mineral exploration.

The main objective was to determine the mineralization potential of IRGS using petrographic study, analysis of multi-elemental rock in encasing rock and in veinlets (where we worked with 75 elements giving relevance to the Magnetite-Ilmenite Series typical of this type of deposit); in addition, we carried out studies of spectral analysis—SWIR spectral indexes, where the zone of hydrothermal alteration was estimated giving, as a result, the predominance of white micas, 2205-2230 (Phengite to Muscovite).

Finally the study gave a result of an important mineralized target and thus part of the mineralization model of the gold systems related to intrusions, sheeted veins, and epithermal-style veins and stockwork Au, Ag, As, Sb, Pb, Hg.

The place called San Antonio de Pupuja is located near a ranch with a population of approximately 100 people, where the main activity is livestock; therefore, responsible discovery and mining in this 21st century would be relevant and important, respecting the environment as the main objective.

Ore Mineralogy and U-Pb and Re-Os Geochronology in the Ikkari Gold Camp, Central Lapland Belt, Northern Finland

Jukka-Pekka Ranta[1], Wäinö Wacklin[1], Charlotte Seabrook[2], Holly Stein[3], Ramiro Lenarduzzi[1, 2], Matti Kurhila[5]
1. Oulu Mining School, University of Oulu, Oulu, Finland, 2. Rupert Resources Ltd, Toronto, ON, Canada, 3. AIRIE, Applied Isotope Research for Industry and Environment, Fort Collins, CO, USA, 4. Department of Geosciences, University of Oslo, Oslo, Norway, 5. Geological Survey of Finland, Espoo, Finland

The Paleoproterozoic Central Lapland belt, part of the Fennoscandian Shield, has proven to be the most prospective area for large mineral deposit discoveries during the last century. One of the recent discovery highlights is the Ikkari gold deposit, part of the more extensive Ikkari Gold Camp (IGC), comprising multiple gold deposits and occurrences hosted by intensely strained metasedimentary rocks and metakomatiites. The most recent combined resource estimation yields >4 Moz Au with potential for growth. In this study, we present mineralogical relationships along with geochronology using in situ LA-SC-ICP-MS U-Pb dating of monazite and xenotime, and Re-Os dating of molybdenite related to the mineralization in the Ikkari and Heinä Central gold deposits within the IGC. At Ikkari, gold is strongly associated with pyrite within carbonate-rich intervals and occurs as native gold, filling fractures in pyrite or forming adjacent to pyrite. Locally, mineralized rocks show elevated Co content where common Co-(and Ni)-rich minerals enclosed by pyrite include siegenite with millerite exsolution lamellae. In contrast to Ikkari, the Heinä Central deposit is characterized by abundant pyrrhotite and chalcopyrite in carbonate-rich breccias. Pyrite is subordinate and focused within local pyrite-rich intervals. Gold is found as native grains associated with pyrrhotite, Bi-Te phases, or intergrown with molybdenite in sulfide-carbonate-quartz breccias.

At Ikkari, hydrothermal monazite and xenotime from Au-rich mineralized intervals show a distribution of ages between ca. 1.83 and 1.75 Ga. Similarly at Heinä Central, hydrothermal xenotime related to gold-rich intervals shows ages between ca. 1.83 and 1.75 Ga. In addition, Re-Os dating of molybdenite intergrown with native gold yields ages of 1802 ± 7 Ma. These results suggest a prolonged mineralizing system during the late-orogenic stages of the Svecofennian orogeny and reflect complex periods of re-activation of major fluid-channeling structures in the Central Lapland belt.

Recent Advances in Hard-Rock Seismic Technology Applications Unlock the Deep

Leigh Rigg[1], Greg Turner[1], Nigel Miller[2]
1. HiSeis Pty Ltd, Subiaco, WA, Australia, 2. HiSeis Pty Ltd, Subiaco, United Kingdom

Resourcing the future would be easier if we could see underground. We could target higher grades at greater depths and optimise surface operations to minimise our environmental footprint and energy consumption across the life of mine. We could make mining safer. The minerals industry needs new tools to better characterise geology in hard-rock environments for the efficient and cost-effective detection, delineation, and extraction of metals that are important for a sustainable future. The seismic reflection technique offers high-resolution insights into the subsurface at depths down to and beyond economic mining limits, opening new frontiers in in situ locations. We believe the uptake of this technology has been held back in part by the complexity of the data sets, which have required specialist interpreters and software to decipher—until now. Cloud computing and cognitive technologies have enabled customised, data-driven approaches to extract more geologically relatable products from the seismic data that are easier to interpret and easier to use in standard mining workflows. This paper will showcase the latest suite of seismic-derived geological products, focused on key elements of the minerals system, and discuss the associated benefits of how these allow for a deeper and more targeted approach to exploration drilling. Answer products include detailed 3D fault framework models and 3D solid geology models that allow mapping of key rock types and intrusive geobodies several kilometres below the limits of traditional geophysical methods. Further product development beyond exploration is made possible as an increasing volume and variety of geoscience data is integrated within the 3D seismic container, providing critical new insights through different lenses across the life of mine. We believe seismic reflection technology is a vehicle for change in orebody characterisation, helping miners see more underground for a smarter, deeper, and cleaner mine of the future.

Gold Exploration Through Integrating Machine Learning, Remote Sensing, Aeromagnetic, and Spectrometric Gamma-Ray Data. A Case Study from the Egyptian Eastern Desert

Ali Shebl[1, 2], Árpád Csámer[1]
1. 1. Department of Mineralogy and Geology, University of Debrecen, Hungary., Debrecen, Hungary, 2. 2. Department of Geology, Tanta University, Egypt, Tanta, Egypt

Mineral exploration is the initial stage that constitutes the basic framework for more expensive exploration and exploitation operations. Consequently, adequate preliminary explorations help in saving time, effort, and money in the whole mining industry process. The current research is an attempt to make the best use of integrated data to raise the chance of productive mineralized zones. This is achieved by applying multi-sensor remote sensing data, airborne geophysical data, including aeromagnetic and spectrometric gamma-ray data, machine learning algorithms, and field observations for accurate lithological discrimination and precise hydrothermal alteration delineation that is mostly associated with orogenic gold deposits in the central Egyptian Eastern Desert. Instead of using remote sensing data or gamma-ray data independently, an excellent lithological map for the examined terrain is created by noting the variable concentrations of K, Th, and U and the distinctive spectral signatures of the rock units. Additionally, the concentrations of K, Th, and U have been used for the delineation of hydrothermal alteration zones, which is confirmed by informative remote sensing combinations and field observations. Thorough structural delineation is required for adequate mineral potentiality mapping as it controls the pathways through which hydrothermal solutions flow and plays a role in concentrating mineral deposits. Thus, automatic lineament extraction and the center for exploration targeting (CET) technique were utilized to create accurate maps of structural complexity by analyzing aeromagnetic information. Our results distinguish various types of alteration zones, including files, argillic, and propylitic. Spatial overlay analysis of the detected structural features and hydrothermal alteration zones highlighted higher potential zones of gold mineralization within the study area. Our field investigations have confirmed the outcomes of the current research, and several random mining sites were observed during our field verification. Thus, we strongly recommend adopting the utilized approach for further investigation and gold exploration in similar terrains.

Disseminated Orogenic Gold Mineralization in High-Grade Metamorphic Turbidites of the Central Zone of the Damara Belt, Namibia: Controls and Characteristics

Josia Shilunga, Alexander Kisters
Stellenbosch University, Stellenbosch, South Africa

The recently discovered Twin Hills deposit is a large (>12-km strike length), disseminated, low-grade mineralized gold-sulfide (pyrrhotite > arsenopyrite » pyrite) system hosted by amphibolite-grade metaturbidites of the Pan-African Damara Belt of central Namibia. The deposit comprises distinct clusters of economic-grade gold mineralization, measuring several hundred meters along strike, separated by several kilometer-long intervals of only sporadic and sub-economic gold grades. Individual higher-grade clusters are controlled by the combination and interplay of two main factors, namely lithological contacts and subtle deflections of bedding of the highly-strained, subvertical metaturbidites. The ENE-trending mineralized corridor of the Twin Hills deposit is largely confined to imbricated and/or tightly folded packages of interbedded metapsammites-metapelites within the otherwise metapelite-dominated sequence. It is these units that promoted fracturing, fluid flow, and the development of fine-scale vein networks as a result of pronounced strain partitioning between schist units (ductile) and interbedded metapsammites (brittle) during regional deformation (flexural-slip folding). Clusters (shoots) of higher-grade mineralization (>1.5 g/t Au) follow these main lithological contacts, but are laterally confined to subtle (5°–15°) clockwise deflections of bedding (dilational jogs) from regional ENE trends. Outside these deflections, gold mineralization is patchy and only sub-economic. On a regional scale, the position of the Twin Hills deposit coincides with the inflection of the vergence direction of regional-scale first-order folds. The structural and lithological controls on the mineralization at Twin Hills closely resemble those of orogenic gold deposits, but the fine-grained and dispersed textures of gold and associated sulfide mineralization are more reminiscent of those found in disseminated gold deposits.

Geochemistry of Felsic Rocks in the Yanıklı Prospect, Artvin District, Eastern Pontides, Turkey: Fertility and Implications for Mineral Exploration

Şafak U. Sönmez[1], Robert Moritz[1], Serdar Keskin[2], François Turlin[1], Alexey Ulianov[3], Ümit Aydın[2]
1. Department of Earth Sciences, University of Geneva, Geneva, Switzerland, 2. General Directorate of Mineral Research and Exploration of Turkey, Department of Mineral Research and Exploration, Ankara, Turkey, 3. Institute of Earth Sciences, University of Lausanne, Lausanne, Switzerland

Late Cretaceous felsic volcanic and volcaniclastic rocks in the Eastern Pontides are well-endowed in volcanogenic massive sulfide (VMS) and epithermal deposits/occurrences. The Artvin district is one of the most representative regions to examine numerous felsic rock types related to a bimodal volcanic and extensional teconic setting within the easternmost Eastern Pontides. The purpose of this study is to determine the fertility of Late Cretaceous dacitic/rhyolitic rocks in the Yanıklı prospect within the Artvin district and compare their geochemical characteristics with those of felsic rocks from the major Late Cretaceous Murgul, Tunca, and Kutlular VMS deposits/occurrences of the Eastern Pontides.

Felsic rocks at Yanıklı have been subdivided into four main units, from oldest to youngest, namely (i) feldspar porphyry dacite (KDa), (ii) fine-grained flow-banded rhyolite-rhyodacite (KRDa-1), (iii) flow banded-columnar rhyodacite-dacite (KRDa-2), and (iv) flow-banded feldspar-rich rhyolite-rhyodacite (KRDa-3).

Trace element signatures for the felsic rocks indicate that they have been formed in a subduction setting. The chondrite-normalized La/Yb ratios vs. Yb data of the felsic rocks from the Yanıklı prospect are very distinct. The oldest KDa unit presents low $La/Yb_{(CN)}$ (up to 4.98) and high $Yb_{(CN)}$ (up to 28.8) values. The immediate felsic host rock units of the three major Murgul, Tunca, and Kutlular VMS deposits/occurrences share similar geochemical compositions. These felsic rocks have tholeiitic to calc-alkaline rhyolitic compositions, which are generally favourable settings for VMS deposits/occurrences. In contrast, the mainly calc-alkaline rhyodacitic rocks of the younger KRDa-1 and KRDa-2 units at Yanıklı have high $La/Yb_{(CN)}$ (up to 8.3 and 12.9, respectively) and low $Yb_{(CN)}$ (up to 6.6 and 6.3, respectively) values. These units are mostly barren in terms of VMS orebodies, but they host epithermal-type polymetallic-Au veins/veinlets. The youngest KRDa-3 unit has alkaline to calc-alkaline compositions, which is commonly barren in terms of VMS and epithermal potential.

Geology and Gold Mineralization of the La Debo Permit, Southwestern Côte D'ivoire, West Africa

G. Doma Sorho[1], Sarah F. Pein[1], Vagne Kouakou[1], Sihle Gqoboka[2], Geoffrey Howarth[2], Robert Kiefer[1], Lynnette Greyling[1]
1. Goldrange Resources, Toronto, ON, Canada, 2. University of Cape Town, Cape Town, South Africa

Gold mineralization on the La Debo exploration permit is located within the Birimian Soubré greenstone belt in the Goh-Nawa region of southwestern Cote D'Ivoire in West Africa. Several mineralized occurrences are found within a 30-km NE-trending structural corridor, which are the focus of an ongoing research project and exploration program. The gold-bearing orebodies are located off major structures, commonly in areas of second-order shears or regional folds, or on extensive dilation jogs. The focus of this study is on the genesis of two NE-trending orebodies, G3 north and G3 south. These orebodies have been the subject of an earlier PEA, where an inferred resource of 500k ounces Au is defined. Reinterpretation of structural and geological models from current and historical drilling indicate the presence of several mineralized shear structures hosting a number of ore zones. Preliminary petrography results show a mineral assemblage of sericite-chlorite-biotite and minor pyrite, suggesting that the host rocks underwent propylitic alteration. Gold mineralization is broadly linked to post-peak metamorphism and is closely associated with sulfides, mainly pyrite and pyrrhotite and, to a lesser extent, chalcopyrite. Additionally, the presence of tourmaline and chlorite may indicate that the rocks were affected by intense shearing and associated boron-rich fluid circulation. The updated structural model shows that mineralization is likely related to wrench tectonics associated with shallow, NE-plunging regional fold hinges. This study will also delve into the physical and chemical explanations for the source and location of the mineralization, thereby improving the understanding of the auriferous potential of the orebodies.

Measure Twice – Rochford-Homestake Gold, South Dakota, USA

Holly J. Stein[1, 2], Zeynep Baran[3], Gang Yang[1]
1. AIRIE Program, Applied Isotope Research for Industry and Environment, Fort Collins, CO, USA, 2. Department of Geosciences, University of Oslo, Oslo, Norway, 3. South Dakota School of Mines and Technology, Rapid City, SD, USA

Re-Os dating of sulfides is now a widely employed analytical tool for providing radiometric ages for ore deposits, yet bold data interpretation and provocative new models lag behind the technology. An Re-Os study of the Rochford Au district, 20 miles south of the renowned Homestake Au mine, affirms widespread Proterozoic Au mineralization in the region. Structural features highlighting ductile-brittle deformation relative to sulfide-forming events are present in the analyzed drill cores. Re-Os dating of unique targeted sulfide occurrences relative to metamorphic fabric in the host rock present a new model for Au formation.

Using a double Os spike, we show *two* distinct generations of arsenopyrite at Rochford (isochrons and model ages). A previously unknown but significant older generation yields a Model 1 age of 1757.8 ± 1.7 Ma (MSWD = 1.7, n = 12, $^{187}Os/^{188}Os \sim 0$). A younger arsenopyrite generation, which includes three pyrrhotite analyses, yields an isochron age of 1735.5 ± 6.2 Ma (MSWD = 2.6, n = 6, $^{187}Os/^{188}Os$ = 0.0006 ± 0.0013, again essentially zero). However, another pyrrhotite analyzed plots precisely on the older arsenopyrite isochron, relating pyrrhotite to both gold generations. These two well-constrained events, both with Os initial ratios of zero, are interpreted as follows: (1) models for Au generation from older continental crust or reconstitution of pre-existing sulfide are conceptually difficult as they are unlikely to yield Os initial ratios of zero, or (2) the presence of Au-rich hydrothermal systems may force exclusion of Os in arsenopyrite in favor of Au. The latter offers a new pathfinder for Au-mineralized systems. That is, if there is common Os in the arsenopyrite structure, significant Au may be absent in the system. Further, this study shows that (some) pyrrhotites can be dated using Re-Os, in contrast to previous suggestions that pyrrhotite Re-Os systematics are easily disturbed.

Gold Fertility of the Inner Lesser Himalayas: Insights from Textural and Mineralogical Investigation of Pyritiferous Mandhali Black Shales, Uttarakhand, India

Abdul Latheef T P, Rajarshi Chakravarti
Department of Earth Sciences, Indian Institute of Technology, Roorkee, Roorkee, UK, India

Pyritiferous black shales are known to host several world-class orogenic/sediment-hosted gold deposits. Several research articles have also discussed the importance of pyritiferous black shales as a suitable source rock for orogenic gold formation, such as in the Sukhoi log, Bendigo, and majority Phanerozoic orogenic gold deposits. Notwithstanding, the gold fertility of laterally extensive pyritiferous black shales along the inner Lesser Himalayas (Mandhali Formation) have been overlooked. To this end, Neoproterozoic black shales of the inner Lesser Himalayas have been investigated here. Textural analysis reveals the presence of syngenetic, diagenetic, and metamorphic/hydrothermal pyrite, along with gold, either occurring as native grains or as micro-inclusions in pyrite. The sub-microscopic as well as native gold grains occur in texturally late paragenetic positions, thereby indicating a post-depositional hydrothermal control on mineralization. While this may be a sub-economic mineralization, previous studies recognizes that a threshold whole-rock content of 5 ppb Au and 250 ppb Au in diagenetic pyrite qualify a black shale as a suitable source rock for orogenic gold formation. Our results reveal whole-rock Au concentrations ranging between 6 and 66 ppb and Au content in pyrite ranging up to 4,300 ppm (mean 2,300 ppm, EPMA X-ray mapping) with As enrichment in diagenetic pyrite and depletion in later hydrothermal/metamorphic pyrite.

Further, the studied black shales exhibit effects of lower-greenschist facies metamorphism, contact metasomatism due to spatially associated sills, mild deformation, and are structurally overlain, at least in the studied area, by impermeable quartzites of the Berinag thrust sheet, thus satisfying the prerequisites for orogenic/sediment-hosted gold mineralization. Our findings, thus provide a starting point for; (a) studying the hitherto-unknown black shale-hosted gold mineralization in the inner Lesser Himalayas, and (b) assess the gold fertility of the inner Lesser Himalayas in context of a carbonaceous sedimentary gold reservoir.

Operating on the Frontier of Discovery: Utilisation of a Multi-Disciplinary Exploration Approach at Hizarliyayla, Northeast Turkey

Berkin Ugurlu[1], Ahmet Kerim Sener[2], Izak J. Van Coller[2]
1. Galata Mineral Madencilik San. ve Tic. A.Ş., Ankara, Turkey, 2. Ariana Resources plc, London, United Kingdom

Exploration for mineral deposits is the art of decision-making through the use of various scientific disciplines tempered by consideration of the time and cost of making a discovery. In addition, discoveries are not limited to objectively determinable factors such as location, grade, tonnage, or impact on local community or environment, but also the time of discovery, an aspect which reflects on the technology, metal prices, and politics of the day. Only when all factors align can conclusions be drawn regarding whether a mineral deposit is likely to be economic. Through early implementation of responsible exploration habits and particularly through the involvement of local communities, the progression of an exploration project can be swift, benefiting society as a whole.

The Hizarliyayla Project is located in the Eastern Pontides magmatic belt in northeast Turkey, situated within the well-known Hod Gold Corridor. This northeast trending, 110-km-long tectonic corridor hosts several potentially economic porphyry, epithermal, carbonate replacement, and skarnoid deposits of likely Eocene age, based on recent studies undertaken at the nearby Salinbas and Ardala deposits. Recent studies have revealed the potential for an intermediate sulfidation style of mineralisation at the Hizarliyayla Project, linked to Eocene magmatism.

The exploration methods employed for the assessment of this brownfield district included a review of historic studies from the last 30 years, remote-sensing analysis, 1:5,000-scale geological and alteration mapping supported by petrographical studies, soil and rock sampling for multi-element geochemical analysis, multi-sensor sample scanning, and an advanced statistical evaluation of the geochemistry. In addition, an environmental impact base-line study and local community engagement is continuing. This holistic approach and early application of multi-disciplinary subjects supports the development of a coherent geological and mineralisation model, which contributes to further project development in a socially and environmentally responsible manner.

Shear-Zone Hosted Gold Exploration, British Columbia: Unmanned Aerial Vehicle Photogrammetry Mapping, and EasyMineXR Application to Inaccessible Steep Terrains

Ronald Uken[1], Sheila Ulansky[1], Emre Onsel[1], Venessa Bennett[2], Thomas Monecke[3], Randell Karcher[3], T. James Reynolds[3], Manuele Lazzarotto[4], Russell Ashton[4], Bill Chornobay[4]
1. SRK Consulting (Canada) Inc, Vancouver, BC, Canada, 2. Drone North, Vancouver, BC, Canada, 3. Colorado School of Mines, Golden, CO, USA, 4. Goliath Resources, Toronto, ON, Canada

High-resolution unmanned aerial vehicle photogrammetry with survey-grade spatial accuracy, generated by Drone North in collaboration with Goliath Resources, together with SRK Consulting (Canada)'s newly developed structural mapping software (EasyMineXR) were effectively used to improve structural data coverage of the Surebet Zone discovery. A multidisciplinary approach using structural data, oriented core, and assay data, and petrography and radiometric ages in collaboration with Colorado School of Mines, was used to advance the understanding of mineralisation controls, structural geometries, and timing relationships.

Digital outcrop mapping with EasyMineXR focussed on the auriferous Surebet Shear Zone exposed along an inaccessible 1-km cliff face. Mapping included bedding orientations, form lines and fold criteria of the host Hazelton Group sedimentary succession, interpretation of the crosscutting auriferous shear zone system, dikes, and brittle faults. Regionally, mineralisation is situated on the northern margin of the Skeena arch, a zone of repeated tectonomagmatism from the Triassic to Eocene transecting the Stikine terrane. Hydrothermal activity was likely associated with Eocene deformation and calc-alkaline magmatism. Shearing was initiated as a thrust system with top-to-NE transport, subsequent to but coaxial to the NE-SW–directed contraction that produced upright to slightly inclined folds. Mineralization formed from auriferous fluids that were CO_2 rich and of comparably low salinity (3.5–5 wt % NaCl). Fluids are interpreted to be of magmatic origin. Two vein quartz types are present, early quartz formed at lithostatic pressures with a later quartz generation formed at hydrostatic conditions. Native gold and sulfide minerals including pyrrhotite, pyrite, sphalerite, galena, chalcopyrite, and tetrahedrite are paragenetically associated with the second quartz. Mineralization postdated peak metamorphism within the contact aureole andalusite zone, at mid- to upper-crustal level (>5–6 km, 350°–400°C). The Surebet discovery adds to the plethora of established mineralisation styles within this exceptional gold- and metal-rich camp.

Lithological-Structural Setting Mineralisation Styles of the Gilar Epithermal Deposit in the Lök-Qarabağ Volcanic Arc, the Lesser Caucasus, Azerbaijan

Anar Valiyev, Stephen J. Westhead, Aydin Bayramov, Sabuhi Mammadov, Javid Ibrahimov, Shakir Gadimov, Samir Mursalov
Azerbaijan International Mining Company (subsidiary of Anglo Asian Mining), Gadabay, Azerbaijan

The Gilar deposit belongs to the Gədəbəy ore district of the Shamkir uplift, which is part of the Lök-Qarabağ volcanic arc—one of the main structural formations of the Lesser Caucasus. The deposit is located over the East flank of the Böyük-Qalaça local deep fault. Gilar was discovered during geological exploration work of the northeast flank of the Gedabek gold-copper mine by the Gedabek Exploration Group (GEG) of Azerbaijan International Mining Company in 2019. Gilar will be the 4th deposit discovered that will be constructed to a mine. The Gilar deposit is located between two systems: the Maarif porphyry and the Ərtəpə mineral occurrences. The rocks are Upper Bathonian extrusive, represented by mostly felsic and intermediate composition rhyolite-dacites, andesite-porphyry, andesites and their tuffs distributed from the surface to depth. Ore minerals are hosted in the metasomatic rhyolite-porphyry of the Upper Bajocian age, at depths ranging from about 130 to more than 400 m from the surface. Preliminary field mapping and outcrop sampling identified a continuous epithermal quartz vein, hosted in a rhyolite volcanic in the northern Gilar area. To the south of the vein system, significant massive mineralisation has been discovered. Mineral content is visual; however, further petrographic studies by polished thin section were carried out on samples taken from outcrop and drill core. The majority of core samples from Gilar contain high-grade gold, copper, and zinc. According to these observations and studies, pyrite is the main mineral which is observed in all drill core, and nearby outcrop samples exhibit crystal structures: massive, disseminated, euhedral, anhedral, etc. with associated chalcopyrite, sphalerite, and other sulfide minerals. Jasper, magnetite, barite, and limonite are non-sulfide minerals present. The mineral composition, textures, relationships, and chemical composition suggest a high-sulphidation type of epithermal system for the Gilar deposit.

The Singular Polymetallic Occurence in Zona Basal, Quadrilátero Ferrífero, Brazil: Mineralogical and Geochemical Characterization and Genetic Implications

Felipe Emerson André Alves[1, 2], Atlas Vasconcelos Correa Neto[3], Mariana Brando Soares[4], Reiner Neumann[2], Gabriel Machado da Silva[1], Gabriela Pereira da Silva[1], Ana Clara Varca Pinheiro[1], Piero Azevedo Berquó de Sampaio[1], Vitor Diniz Silveira[5]

1. Programa de Pós-Graduação em Geologia (PPGL) - UFRJ, Rio de Janeiro, RJ, Brazil, 2. Centre for Mineral Technology (CETEM), Rio de Janeiro, RJ, Brazil, 3. Institute of Geology, Federal University of Rio de Janeiro, Rio de Janeiro, RJ, Brazil, 4. Geology Institute, Rio de Janeiro State University (Uerj), Rio de Janeiro, RJ, Brazil, 5. Jaguar Mining, Inc., Belo Horizonte, MG, Brazil

The Quadrilátero Ferrífero (QF) mining district, one of the main Au-Fe metallogenic provinces in Brazil, holds historic, strategic, and economic importance. Located in the southern São Francisco craton, it comprises gold deposits discovered as early as in the 17th century. The Turmalina area deposits are amongst those, and are modernly known to be hosted in amphibolite-facies rocks of the Archean Pitangui Greenstone Belt (PGB) in the northwestern portion of the QF. While the most widely known deposits in the QF are regarded as classic orogenic gold type, those situated in the PGB (such as Satinoco—part of the Turmalina area—and São Sebastião) show hypozonal orogenic gold characteristics. The recently discovered Zona Basal (ZB) gold prospect in the Turmalina area shows even further differentiation, as it displays a singular Au-Ag-Cu-Zn-Pb-Sb association. Since base metal occurrences in the QF are scarce, geochemical, mineralogical, and textural analyses of ZB deposit samples are key to revealing the deposit's genetic conditions. So far, geochemical and mineralogical characterization unveiled two stages of mineralization: an Au-W-As late- to post-tectonic stage, displaying arsenopyrite-pyrrhotite-pyrite assemblages associated with gold and scheelite; and the following Ag and base metals stage, with its strictly post-tectonic pyrrhotite-pyrite-sphalerite-chalcopyrite-galena-ullmanite±meneghinite± fahlore± miargyrite± pyrargyrite assemblage. While the first stage crystallized from ~491° to 404°C, during the transition from peak greenschist facies to retrograde metamorphism, the second stage crystallized from ~400° to <170°C. Possible mechanisms that may elucidate the chemical changes between both stages are participation of magmatic-hydrothermal, relatively low-temperature fluids, associated with syn- to late-tectonic granite intrusion, or localized re-concentration of base metals caused by fluid immiscibility after the transition from ductile to brittle conditions, generating fluids with contrasting salinities. Those results point to the necessity of understanding the nature of reactions of mineral re-equilibrium in complex ore formation.

A Study of the Gold-To-Sulphide Ratio in the Cononish Au-Ag Deposit, Tyndrum, Scotland

Amy L. Walsh
The University of St Andrews, Ashtead, Surrey, United Kingdom

The Cononish Au-Ag deposit is Scotland's only operational gold mine, located near the historic mining town of Tyndrum in the Loch Lomond and Trossachs National Park. Gold at Cononish is not visible at the face- or hand-specimen scale, but it is associated with abundant, easily visible sulphide phases. The project aim is to explore the paragenetic and visual relationship between the occurrence of sulphides, their textures, and the presence of gold at the Cononish deposit.

The gold-to-sulphide ratio will be studied qualitatively through textural analysis on both a micro- and macroscopic scale through field- and labratory-based data collection. Assayed samples from the Cononish vein with variable gold-to-sulphide ratios will be examined using polished thin sections and SEM-EDS analysis. The ratio will also be studied quantitively through statistical data analysis of screen fire assay results from the Scotgold Resources Ltd. (the deposit owners) database of face mapping and channels samples. The association with, and abundance of, penalty elements (Pb and Zn) will also be examined.

The findings will be used to develop a visual guide of sulphide patterns to assist geologists to better estimate gold concentrations. This research aims to contribute to the understanding of gold and sulphide relationships in vein-hosted Au-Ag deposits by studying Cononish, which is the best characterised of the Dalradian gold deposits. By focusing on the Cononish deposit, this study will provide valuable insights into the significance of the gold-to-sulphide ratio and its implications for the mineralisation process. This work could also provide insights into mineralisation across the Dalradian gold belt and the Grampian Terrane of the Caledonian orogenic belt, having applications for regional projects in the vicinity of Cononish to the national scale of Scotland as a whole, and internationally to Dalradian gold in Ireland, and possibly even Canada.

Calcatreu Project 3D Vein Modelling, Río Negro Province, Argentina

Matías S. Wernert[1], Darío Vera[2], Marcelo Paolini[2], Sebastián M. Jovic[1], Diego M. Guido[1]
1. Instituto de Recursos Minerales, Ciudad de La Plata, Argentina, 2. Patagonia Gold Corp, Buenos Aires, Argentina

The Calcatreu Project is a low-sulphidation epithermal Au-Ag deposit located in Río Negro Province, Argentina, currently undergoing advanced exploration stages. Indicated resources are estimated at 746 kOz, with ore grades of 2.36 g/t AuEq.

The district is located in the Somún Curá Massif, a geological province characterized by extensive Jurassic vulcanism related to extensional events linked to the Gondwanide Orogeny. This sub-explored region bears similar geological features and mining potential to the important low-sulphidation Deseado Massif province.

The deposit hosts an array of NE-SW–trending quartz-carbonate veins, hosted by Jurassic andesites and tuffs. Furthermore, the district is intersected by the Gastre Fault Zone, a Jurassic NW-SE regional megastructure that controlled vein disposition. The main gold ore is electrum, mostly linked to an open-space filling pulse composed of silica, carbonate, adularia, and ginguro bands.

The main exploration targets are the Nelson E, Nelson W, and 49 veins, where extensive mapping, drilling, and geophysical surveys were conducted. This data was used to create a 3D geological model of the orebodies by employing Leapfrog software.

The mineralization took place under $\sigma 1$ stresses of an approximate 60°N orientation. The orebodies follow an overall NE-SW orientation and a 60° to 80° SE dip. Specifically, the main vein trends vary between 40° to 80° N, whereas related splays and offshoots fall between 0° and 30°N ranges.

Vein 49 features a well-defined 1-km-long strike with few secondary ramifications. Normal faulting with varying shear stresses occurred during mineralization, as evidenced by slickenlines with rake angles between 40° and 80°E. Post-mineral faulting is also present, as suggested by common vein displacements. A major E-O structure limits Vein 49 with the southern Nelson targets, which consists of a 2-km-wide vein swarm that defines a mesh-type architecture.

Carlin-Type Gold Deposits in Youjiang Basin, Southwest China: Key Characteristics and Genetic Model

Zhuojun Xie[1], Yong Xia[1], Qinping Tan[1], Jianzhong Liu[2]
1. State Key Laboratory of Ore Deposit Geochemistry, Institute of Geochemistry, Chinese Academy of Sciences, Guiyang, China, 2. Guizhou Bureau of Geology and Mineral Exploration & Development, Guiyang, China

More than 100 Carlin-type gold deposits/occurrences have been discovered in the Youjiang Basin, southwest China (hereinafter referred to as Guizhou deposits). Currently, the Au endowment of Guizhou deposits is more than 900 tons of Au, the second largest Carlin-type gold concentration in the world (after the Nevada deposits).

Key characteristics for the Guizhou deposits include the following: gold is mainly ionically bound in trace element-rich, subhedral-euhedral arsenian pyrite, with minor Au in arsenopyrite; alteration includes dissolution of calcite, jasperoid replacement of calcite, sulfdation of Fe in Fe-dolomite/Fe-calcite to form Au-bearing pyrite and dolomite, minor illite replacement of K-feldspar/muscovite, and formation of minor quartz veins; late-ore-stage stibnite, realgar, calcite, quartz, and lesser orpiment and cinnabar precipitated in open space within veins. Fuid inclusion studies reflect these deposits formed at depths between 2 and 8 km and at temperatures of 190° to 300°C from low-salinity (<5 wt % NaCl equiv.) and CO_2-rich reduced fluids.

Combining geologic characteristics with isotope evidences (S, Hg, Mg, and He-Ar) supports a magmatic-hydrothermal origin for the ore fluids. The thick Devonian-Triassic sedimentary sequence and weak extension following sedimentation prevented igneous activity from reaching the surface. However, buoyant ore fluids released from the deep-seated intrusions have ascended along basement-penetrating faults and eventually produced the Guizhou deposits at shallow levels.

Structural Controls on Carlin-Type Gold Deposits at the Huijiabao Trend, Southwestern Guizhou Province, China

Chengfu Yang[1], Jianzhong Liu[2], Zepeng Wang[1], Faen Chen[1], Wenxing Tai[1]
1. No. 105 Geological Team, Guizhou Bureau of Geology and Mineral Exploration and Development, Guiyang, China, 2. Guizhou Bureau of Geology and Mineral Exploration and Development, Guiyang, China

The Huijiabao Gold Mining District has a total of 370 tonnes of proven gold resources and is expected to exceed 400 tonnes in the future. It is the most important gold-forming area in the "Golden Triangle" region of Yunnan, Guizhou, and Guangxi provinces, China, and is mainly controlled by the Huijiabao anticline and its axial or flank thrust faults.

The representative deposit is the giant Shuiyindong gold deposit, which is located on the side of the shallow-water carbonate rock distribution area at the boundary of the platform basin, and the contractional structural strain is weaker than that in the basin. The ore-controlling structure of the Shuiyindong gold deposit is mainly composed of folds and axial faults. In the Lower Triassic Yelang Formation, the orebodies are mainly controlled by the axial thrust fault on the limbs of the Huijiabao anticline and are hosted in the joint drag kink bands of the thrust fault; in the Upper Permian Longtan Formation, the orebodies are mainly controlled by the anticline core and "X"-type shear faults (back thrust faults) on the two limbs of the anticline, and the faults often show the characteristics of high-angle compressional thrust faults. Structural stripping zones in the anticline core between the carbonate rock layers form saddle-shaped orebodies, and thick and large lenticular orebodies are mostly hosted at the intersections of the carbonate rock layers and faults, especially in the third member of the Upper Permian Longtan Formation. There is a regional detachment zone between the Middle Permian Longtan Formation and the Upper Permian Longtan Formation, which acts as a channel for hydrothermal fluid migration and hosts plate-shaped orebodies.

Ore Mineral Textures of the Midas Low-Sulfidation Epithermal Deposit: Implications for Ore-Forming Processes

Lauren R. Zeeck[1], Thomas Monecke[1], T. J. Reynolds[1, 2], Katharina Pfaff[1], Nigel M. Kelly[3]
1. Colorado School of Mines, Golden, CO, USA, 2. FLUID INC., Denver, CO, USA, 3. Bruker Nano Analytics, Kennewick, WA, USA

Ore and gangue mineral textures of high-grade quartz vein material from the middle Miocene Midas Au-Ag deposit near Elko, Nevada, were studied to identify the processes that resulted in precious metal enrichment at this low-sulfidation epithermal deposit. Ore minerals in high-grade samples occur as dendritic aggregates within distinct ginguro bands. The delicate ore mineral dendrites are hosted by mosaic quartz that appears to have formed through recrystallization of a microspherical silica precursor. Based on the comparison to modern geothermal systems, this microspherical silica precursor is interpreted to originally have been gel-like opal-AG. The occurrence of ore mineral dendrites and microspherical opal-AG suggests that mineral precipitation at Midas occurred at far-from-equilibrium conditions. The findings of this study are consistent with previous models linking high-grade precious metal enrichment in low-sulfidation epithermal deposits to episodic flash vaporization of the hydrothermal liquids within hundreds of meters from the paleosurface.

Identification of Ore-Finding Targets Using the Anomaly Components of Ore-Forming Element Associations Extracted by BEMD and PCA in the Jiaodong Gold Clusters, Eastern China

Aoyue Zheng, Yongqing Chen
School of Earth Sciences and Resources, China University of Geosciences, Beijing, China

The Jiaodong gold concentrated luster area, with approximate gold reserve of 5,000 t, located at the southeastern edge of the North China Craton and to the east of the Tan-Lu Fault, is the third largest gold concentrated cluster area in the globe. Multi-stage geological processes (i.e., magma and hydrothermal processes) produce generally regional and local element concentration anomalies of ore-forming elements. Bi-empirical mode decomposition (BEMD) can be applied to establish various types of ore-forming element associations, which can be further divided into two different anomalous components with the help of BEMD. One can be defined as a regional geochemical anomaly generated by geological processes (i.e., magmatism); and the other can be regarded as a local geochemical anomaly created by ore-forming events. Three aspects are illustrated in this paper. (a) Two types of ore-forming element associations are established by PCA: Th-U-Y-Nb-Mo-La and Ag-Cd-Au. The former characterizes molybdenum polymetallic mineralization; the latter does gold polymetallic mineralization. (b) The geochemical anomaly components created by both magmatism and hydrothermal ore-forming processes are respectively identified by BEMD from the ore-forming element associations. (c) The regional and local geochemical anomalies are delineated using the geochemical anomaly components corresponding to different eigenspaces. The PCA and BEMD methods were applied to gold polymetallic ore-prospecting target areas and molybdenum polymetallic ore-prospecting target areas in the Jiaodong gold cluster area, eastern China, by delineating their regional and local geochemical anomalies. The results show that BEMD can extract not only regional anomalies associated with magmatism, but also local anomalies related to the hydrothermal processes from multi-element associations produced by PCA.

Application of BEMD in Extracting Gravity Anomaly Components Showing Deep Ore-Forming Geo-Dynamic Background of Jiaodong Giant Gold Cluster Region, Eastern China

Aoyue Zheng, Yongqing Chen
School of Earth Sciences and Resources, China University of Geosciences, Beijing, China

The Jiaodong gold cluster region, with approximate gold reserve of 5,000 t and located at the southeastern edge of the North China Craton, is the third largest gold cluster region in the world. BEMD (Bi-empirical mode decomposition) is applied in extracting gravity anomaly components showing deep ore-forming geodynamics at different depths. (a) At a depth of about 30 km, there are three tectonic units: the mantle uplift (I) with gravity values ranging from 2 to 14 μm/s^2, a mantle depression (II) with gravity values varying from 0 to -13 μm/s^2, and a mantle flat (III) with gravity values ranging from -2 to 2 μm/s^2. All giant gold deposits are distributed within the mantle depression. It may illustrate that mantle uplifting triggered hot, ore-forming fluids concentrating into the mantle depression to accumulate, forming giant gold deposits. (b) At about 17.1 to 12.5 km, there three tectonic units: Jiaolai-Jiaobei mantle uplift showing a strong positive gravity anomaly with gravity values ranging from 1.5 to 10 μm/s^2; the Sulu ultra-high pressure metamorphic block displaying a negative gravity anomaly with gravity field values ranging from -10 to -1.5 μm/s^2; and the Jiaoxibei gold cluster region exhibiting gravity background with gravity values varying from -1.5 to 1.5 μm/s^2. (c) At about 8.9 to 5.3 km, there is a series of positive and negative gravity anomalies. Most granites with low density display negative gravity anomalies, among which there are some negative anomalies with positive anomalous edges which contain gold deposits. This may illustrate an ore-forming pattern of granite (negative gravity anomaly) alteration (positive gravity anomaly). It has been illustrated that geological architectures at different depths as well as the giant Jiaodong gold cluster region was formed by the asthenosphere rises triggered by NNW subduction of the Izanagi Plate over a time period from about 200–100 Ma

Melt Inclusion Constraints on the Initial Metal Budget of Suspected Causative Magmas in the Au-Rich Doyon-Bousquet-LaRonde Mining District, Abitibi Greenstone Belt, Québec

Kevin Neyedley[1], Jacob J. Hanley[1], Patrick Mercier-Langevin[2], Zoltan Zajacz[3], Alexandra Tsay[3]
1. Department of Geology, Saint Mary's University, Halifax, NS, Canada, 2. Geological Survey of Canada, Quebec, QC, Canada, 3. Department of Earth Sciences, University of Geneva, Geneva, Switzerland

The synvolcanic Mooshla Intrusive Complex (MIC) in the Doyon-Bousquet-LaRonde mining camp, Abitibi greenstone belt, Canada, may be genetically associated with numerous Au-rich VMS, epizonal intrusion-related Au ± Cu, and shear zone-hosted orogenic Au and/or remobilized VMS deposits. To understand the magmatic evolution of the MIC and its relationship to these Au-rich deposits, analysis of silicate melt inclusions (SMIs) preserved in zircon and quartz provides a means to determine the original major- and trace-element composition and metal tenor of the magmatic system. The use of SMIs is preferred for quantification of the pre-emplacement melt chemistry (in particular, ore metals) compared to bulk-rock compositions commonly modified by magmatic (e.g., degassing) and post-solidus processes. Characterization of SMI chemistry from different lithologies of the MIC (gabbro, quartz diorites, tonalites, and trondhjemites) was done by integrating petrography with SEM-EDS, EMPA, and LA-ICP-MS analyses. Data presented constitute the first SMI study in Archean-age rocks within Canada.

Melt inclusions, trapped at depths of ~18.5 to 40 km (based on thermobarometry in a parallel study) are sub-alkaline to alkaline and rhyolitic in composition from all lithologies. Importantly, within individual lithologies, SMIs span from sub-alkaline to alkaline compositions, indicating melt entrapment at various stages of magmatic evolution, requiring contamination and/or magma mixing. Present in both zircon- and quartz-hosted SMIs are accidentally trapped sulphide phases (enriched in Au, Ag, Cu, Bi), indicating the melts were sulphide-saturated at the time of SMI entrapment. In quartz-hosted SMIs with no accidentally trapped sulphide, Au concentrations range from 10 to 120 ppb (70 ± 50 ppb; n = 6). Enrichment of SMIs in Au and other metals (e.g., Ag, Cu, Sb, Bi) typical of porphyry-epithermal environments in arc settings reinforces that consideration must be made for the direct involvement of magmatically sourced metals transferred into shallow ore-forming hydrothermal systems through deep magmatic devolatilization.

Society of Economic Geologists (SEG)
7811 Shaffer Parkway
Littleton, CO 80127-3732
USA

ISBN 978-1-7138-7964-0
9 781713 879640

CURRAN ASSOCIATES INC.
proceedings
.com

Design & Operation of Passenger Ships 2019

London, United Kingdom
1 May 2019

The Royal Institution of
Naval Architects